WORKBOOK

S N A P

Student Notes and Problems

MATH 6 Alberta

CASTLE ROCK
RESEARCH CORP

Rao, Gautam, 1961 –
STUDENT NOTES AND PROBLEMS – Math 6 Workbook
Alberta
(Second Edition)

1. Science – Juvenile Literature. I. Title

Published by
Castle Rock Research Corp.
2340 Manulife Place
10180 – 101 Street
Edmonton, AB T5J 3S4

1 2 3 FP 13 12 11

Publisher
Gautam Rao

Contributors
Amanda Brouwer
Phyllis Kozak
Sonya Witzman

Reviewers
Adam Boothe
Samantha Frampton
Pam Mosen

Dedicated to the memory of Dr. V. S. Rao

STUDENT NOTES AND PROBLEMS WORKBOOKS

Student Notes and Problems (SNAP) workbooks are a series of support resources in mathematics for students in grades 3 to 12 and in science for students in grades 9 to 12. SNAP workbooks are 100% aligned with curriculum. The resources are designed to support classroom instructions and provide students with additional examples, practice exercises, and tests. SNAP workbooks are ideal for use all year long at school and at home.

The following is a summary of the key features of all SNAP workbooks.

UNIT OPENER PAGE

- summarizes the curriculum outcomes addressed in the unit in age-appropriate language
- identifies the lessons by title
- lists the prerequisite knowledge and skills the student should know prior to beginning the unit

LESSONS

- provide essential teaching pieces and explanations of the concepts
- include example problems and questions with complete, detailed solutions that demonstrate the problem-solving process

NOTES BARS

- contain key definitions, formulas, reminders, and important steps or procedures
- provide space for students to add their own notes and helpful reminders

PRACTICE EXERCISES

- include questions that relate to each of the curriculum outcomes for the unit
- provide practice in applying the lesson concepts

REVIEW SUMMARY

- provides a succinct review of the key concepts in the unit

PRACTICE TEST

- assesses student learning of the unit concepts

ANSWERS AND SOLUTIONS

- demonstrate the step-by-step process or problem-solving method used to arrive at the correct answer

Answers and solutions for the odd-numbered questions are provided in each student workbook. A *SNAP Solutions Manual* that contains answers and complete solutions for all questions is also available.

CONTENTS

Place Value

Place Value and Large Numbers 2
Practice Exercises 4
Place Value and Decimal Numbers 5
Practice Exercises 7
Reading and Writing Numbers 8
Practice Exercises 13
Review Summary 14
Practice Test 15

Working with Decimals

Multiplication of Decimal Numbers by Multipliers from 0 to 9 20
Practice Exercises 21
Division of Decimal Numbers by Divisors from 1 to 9 22
Practice Exercises 24
Estimation Strategies for Multiplication and Division of Decimal Numbers 25
Practice Exercises 28
Review Summary 29
Practice Test 30

Integers

Understanding Integers 34
Practice Exercises 37
Integers on a Number Line 38
Practice Exercises 41
Comparing Integers 42
Practice Exercises 44
Review Summary 45
Practice Test 46

Multiples and Factors

Multiples 50
Practice Exercises 55
Prime and Composite Numbers 57
Practice Exercises 61

PLACE VALUES

When you are finished this unit, you will be able to…

- Understand the patterns of the place value system and the repetition of ones, tens, and hundreds within each period
- Understand how the place value system works with decimals
- Use the place value system to read and write numbers of any magnitude

Lesson	Page	Completed on
1. Place Value and Large Numbers	2	
2. Place Value and Decimal Numbers	5	
3. Reading and Writing Numbers	8	
Review Summary	14	
Practice Test	15	
Answers and Solutions	at the back of the book	

PREREQUISITE SKILLS AND KNOWLEDGE

Prior to starting this unit, you should be able to…

- Represent and describe whole numbers to 1 000 000
- Represent and describe decimal numbers as small as thousandths

Lesson 1 PLACE VALUE AND LARGE NUMBERS

NOTES

The **place value system** makes it possible to read and write numerals for numbers of any size.

Large numbers are organized in groups of three digits from right to left. These groups are called **periods**. The periods are separated by spaces.

A digit is any number from 0 to 9.

The three digits in a period have place values of ones, tens, and hundreds, moving from right to left. The ones digit of each period gives the period its name and unit value. By looking at a digit in the place value system, you can determine its value in a period.

The value of each digit is determined by its location within the period. As you move from right to left in the place value system, the values increase by a factor of ten. This means that you multiply the place value on the right by 10 to get the next place value immediately to the left.

The relationship between place values can be shown using a place value chart.

Thousands			Ones		
H	**T**	**O**	**H**	**T**	**O**
100 000	10 000	1 000	100	10	1

× 10 ↖ × 10 ↖ × 10 ↖ × 10 ↖ × 10 ↖ × 10

The first period is called the ones. It consists of ones, tens, and hundreds.

The second period is called the thousands. It consists of one thousands, ten thousands, and hundred thousands.

The third period is called the millions. It consists of one millions, ten millions, and hundred millions.

The fourth period is called the billions. It consists of one billions, ten billions, and hundred billions.

This system of repetition of ones, tens, and hundreds continues as you extend the place value chart to the left.

NOTES

Example

Show the number 912 673 950 824 in a place value chart, and give the value of each digit.

Solution

Billions			Millions			Thousands			Ones		
H	**T**	**O**	**H**	**T**	**O**	**H**	**T**	**O**	**H**	**T**	**O**
9	1	2	6	7	3	9	5	0	8	2	4

The digit 9 is in the hundred billions place. The value is $9 \times 100\ 000\ 000\ 000$, or 900 000 000 000.

The digit 1 is in the ten billions place. The value is $1 \times 10\ 000\ 000\ 000$, or 10 000 000 000.

The digit 2 is in the one billions place. The value is $2 \times 1\ 000\ 000\ 000$, or 2 000 000 000.

The digit 6 is in the hundred millions place. The value is $6 \times 100\ 000\ 000$, or 600 000 000.

The digit 7 is in the ten millions place. The value is $7 \times 10\ 000\ 000$, or 70 000 000.

The digit 3 is in the one millions place. The value is $3 \times 1\ 000\ 000$, or 3 000 000.

The digit 9 is in the hundred thousands place. The value is $9 \times 100\ 000$, or 900 000.

The digit 5 is in the ten thousands place. The value is $5 \times 10\ 000$, or 50 000.

The digit 0 is in the one thousands place. The value is 0. You still need to put the 0 in the chart to keep the other digits in the right place.

Remember, zero (0) is used to maintain proper place value positioning.

The digit 8 is in the hundreds place. The value is 8×100, or 800.

The digit 2 is in the tens place. The value is 2×10, or 20.

The digit 4 is in the ones place. The value is 4×1, or 4.

Time to Try 1

Show the number 28 412 567 103 in a place value chart, and give the value of each digit.

PRACTICE EXERCISES

Represent the following numbers using place value charts, and give the value of each digit.

1. 23 403 796

2. 514 372

3. 817 203 986 014

4. 410

Lesson 2 PLACE VALUE AND DECIMAL NUMBERS

NOTES

A decimal point separates whole numbers from numbers that are less than 1. The numbers to the right of the decimal show parts of a whole.

The relationship between place values is the same in decimal numbers as it is in whole numbers. As you move from right to left in the place value chart, multiply the place value on the right by 10 to get the place value immediately to the left.

Ones	**.**	**Parts of a Whole**				
O	**.**	**Tth**	**Hth**	**Thth**	**T Thth**	**H Thth**
1	.	0.1	0.01	0.001	0.0001	0.000 01

× 10 ↖ × 10 ↖ × 10 ↖ × 10 ↖ × 10 ↖

Abbreviations are used for decimals in a place value chart:
Tth—tenths
Hth—hundredths
Thth—thousandths
T Thth—ten-thousandths
H Thth—hundred-thousandths

This system of decimals can be continued if you extend the place value table to the right.

Example

Show the number 2.315 86 in a place value chart, and give the value of each digit.

Solution

Ones	**.**	**Parts of a Whole**				
O	**.**	**Tth**	**Hth**	**Thth**	**T Thth**	**H Thth**
2	.	3	1	5	8	6

The digit 2 is in the ones place. The value is 2 × 1, or 2.
The digit 3 is in the tenths place. The value is 3 × 0.1, or 0.3.
The digit 1 is in the hundredths place. The value is 1 × 0.01, or 0.01.
The digit 5 is in the thousandths place. The value is 5 × 0.001, or 0.005.
The digit 8 is in the ten-thousandths place. The value is 8 × 0.0001, or 0.0008.
The digit 6 is in the hundred-thousandths place. The value is 6 × 0.000 01, or 0.000 06.

NOTES

Time to Try 1

Show the number 4.901 28 in a place value chart, and give the value of each digit.

PRACTICE EXERCISES

Show the following numbers in a place value chart, and give the value of each digit.

1. 2.4016

2. 41.87

3. 8.632 41

4. 0.512

Lesson 3 READING AND WRITING NUMBERS

NOTES

READING AND WRITING WHOLE NUMBERS

When reading whole numbers out loud, work from left to right, and read each period separately. After you read each period, state the name of the period.

After the billions period, include the word *billion* before reading the millions period.

After the millions period, include the word *million* before reading the thousands period.

After the thousands period, include the word *thousand* before reading the ones period.

It is not necessary to include the word *ones* after the ones period.

Always put a comma between each period, but use this strategy only when reading whole numbers.

Example

In 2008, the population of Canada was estimated to be 33 311 400 people. Read the number 33 311 400, and write it using words.

Use a hyphen when writing the compound numbers from twenty-one to ninety-nine.

Remember to put a comma between each period. Never use the word *and* when reading and writing whole numbers.

Solution

Work from left to right.

First, read the millions period. This is written in words and read as *thirty-three*, followed by the word *million.*

Next, read the thousands period. This is written in words and read as *three hundred eleven*, followed by the word *thousand.*

Finally, read the ones period. This is written in words and read as *four hundred.*

The number 33 311 400 is read and written as *thirty-three million, three hundred eleven thousand, four hundred.*

Time to Try 1

Read the number 51 129 022 439, and write it using words.

NOTES

EXPRESSING WHOLE NUMBER WORD FORMS AS NUMERALS

When converting whole numbers from word form to numeral form, work from left to right. Leave a space (not a comma) between each period. Remember to use a digit for each place value inside the number. A zero will sometimes be required as a placeholder.

The words that represent the digits in the millions period are located to the left of the word *billion.*

The words that represent the digits in the millions period are located in between the words *billion* and *million.*

The words that represent the digits in the thousands period are located in between the words *million* and *thousand.*

The words that represent the digits in the ones period are located to the right of the word *thousand.*

Example

Shaddy's high score in his favourite video game is twenty-eight million, five hundred fifteen thousand, twenty-two. Write Shaddy's high score in numeral form.

Solution

Work from left to right.

Start by writing the millions period in numeral form.
twenty-eight million → 28

Leave a space, then write the thousands period in numeral form.
five hundred fifteen thousand → 515

Leave a space, then write the ones period in numeral form.
twenty-two → 022

Written in numeral form, Shaddy scored 28 515 022 points.

Remember, zero (0) is used to maintain proper place value positioning.

Time to Try 2

Write the number sixteen billion, one hundred forty-five million, sixty-two thousand, nine hundred two in numeral form.

NOTES

READING AND WRITING DECIMAL NUMBERS

A decimal point separates whole numbers from numbers that are less than 1. The numbers to the right of the decimal show parts of a whole.

When reading decimal numbers, read the digits from left to right. Start by reading the whole numbers, followed by the word *and.* Then, read the digits of the decimal portion as if they were a whole number, followed by the place value of the last digit.

If there is one digit after the decimal, include the word *tenths* at the end.

If there are two digits after the decimal, include the word *hundredths* at the end.

If there are three digits after the decimal, include the word *thousandths* at the end.

If there are four digits after the decimal, include the words *ten-thousandths* at the end.

If there are five digits after the decimal, include the words *hundred-thousandths* at the end.

Example

Read the number 76.052 31, and write it using words.

Solution

Work from left to right.

First, read the whole numbers. These are read and written as *seventy-six*, followed by the word *and.*

Next, read the decimal numbers. These are written in words and read as *five thousand, two hundred thirty-one*, followed by the place value *hundred-thousandths*.

The number 76.052 31 is read and written as *seventy-six and five thousand, two hundred thirty-one hundred-thousandths*.

In casual speech, the word *and* is often replaced with the word *decimal* or *point*. (As in seventy-six *point* five hundred twenty-three ten-thousandths.) However this is not formally correct language.

NOTES

If the whole number portion of a decimal number is zero, it is not necessary to write *zero and* before the decimal portion of the number.

Example

Read the number 0.305, and write it using words.

Solution

Work from left to right.

The whole number portion of the number is zero.

Read the decimal numbers, followed by the place value of the last digit. The number 0.305 is read and written as *three hundred five thousandths*.

Time to Try 3

Read each of the following decimal numbers. Write each number using words.

a) 631.02

b) 0.951 62

EXPRESSING DECIMAL NUMBER WORD FORMS AS NUMERALS

When converting a decimal number from word form to numeral form, work from left to right. First, write the digits representing the whole number portion of the number. Then, write a decimal point, followed by the digits representing the decimal portion of the number.

When writing numerals, remember to use a digit for each place value and to leave a space (not a comma) between each period.

The words that represent the whole number digits are located to the left of the word *and.*

The words that represent the decimal number digits are located to the right of the word *and.*

NOTES

Example

Pia measured the height of her ice-cream cone. She found that it measured nine and thirty-seven hundredths of a centimetre. Written as a numeral, how tall was Pia's ice-cream cone?

Solution

Work from left to right.

Start by writing the whole numbers in numeral form.
nine → 9

Write a decimal point, followed by the decimal numbers in numeral form.
thirty-seven hundredths → .37

Written in numeral form, Pia's ice-cream cone was 9.37 cm tall.

If the word *and* does not appear in the word form of a decimal number, the whole number portion of the decimal number is zero.

Example

Write the number eighty-eight ten-thousandths in numeral form.

Solution

Work from left to right.

Start by writing the whole numbers in numeral form.
The whole number is zero → 0

Write a decimal point, followed by the decimal numbers in numeral form.
eighty-eight ten-thousandths → .0088

Written in numeral form, the number is 0.0088.

Be careful when reading the place value of the decimal number. For example, eighty-eight ten-thousandths (0.0088) is not the same as eighty-eight hundredths (0.88).

Time to Try 4

Write each of the following decimal numbers in numeral form.

a) Two hundred five thousandths

b) Seventy-three and five ten-thousandths

PRACTICE EXERCISES

Read each of the following numbers. Write each number using words.

1. 261 005 065 023

2. 72.082

3. 0.702 33

Write each of the following in numeric form.

4. Six billion, twenty-two million, four thousand, seven hundred forty-eight

5. Five hundred sixty-seven and forty-five hundredths

6. Twelve ten-thousandths

REVIEW SUMMARY

- The relationship between place values in any whole number or decimal number can be shown using a place value chart.
- Moving from right to left in a place value chart, the values increase by a factor of ten.
- When using a place value chart, remember that each digit tells you something about the value.
- The digits after a decimal point are parts of a whole number.
- The system of repetition of ones, tens, and hundreds continues as you extend the place value chart to the left.
- The system of decimals can be continued if you extend the place value table to the right.
- When reading and writing numbers, always work from left to right.
- When writing whole numbers in word form, put a comma between each period.
- When writing whole numbers in numeral form, put a space between each period.
- Read decimal numbers by naming the place value of the final digit.

PRACTICE TEST

Represent the following numbers using place value charts, and give the value of each digit.

1. 6.597 13

2. 281.735

3. 62 403 972 185

4. Milo used a digital scale to weigh his pet turtle. He found that his turtle weighed 762.25 grams. In word form, how much does Milo's pet turtle weigh?

5. In the year 2005, the population of Baltimore was 635 815 people. Read the number 635 815, and write it using words.

Read each of the following numbers. Write each number using words.

6. 45 260 088 201

7. 23.11

8. 0.0903

Write each of the following in numeric form.

9. Twenty billion, four hundred thirty-one million, eighteen thousand, two hundred

10. Five thousand, thirty-nine hundred-thousandths

11. Six hundred sixty-eight and fourteen hundredths

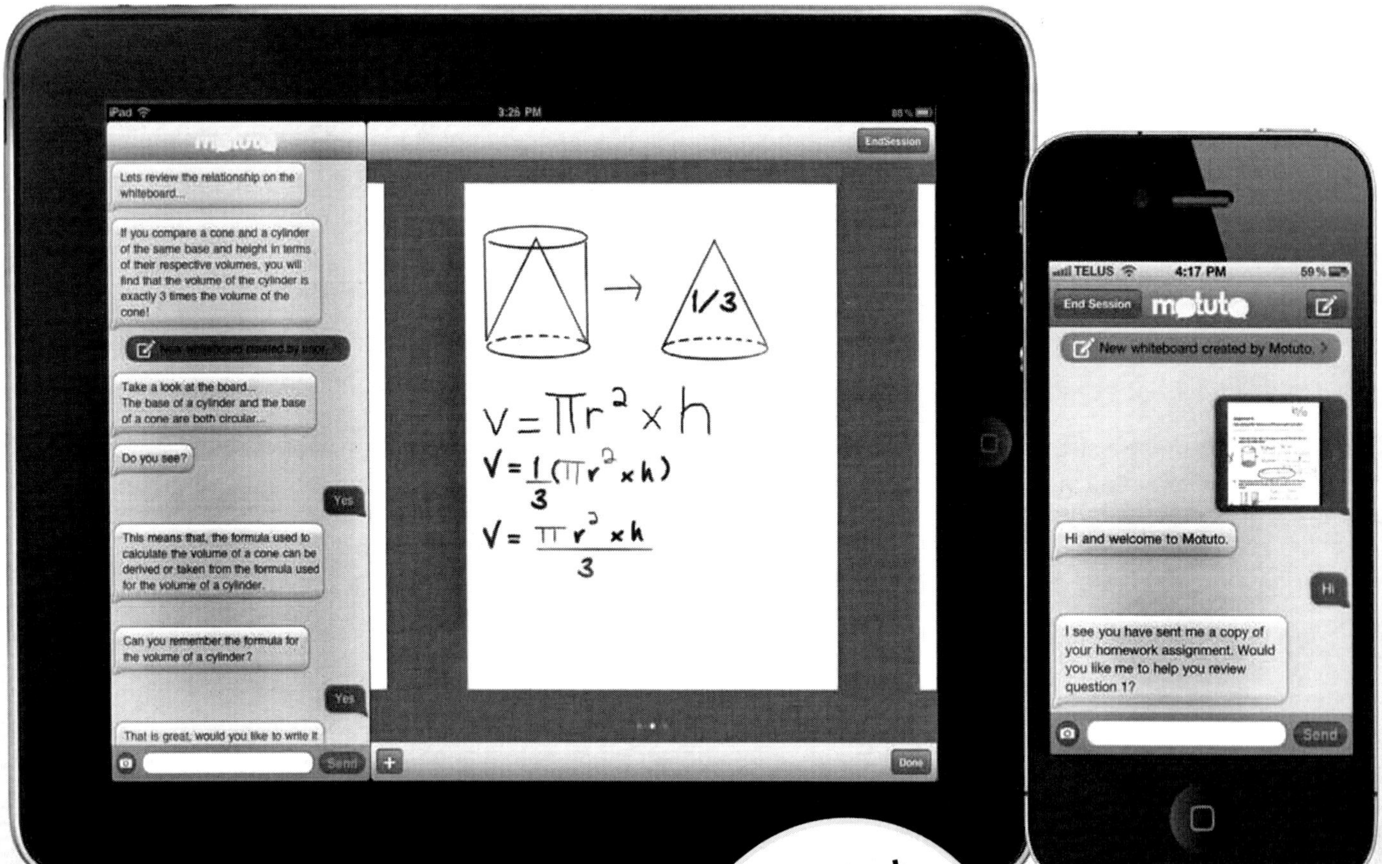

3:26 PM
EndSession
Lets review the relationship on the whiteboard...
If you compare a cone and a cylinder of the same base and height in terms of their respective volumes, you will find that the volume of the cylinder is exactly 3 times the volume of the cone!
Take a look at the board... The base of a cylinder and the base of a cone are both circular...
Do you see?
Yes
This means that, the formula used to calculate the volume of a cone can be derived or taken from the formula used for the volume of a cylinder.
Can you remember the formula for the volume of a cylinder?
Yes
That is great, would you like to write it
Send
Done
1/3
V = πr² x h
V = 1/3 (πr² x h)
V = πr² x h / 3
TELUS
4:17 PM
59 %
End Session
motuto
New whiteboard created by Motuto.
Hi and welcome to Motuto.
Hi
I see you have sent me a copy of your homework assignment. Would you like me to help you review question 1?
Send

WORKING WITH DECIMALS

When you are finished this unit, you will be able to…

- Multiply decimal numbers by one-digit whole numbers
- Divide decimal numbers by one-digit whole numbers
- Use estimation strategies to predict products and quotients of decimal numbers

Lesson	Page	Completed on
1. Multiplication of Decimal Numbers by Multipliers from 0 to 9	20	
2. Division of Decimal Numbers by Divisors from 1 to 9	22	
3. Estimation Strategies for Multiplication and Division of Decimal Numbers	25	
Review Summary	29	
Practice Test	30	
Answers and Solutions	at the back of the book	

PREREQUISITE SKILLS AND KNOWLEDGE

Prior to beginning this unit, you should be able to…

- Select and use an estimation strategy (front-end rounding, compensation, compatible numbers) for a given problem
- Select and use mental strategies and number properties to solve basic multiplication and division facts
- Use mental strategies to solve multiplication problems involving whole numbers
- Multiply two-digit whole numbers by two-digit whole numbers
- Divide three-digit whole numbers by one-digit whole numbers
- Add and subtract decimal numbers to thousandths

Lesson 1 MULTIPLICATION OF DECIMAL NUMBERS BY MULTIPLIERS FROM 0 TO 9

NOTES

Multiplying decimals by whole numbers is similar to multiplying whole numbers by whole numbers. Follow the same procedure used for multiplying whole numbers. Then, put the decimal point in its correct place in the solution.

To determine where the decimal point should be placed, count the number of digits to the right of the decimal point in the number you are multiplying. The solution should have the same number of digits to the right of the decimal point as there are in the decimal number being multiplied.

Example

The *product* is the result of multiplying two numbers together.

Calculate the product.

$$\begin{array}{r} 24.43 \\ \times \quad 2 \\ \hline \end{array}$$

Solution

Calculate the product as a whole number.

$$\begin{array}{r} 24\,43 \\ \times \quad 2 \\ \hline 48\,86 \end{array}$$

There are two digits to the right of the decimal point in 24.43. Therefore, there will be two digits to the right of the decimal point in the product.

4886

$$\begin{array}{r} 24.43 \\ \times \quad 2 \\ \hline 48.86 \end{array}$$

Time to Try 1

Calculate the product.

$$\begin{array}{r} 5.2213 \\ \times \quad 3 \\ \hline \end{array}$$

PRACTICE EXERCISES

Calculate the product.

1. $\begin{array}{r} 64.33 \\ \times \quad 3 \\ \hline \end{array}$

2. $\begin{array}{r} 200.1 \\ \times \quad 9 \\ \hline \end{array}$

3. $\begin{array}{r} 0.852 \\ \times \quad 2 \\ \hline \end{array}$

4. $\begin{array}{r} 3.1641 \\ \times \quad 7 \\ \hline \end{array}$

Use the following information to answer the next question.

> Sue solved a problem by multiplying 105.071 by 3: $105.071 \times 3 = 315213$

5. Place the decimal point in the product where it belongs. Explain your answer.

Use the following information to answer the next question.

> Tristan bought three chocolate bars at the school canteen. Each chocolate bar cost $1.35.

6. Use multiplication to calculate how much money Tristan spent on the chocolates.

Lesson 2 *DIVISION OF DECIMAL NUMBERS BY DIVISORS FROM 1 TO 9*

NOTES

Recall that each number in a division has a label.

```
                 1.4  ←—— Quotient
Divisor ——→  3)4.2  ←—— Dividend
                 3↓
                 1 2
                 1 2
                   0
```

Dividing decimal numbers by whole numbers is similar to dividing a whole number by a whole number. Follow the same procedure used when dividing whole numbers. Then, put the decimal in its correct place in the quotient

The decimal point in the quotient should be placed directly above the decimal point in the dividend. The solution should have the same number of digits to the right of the decimal point as there are in the number being divided.

Example

Calculate the quotient.

$3\overline{)37.68}$

Solution

Calculate the quotient as a whole number.

```
   12 56
3)37 68
  3
  07
   6
   1 6
   1 5
     18
     18
      0
```

NOTES

There are two digits to the right of the decimal point in 37.68. Therefore, there will be two digits to the right of the decimal point in the quotient.

1256

$$\begin{array}{r} 12.56 \\ 3\overline{)37.68} \end{array}$$

Time to Try 1

Calculate the quotient.

$3\overline{)7.851}$

PRACTICE EXERCISES

Calculate the quotient.

1. $6\overline{)15.06}$

2. $4\overline{)24.564}$

3. $8\overline{)205.6}$

4. $3\overline{)10.65}$

Use the following information to answer the next question.

Bob calculated $106.2 \div 9$ and got the quotient of 1.18.

5. Did Bob put the decimal point in the correct place? Justify your answer.

Use the following information to answer the next question.

Eight bags of gravel have a total mass of 33.12 kg. Each bag has the same mass.

6. Use division to calculate the weight of each bag of gravel.

Lesson 3 ESTIMATION STRATEGIES FOR MULTIPLICATION AND DIVISION OF DECIMAL NUMBERS

NOTES

Estimating the product or the quotient of a decimal number and a whole number is similar to estimating the product or the quotient of two whole numbers. Round the numbers to make the calculation easier and to get an approximate solution. It is then possible to describe the actual solution in terms of the estimated solution.

DROPPING THE DECIMAL DIGITS

One way to estimate a product or a quotient is to drop the decimal point and the digits to the right of the decimal point for the decimal number.

The estimated solution will always be a little less than the actual solution because the decimal number being divided or multiplied has been slightly reduced.

Example

Use the strategy of dropping the decimal digits to estimate the product.

$$\begin{array}{r} 18.234 \\ \times \quad 3 \\ \hline \end{array}$$

Solution

Drop the decimal point and the digits to the right of the decimal point.

$18.234 \rightarrow 18$

Perform the multiplication.

$$\begin{array}{r} 18 \\ \times\ 3 \\ \hline 54 \end{array}$$

The actual product of 18.234×3 is greater than 54.

Since 18.234 is greater than 18, the actual product is greater than the estimated product.

Time to Try 1

Use the strategy of dropping the decimal digits to estimate the quotient.

$2\overline{)36.64}$

NOTES

FRONT-END ESTIMATION

Another way to estimate the product is to use the front-end digit for the whole number part of the decimal number. Keep the digit with the highest place value and replace all the other digits with zero. Drop the decimal point and the digits to the right of the decimal point.

The estimated solution will usually be quite a bit lower than the actual solution, because the first number is often greatly reduced.

Example

Use the strategy of front-end estimation to estimate the quotient.

$4\overline{)124.073}$

Solution

Use the front-end digit of the decimal number.

124.073 → 100

Perform the division.

$$\begin{array}{r} 25 \\ 4\overline{)100} \\ \underline{8} \\ 20 \\ \underline{20} \\ 0 \end{array}$$

The actual quotient of 124.073 ÷ 4 is greater than 25.

Since 124.073 is greater than 100, the actual quotient will be greater than the estimated quotient.

Time to Try 2

Use the strategy of front-end estimation to estimate the product.

$$\begin{array}{r} 252.36 \\ \underline{\times \quad\quad 4} \end{array}$$

COMPATIBLE NUMBERS

A third estimation strategy involves changing the decimal number to a whole number that will be easier to multiply or divide. The new compatible number should be as close as possible to the original decimal number.

If the new compatible number is greater than the original number, the estimated solution will be greater that the actual solution.
If the new compatible number is less than the original number, the estimated solution will be less than the actual solution.

NOTES

For division, the compatible numbers [illegible] divide evenly [illegible] each other.

Example

Use the strategy of compatible numbers to estimate the product.

$$\begin{array}{r} 9.73 \\ \times \quad 5 \\ \hline \end{array}$$

Solution

Change the decimal number to a whole number that is compatible with multiplication by 5.

One possible method is to increase 9.73 to 10, which can easily be multiplied by any number.
$9.73 \rightarrow 10$

Perform the multiplication.

$$\begin{array}{r} 10 \\ \times 5 \\ \hline 50 \end{array}$$

The actual product of 9.73×5 is less than 50.

Since 9.73 is less than 10, the actual product is less than the estimated product.

Time to Try 3

Use the strategy of compatible numbers to estimate the quotient.

$3\overline{)11.65}$

PRACTICE EXERCISES

Use the strategy of dropping the decimal digits to estimate each solution.

1. $\begin{array}{r} 201.54 \\ \times \quad 6 \\ \hline \end{array}$

2. $7\overline{)56.951}$

Use the strategy of front-end estimation to estimate each solution.

3. $\begin{array}{r} 28.512 \\ \times \quad 9 \\ \hline \end{array}$

4. $4\overline{)8\,191.2}$

Use the strategy of compatible numbers to estimate each solution.

5. $\begin{array}{r} 41.56 \\ \times \quad 8 \\ \hline \end{array}$

6. $5\overline{)24.97}$

REVIEW SUMMARY

- Multiplying and dividing decimal numbers by one-digit whole numbers is similar to dividing whole numbers by whole numbers.
  - Ignore the decimal point and perform the calculation as with whole numbers.
  - Insert the decimal point into the solution. The solution should have the same number of digits to the right of the decimal point as there are in the decimal number being multiplied or divided.
- Three commonly used estimation strategies for predicting products and quotients are
  - dropping the decimal digits
  - using front-end estimation
  - using compatible numbers
- When estimating by dropping the decimal digits, drop the decimal point and the digits to the right of the decimal point to create a whole number.
- When estimating using front-end estimation, keep only the digit with the highest place value and replace all other digits with zero. Drop the decimal point and digits to the right.
- When estimating with compatible numbers, look for numbers that are related in some way and are easier to multiply or divide mentally. For division, the compatible numbers must divide evenly into each other.
- An actual product or quotient can be described in terms of the estimated product or quotient.

PRACTICE TEST

Calculate the product.

1. $$\begin{array}{r} 43.12 \\ \times \quad 3 \\ \hline \end{array}$$

2. $$\begin{array}{r} 1.615 \\ \times \quad 5 \\ \hline \end{array}$$

Calculate quotient.

3. $2\overline{)32.58}$

4. $4\overline{)9.124}$

5. Use the strategy of dropping the decimal digits to estimate the solution.

$$\begin{array}{r} 403.92 \\ \times \quad 7 \\ \hline \end{array}$$

6. Use the strategy of front-end estimation to estimate the solution.

$$\begin{array}{r} 7\,618.3 \\ \times \quad 5 \\ \hline \end{array}$$

7. Use the strategy of compatible numbers to estimate the solution.

$6\overline{)73.014}$

Use the following information to answer the next question.

Jacob solved a problem by multiplying 152.36 by 4: $152.36 \times 4 = 60944$

8. Place the decimal point in the product where it belongs. Explain your answer.

Use the following information to answer the next question.

Alesha calculated $29.4 \div 2$ and got the quotient of 1.47.

9. Did Alesha put the decimal point in the correct place? Justify your answer.

Use the following information to answer the next question.

Heather decided to buy seven binders for school. Each binder cost $3.91 including tax. She estimated the total cost of the binders and decided that $28.00 would be enough to pay for all seven binders.

10. Compare Heather's estimate to the actual cost of the seven binders. Did Heather make a reasonable estimate?

NOTES

INTEGERS

When you are finished this unit, you should be able to…

- Demonstrate and explain the meaning of integers concretely, pictorially, and symbolically

Lesson	Page	Completed on
1. Understanding Integers	34	
2. Integers on a Number Line	38	
3. Comparing Integers	42	
Review Summary	45	
Practice Test	46	
Answers and Solutions	at the back of the book	

PREREQUISITE SKILLS AND KNOWLEDGE

Prior to beginning this unit, you should be able to…

- Count by whole numbers
- Order numbers on a number line
- Understand the difference between negative and positive numbers

Lesson 1 UNDERSTANDING INTEGERS

NOTES

A whole number is a number that does not contain a decimal or a fraction.

An **integer** is any whole number that is positive or negative, including zero. Zero is neither negative nor positive. It is neutral.

A **positive integer** is a whole number that is greater than zero. Positive integers are usually marked with a (+) sign. If there is no sign in front of a number, it is assumed to be positive. For example, 2 is the same as +2.

A **negative integer** is a whole number that is less than zero. Negative integers are always marked with a (–) sign.

Integers are used often in everyday life.

TEMPERATURE

One of the most common uses of integers is to indicate temperature. Positive integers represent a temperature above the freezing point (above zero), while negative integers indicate a temperature below the freezing point (below zero).

A thermometer is an example of a number line that uses positive and negative integers.

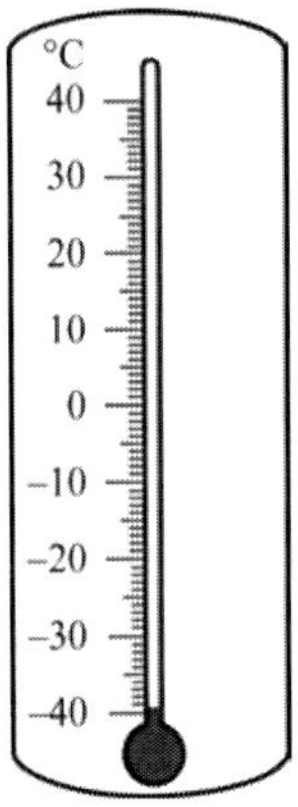

The positive integers are at the top of the thermometer, representing the warmer temperatures. The negative integers are at the bottom of the thermometer, representing the colder temperatures.

NOTES

Example

Maria is baking a cake. Would the temperature inside the oven be represented by a positive integer or a negative integer? Explain your answer.

Solution

When baking a cake, the oven would have to be very hot, so the temperature would be above zero.

Therefore, the temperature inside the oven would be represented by a positive integer.

Time to Try 1

Ashton is skating on a frozen lake. Would the temperature of the lake be represented by a positive integer or a negative integer? Explain your answer.

ELEVATION

Another common use of integers is to describe the elevation of an area as either above or below sea level.

If an area has an elevation of zero (0) it is *at sea level*. An elevation represented by a positive number is said to be *above sea level*. An elevation represented by a negative number is said to be *below sea level.*

Example

The Caspian Sea in Russia has an elevation of 28 m below sea level. Write an integer to represent this elevation.

Solution

Since the elevation is below sea level, the integer will be negative.

The elevation is represented by the integer –28 m.

NOTES

Time to Try 2

The peak of Windtower Mountain near Canmore, Alberta, has an elevation of 2 697 m above sea level. Write an integer to represent this elevation.

GOLF

It usually takes a certain number of golf strokes to complete a hole on a golf course. This number is called *par*. A score of par is represented by the number zero (0).

If a golfer uses more strokes than par, the golfer is said to be *over par*, and the score is a positive number. If the golfer uses fewer strokes than par, the golfer is said to be *under par*, and the score is a negative number.

Since golfers try to complete each hole in the fewest number of strokes, a negative score (under par) is actually a good score.

Example

Daniel's score at the end of a game of golf was –2. Was Daniel's score above par or below par? Explain your answer.

Solution

Daniel's score is a negative integer. Therefore, his score was two below par.

Time to Try 3

Jackie played golf and scored +1. Was Jackie's score above par or below par? Explain your answer.

PRACTICE EXERCISES

Identify whether each situation would be represented by a positive integer or a negative integer.

1. The elevation of a tall mountain

2. The temperature on a cold winter day

Write an integer to represent each of the following situations.

3. Ten degrees below zero

4. An elevation of 355 m below sea level

5. A golf score of four over par

Lesson 2 INTEGERS ON A NUMBER LINE

NOTES

A **number line** can be used to show the relationship between positive and negative integers.

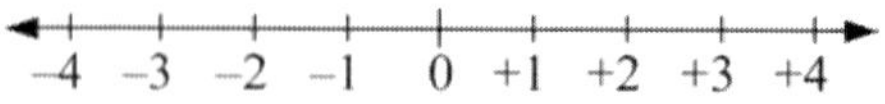

EXTENDING THE NUMBER LINE

The arrowheads on each end of a number line show that the integers continue forever in each direction.

Positive integers are located to the right of the zero on a number line. As you move to the right on a number line, the positive integers become larger.

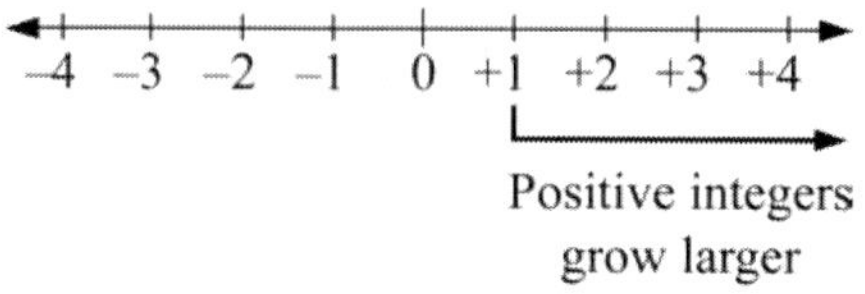

Example

Draw a number line to show the next integer larger than +4.

Solution

The integer +4 is positive. Positive integers get larger as you move to the right on a number line.

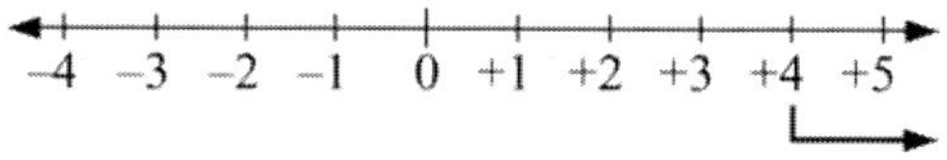

The integer +5 is larger than +4.

Time to Try 1

Draw a number line to show the next two integers larger than +6.

Negative integers are located to the left of the zero on a number line. As you move to the left on a number line, negative integers become smaller.

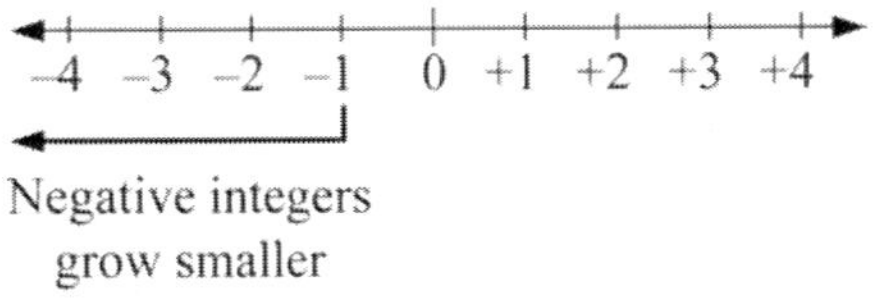

Example

Draw a number line to show the next integer smaller than –4.

Solution

The integer –4 is negative. Negative integers get smaller as you move to the left on a number line.

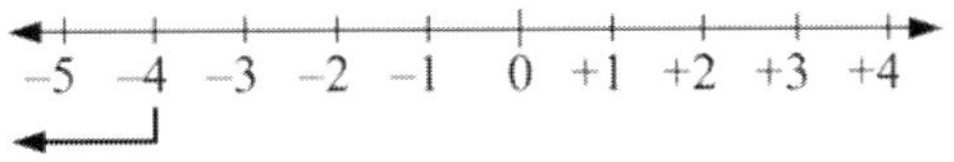

The integer –5 is smaller than –4.

Time to Try 2

Draw a number line to show the next two integers smaller than –6.

ORDERING INTEGERS ON A NUMBER LINE

Positive integers are always larger than negative integers because positive integers are greater than zero. When comparing two negative integers, the number that is closest to zero is the larger integer.

To compare integers using a number line, remember that the number farthest to the right of zero is the largest number. The number farthest to the left of zero is the smallest number.

NOTES

NOTES

Example

Using a number line, arrange the integers +4, –2, +1, –3, and 0 in order from smallest to largest (ascending order).

Solution

Draw a number line, and place the integers on it.

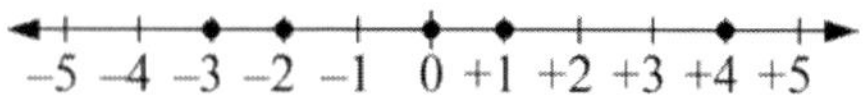

Of the given integers, –3 is the smallest because it is the farthest to the left of 0.

The integer –2 is larger than –3 because it is closer to 0.

The 0 is next. It is not positive or negative.

The integer +1 is larger than 0 because it is to the right of 0.

Of the given integers, +4 is the largest because it is farthest to the right of 0.

The integers in order from smallest to largest are –3, –2, 0, +1, and +4.

Time to Try 3

Using a number line, arrange the integers +1, +6, –3, +3, and –1 in order from smallest to largest (ascending order).

PRACTICE EXERCISES

1. Complete the given number line by filling in the missing integers.

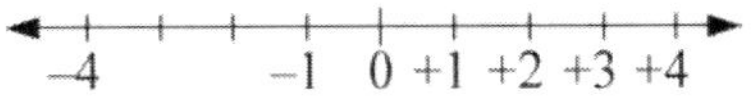

2. Draw a number line to show the next two integers smaller than –8.

Using a number line, arrange each set of integers in order from smallest to largest (ascending order).

3. –7, +3, +1, –1, and +5

4. –5, +4, 0, –3, and +6

5. +2, 0, +7, –2, and –4

Lesson 3 COMPARING INTEGERS

NOTES

There are three symbols commonly used to express the relationship between two integers.

The first integer can have a value greater than the second integer. This relationship can be expressed with a greater than (>) symbol. For example, $+6 > +2$.

The first integer can have a value less than the second integer. This relationship can be expressed with a less than (<) symbol. For example, $+1 < +3$.

The first integer can have the same value as the second integer. This relationship can be expressed with an equal to (=) symbol. For example, $-2 = -2$.

Example

Use the greater than (>), less than (<), or equal to (=) symbols to describe the relationship between each of the following pairs of integers.

a) +7, +6

Solution

When comparing two positive integers, the integer that is farther to the right of zero is the greater number.

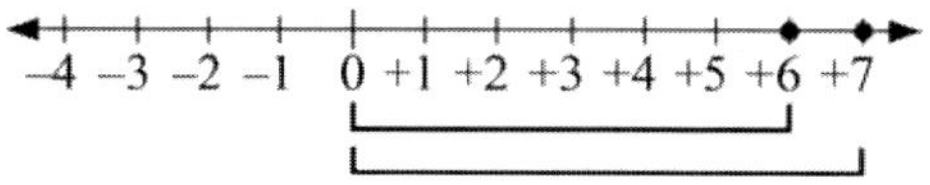

In this case, +7 is farther to the right of 0 than +6, so +7 is greater than +6.
$+7 > +6$

b) –5, –2

Solution

When comparing two negative integers, the integer that is farther to the left of zero is the smaller number.

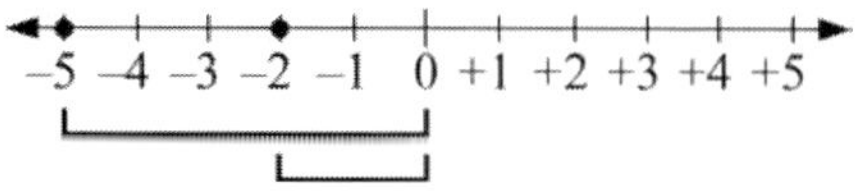

In this case, –5 is farther to the left of 0 than –2, so –5 is less than –2.
$-5 < -2$

NOTES

c) +3, –1

Solution

When comparing a positive integer with a negative integer, the positive integer is always greater than the negative integer.

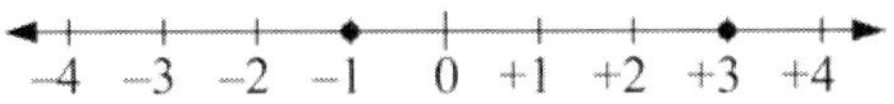

In this case, +3 is positive and –1 is negative, so +3 is greater than –1.

+3 > –1

d) 0, +4

Solution

Positive integers always have a value greater than zero.

Therefore, 0 is less than +4.

0 < +4

Time to Try 1

Use the greater than (>), less than (<), or equal to (=) symbols to describe the relationship between each of the following pairs of integers.

a) –1, –6

b) –3, +2

c) +4, +4

d) –5, 0

PRACTICE EXERCISES

Use the greater than (>), less than (<), or equal to (=) symbols to describe the relationship between each of the following pairs of integers.

1. –1, +4

2. +3, +5

3. –2, –6

4. –3, –3

5. +4, –5

REVIEW SUMMARY

- Integers are whole numbers, both positive and negative, including zero.
- A positive integer, which is represented by a (+) sign, is any whole number that is greater than zero.
- A negative integer, which is represented by a (–) sign, is any whole number that is less than zero.
- Zero is neither positive nor negative. It is neutral.
- The number line is labelled with integers in order of increasing value from left to right.
- Examples of integers that are used in everyday life include temperature, elevation, and golf scores.
- When comparing integers on a number line, the number farthest to the right is the largest.
- When comparing a set of integers, the symbol (>) is used to mean "greater than," the symbol (<) is used to mean "less than," and the symbol (=) is used to mean "equal to."

PRACTICE TEST

1. Complete the given number line by filling in the missing integers.

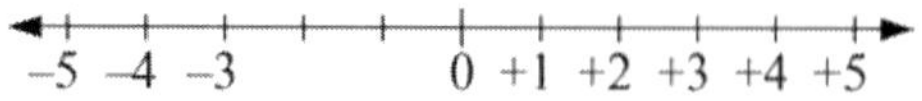

2. Camilla is eating a cherry Popsicle. Would the temperature of the Popsicle be represented by a positive integer or a negative integer? Explain your answer.

3. Write an integer to represent an elevation of 140 m below sea level.

4. Write an integer to represent a golf score of 3 over par.

Using a number line, arrange each set of integers in order from smallest to largest.

5. –3, 0, +5, –6, and –1

6. +4, –2, +1, –4, and +5

Use the greater than (>), less than (<), or equal to (=) symbols to describe the relationship between each of the following pairs of integers.

7. +2, +5

8. –3, –1

9. +1, –4

10. –6, –6

11. –5, +5

Use the following information to answer the next question.

Roger compared the integers –6 and +4. He decided that –6 is greater than +4 because 6 is farther away from zero on the number line than 4.

12. Was Roger's comparison correct? Explain your answer.

MULTIPLES AND FACTORS

When you are finished this unit, you will be able to…

- Explain why given numbers are prime or composite
- Sort prime and composite numbers
- Identify multiples, common multiples, and lowest common multiples of numbers less than 100
- Solve problems involving multiples
- Identify factors, common factors, and greatest common factors of numbers less than 100
- Solve problems involving factors
- Use factor trees and repeated division by prime numbers to determine prime factors of given numbers

Lesson	Page	Completed on
1. Multiples	50	
2. Prime and Composite Numbers	57	
3. Factors	63	
4. Prime Factorization	71	
Review Summary	75	
Practice Test	76	
Answers and Solutions	at the back of the book	

PREREQUISITE SKILLS AND KNOWLEDGE

Prior to beginning this unit, you should be able to…

- Relate multiplication to repeated addition
- Relate division to repeated subtraction
- Skip count up by one or two groups
- Skip count down by one or two groups
- Determine basic multiplication facts to 81 and related division facts

NOTES

Lesson 1 MULTIPLES

A **multiple** is the product of any given number and another whole number, except zero. For example, 8 is a multiple of 2 because $2 \times 4 = 8$.

STRATEGIES FOR INDENTIFYING MULTIPLES

The most common strategies for determining multiples are repeated addition, skip counting, and multiplication.

USING REPEATED ADDITION

Multiples of a given number can be identified by repeatedly adding the number to itself.

Example

Use the strategy of repeated addition to determine the first four multiples of the number 12.

Solution

To find four multiples, start with one 12 and add three more: $12+12+12+12$.

Repeated Addition	Multiples
	12
$12+12=24$	24
$24+12=36$	36
$36+12=48$	48

The first four multiples of 12 are 12, 24, 36, and 48.

Time to Try 1

Use the strategy of repeated addition to determine the first five multiples of the number 20.

NOTES

USING SKIP COUNTING

Multiples of a given number can be identified by skip counting by the given number.

Example

Use the strategy of skip counting to determine the first five multiples of the number 3.

Solution

To find five multiples, start with 3 and skip count by 3s for four more counts.

[3], 4, 5, [6], 7, 8, [9], 10, 11, [12], 13, 14, [15]

The first five multiples of 3 are 3, 6, 9, 12, and 15.

Time to Try 2

Use the strategy of skip counting to determine the first four multiples of the number 4.

USING MULTIPLICATION

Multiples of a given number can be identified by multiplying that given number by any other whole number, except 0. The products of the two numbers are the multiples of the given number.

Example

Use the strategy of multiplication to determine the first five multiples of the number 4.

Solution

Multiply 4 by 1, then by 2, and so on for five counts.

Multiplication	Multiples
$4 \times 1 = 4$	4
$4 \times 2 = 8$	8
$4 \times 3 = 12$	12
$4 \times 4 = 16$	16
$4 \times 5 = 20$	20

The first five multiples of the number 4 are 4, 8, 12, 16, and 20.

NOTES

Time to Try 3

Use the strategy of multiplication to determine the first four multiples of the number 6.

IDENTIFYING COMMON MULTIPLES

A common multiple is a number that is a multiple of two or more numbers.

Example

List the first three common multiples of 2 and 3.

Solution

Step 1

List the multiples of both numbers.

- The multiples of 2 are 2, 4, 6, 8, 10, 12, 14, 16, 18,…
- The multiples of 3 are 3, 6, 9, 12, 15, 18,…

Step 2

Identify the multiples that are common to both lists.

2, 4, [6], 8, 10, [12], 14, 16, [18]

3, [6], 9, [12], 15, [18]

The first three common multiples of the numbers 2 and 3 are 6, 12, and 18.

Time to Try 4

List the first two common multiples of 6 and 9. Show your work.

IDENTIFYING LOWEST COMMON MULTIPLES

The least or lowest common multiple (LCM) of two or more numbers is the common multiple that has the lowest value.

Example

Identify the lowest common multiple of the numbers 4 and 6.

Solution

Step 1

Identify the common multiples of the two numbers.

4, 8, $\boxed{12}$, 16, 20, $\boxed{24}$, 28,…

6, $\boxed{12}$, 18, $\boxed{24}$, 30,…

The common multiples of 4 and 6 are 12 and 24.

Step 2

Identify the lowest common multiple (LCM).

Since $12 < 24$, 12 is the lowest common multiple of 4 and 6 because it has the least value.

Time to Try 5

Identify the lowest common multiple of the numbers 10 and 15.

NOTES

SOLVING PROBLEMS USING MULTIPLES

Some problems can be solved by identifying common multiples or by determining the lowest common multiple.

Example

Jenny and Stella join a running club for the month of September, which has 30 days.

- They both run on the first day of September.
- Jenny then only runs every 3rd day of the month.
- Stella then only runs every 4th day of the month.

On which other days will the two girls both run on the same day?

Solution

This type of problem can easily be solved by determining the common multiples of 3 and 4 up to 30 (the number of days in September).

Multiples of 3: 3, 6, 9, [12], 15, 18, 21, [24], 27, 30,...
Multiples of 4: 4, 8, [12], 16, 20, [24], 28,...

After the first day of September when both girls started running, they will run on the same day only two more times: September 12th and September 24th.

Time to Try 6

One timer in a factory beeps every 15 minutes. Another timer in the same factory beeps every 20 minutes. If both timers are set to start at the same time, when will the two timers first beep at the same time?

PRACTICE EXERCISES

1. Identify the first four multiples of 21. Explain the strategy you used to identify them.

2. Use a different strategy to identify the first four multiples of 25. Explain the strategy you used to identify them.

3. Determine the common multiples of 4 and 12 up to the number 40. Show your work.

4. Identify the lowest or least common multiple (LCM) of the numbers 2 and 6. Show your work.

Use the following information to answer the next question.

Bob and Brett jog around an oval track. They start at the same starting line, at the same time. Bob completes the first lap in 5 minutes, and Brett completes the first lap in 7 minutes.

5. If Bob and Brett continue to jog at the same speeds, how many minutes will it take before they both arrive at the starting point together?

Use the following information to answer the next question.

Marnie is buying fruit bars and granola bars for the school picnic. There are 8 fruit bars in each package and there are 12 granola bars in each package. Marnie wants to buy the same number of fruit bars and granola bars.

6. How many packages of fruit bars and granola bars does Marnie need to buy?

Lesson 2 PRIME AND COMPOSITE NUMBERS

NOTES

Whole numbers greater than 0 can be described as either prime or composite. A number is classified as prime or composite based on the number of factors it has.

A **factor** is a number that is multiplied by another number to get a product.

factor		factor		product
3	×	4	=	12

PRIME NUMBERS

A prime number is any number that has exactly two factors: the number itself and one. A prime number can only be divided equally by itself or the number 1.

Example

Explain why the number 2 is a prime number.

Solution

The number 2 is a prime number because it has only two factors: 1 and itself.

$2 \times 1 = 2$
$1 \times 2 = 2$

It can only be divided equally by 1 or itself.

$2 \div 1 = 2$
$2 \div 2 = 1$

Time to Try 1

Write a different one-digit prime number. Explain why it is a prime number.

NOTES

COMPOSITE NUMBERS

A composite number is any number with more than two factors. The product can result from different combinations of factors being multiplied by each other.

Example

Explain why the number 16 is a composite number.

Solution

The number 16 is a composite number because it has more than two factors.

Think of all the whole numbers that can be multiplied by each other to result in the product of 16.

$1 \times 16 = 16$

$2 \times 8 = 16$

$4 \times 4 = 16$

The number 16 has five factors: 1, 2, 4, 8, and 16.

Time to Try 2

Write a one-digit composite number. Explain why it is a composite number.

ZERO AND ONE—NEITHER PRIME NOR COMPOSITE

The numbers 0 and 1 are neither prime nor composite numbers.

Since 0 does not have any factors, and 1 has only one factor ($1 \times 1 = 1$), 0 and 1 cannot be described as prime (exactly two factors) or composite (more than two factors).

Example

Use the definitions of prime numbers and composite numbers to explain why 1 is neither prime nor composite.

Solution

A prime number is a whole number that has exactly two factors, 1 and the number itself. A composite number is a whole number with two or more factors.

Therefore, the number 1 is neither prime nor composite because it has only one factor, and that is 1.

NOTES

Time to Try 3

Use the definitions of prime and composite numbers to explain why 0 is neither prime nor composite.

SORTING PRIME AND COMPOSITE NUMBERS

Following are some helpful hints when sorting a set of whole numbers as either prime or composite.

- The number 2 is the only even number that is a prime number.
- All even numbers except 2 are composite numbers.
- Composite numbers can be even or odd.
- The numbers 0 and 1 are neither prime nor composite.

Example

A teacher wrote this set of numbers on the whiteboard at the front of the classroom: 2, 5, 9, 12, 13, 21.

Sort the given numbers into two groups: prime numbers and composite numbers. Explain your work.

Solution

Step 1
Identify the prime numbers.

The numbers 2, 5, and 13 are prime numbers because they each have only two factors: 1 and the number itself.

- Factors of 2: 1, 2
- Factors of 5: 1, 5
- Factors of 13: 1, 13

Step 2
Identify the composite numbers.

The numbers 9, 12, and 21 are composite numbers because they each have more than two factors.

- Factors of 9: 1, 3, 9
- Factors of 12: 1, 2, 3, 4, 6, 12
- Factors of 21: 1, 3, 7, 21

NOTES

Time to Try 4

Sort the given set of numbers as prime or composite.
43, 25, 44, 12, 27, 14, 19, 8, 10

Prime Numbers	Composite Numbers

PRACTICE EXERCISES

Identify the following numbers as prime or composite. Explain your ans

1. 49

2. 53

3. Use the definitions of prime and composite numbers to explain why 1 is neit nor a composite number.

4. Write a number between 20 and 30. Explain whether the number you wrote is a prime or composite number.

Use the following information to answer the next question.

14, 17, 36, 41, 28, 57, 23, 7, 62

5. Sort the given set of numbers as prime and composite.

Prime Numbers	Composite Numbers

6. Sort the numbers from 0 to 20 as prime, composite, or neither.

Prime Numbers	Composite Numbers	Neither Prime Nor Composite

Lesson 3 FACTORS

NOTES

A **factor** is a number that is multiplied by another number to get a product.

6	$\times$	2	$=$	12
↑ Factor		↑ Factor		↑ Product

Factors are not considered to be distinct if the order of the factors is changed. The numbers 6 and 2 are factors of 12, regardless of whether the multiplication fact is written as 6×2 or 2×6.

Every number has the number 1 and itself as factors.

Following are two strategies that can be used to determine all the factors of a given number.

USING CONCRETE OR VISUAL REPRESENTATIONS TO DETERMINE ALL FACTORS

An **array** is a concrete or visual representation of a number. An array uses rows and columns in the form of a rectangle to show the factors. Arranging square tiles or drawing squares will show all the factors of a given number.

Example

Use square tiles or draw squares in a rectangular form to illustrate all the factors of the number 12.

Solution

Step 1
Draw or model arrays for the number 12.

Think of all the ways to arrange 12 same-sized squares into a rectangle.

The following arrays show three sets of numbers that result in a product of 12 when multiplied by each other.

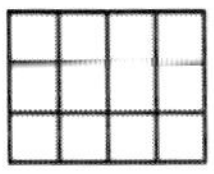
$3 \times 4 = 12$

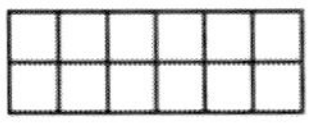
$2 \times 6 = 12$

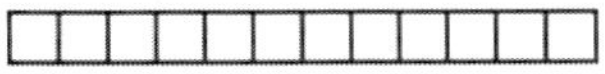
$1 \times 12 = 12$

NOTES

Step 2
Identify all the factors of the number 12.

The number 12 has six factors: 1, 2, 3, 4, 6, and 12.

Time to Try 1

Use square tiles or draw arrays to illustrate all the factors of the number 24.

USING MULTIPLICATION FACTS TO DETERMINE ALL FACTORS

When determining all the factors of a given number, think of all the whole numbers that can be multiplied by each other to result in the given product.

- Remember that factors are not considered to be distinct if the order of the factors has changed.
- Remember that every number has the number 1 and itself as factors.

Example

Use multiplication to determine all the factors of the number 16.

Solution

Step 1
Think of all the factors that result in 16.

Start with 1 and work your way up.
$1 \times 16 = 16$
$2 \times 8 = 16$
$4 \times 4 = 16$

Step 2
List the factors.

The number 16 has five factors: 1, 2, 4, 8, and 16.

Note that if the number is repeated, like 4 (4×4), it is only listed once.

NOTES

Time to Try 2

Use multiplication to determine all the factors of 30.
Show your work.

IDENTIFYING A FACTOR

One way to identify if a given number is a factor of another number is to think division. If the larger number can be divided evenly by the smaller number, then the smaller number is a factor of the larger number.

Example

Is 8 a factor of 40? Use division to explain your answer.

Solution

Yes, 8 is a factor of 40 because 40 can be divided evenly by 8.
$40 \div 8 = 5$

Time to Try 3

Is 3 a factor of 99? Use division to explain your answer.

DETERMINING COMMON FACTORS

Common factors are factors that are the same for two or more numbers. To determine the common factors, follow these steps.

1. List the factors of the first number.
2. List the factors of the second number.
3. Compare the two lists of factors to identify the numbers that appear on both lists.

A visual representation like a Venn diagram is an organized way of showing common factors. A Venn diagram is made up of two or more circles that overlap. It can show the factors of two or more numbers. The part of the circles that overlap shows the common factors of the given numbers. The areas that do not overlap show the factors that are not common to the given numbers.

NOTES

Example

Show the factors of 24 and 28 in a Venn Diagram. What are the common factors?

Solution

Step 1
Determine the factors of the two numbers.

The factors of 24 are 1, 2, 3, 4, 6, 8, 12, 24.
The factors of 28 are 1, 2, 4, 7, 14, 28.

Step 2
Identify the common factors.
[1], [2], 3, [4], 6, 8, 12, 24

[1], [2], [4], 7, 14, 28

The common factors are 1, 2, and 4.

Step 3
Enter the factors into the Venn diagram.

Write the factors of 24 in the left circle. Write the factors of 28 in the right circle. Write the common factors in the overlapped part of the two circles.

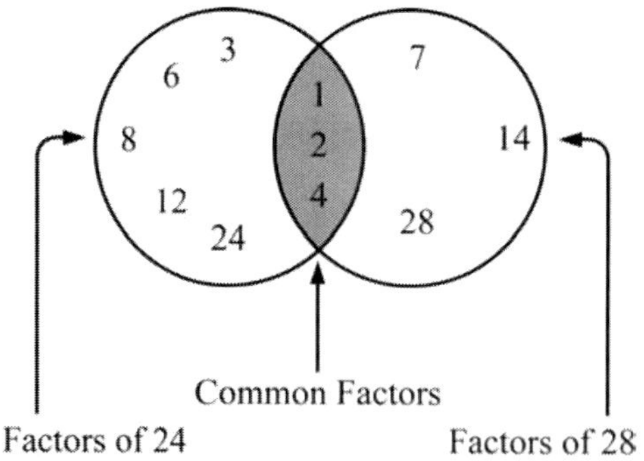

Time to Try 4

Show the factors of 28 and 32 in a Venn Diagram. What are the common factors?

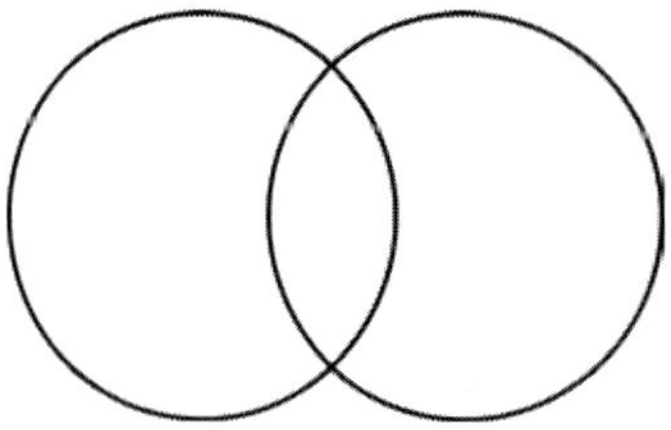

IDENTIFYING THE GREATEST COMMON FACTOR

NOTES

The **greatest common factor** (GCF) is the greatest factor that two or more numbers have in common.

To determine the greatest common factor (GCF), follow these steps:

1. List the factors of the first number.
2. List the factors of the second number.
3. Compare the two lists of factors to identify the numbers that appear on both lists.
4. Select the common factor that has the greatest value.

Example

What is the greatest common factor for the numbers 18 and 24?

Solution

Step 1
Identify the factors of 18.
$1 \times 18 = 18$
$2 \times 9 = 18$
$3 \times 6 = 18$

The factors are 1, 2, 3, 6, 9, 18

Step 2
Identify the factors of 24.
$1 \times 24 = 24$
$2 \times 12 = 24$
$3 \times 8 = 24$
$4 \times 6 = 24$

The factors are 1, 2, 3, 4, 6, 8, 12, 24

Step 3
Identify the common factors.

Compare the two sets of factors to see which factors appear on both lists.

[1], [2], [3], [6], 9, 18

[1], [2], [3], 4, [6], 8, 12, 24

The common factors are 1, 2, 3, and 6.

Step 4
Identify the greatest common factor.

Since $6 > 3 > 2 > 1$, 6 is the greatest common factor of 18 and 24.

NOTES

Time to Try 5

What is the greatest common factor of 15 and 20? Show your work.

SOLVING PROBLEMS INVOLVING FACTORS

Some problems can be solved by determining the greatest common factor. Problems like these usually ask for the greatest number of something, and involve equal groupings.

Example

A florist has 90 red roses and 60 white roses to place into vases for selling. The florist places an equal number of red roses into each vase he uses. He then places an equal number of white roses in with the red roses.

Use the greatest common factor to determine the greatest number of vases the florist can use.

Solution

Step 1
Identify the common factor of 90 and 60.

Factors of 90: [1], [2], [3], [5], [6], 9, [10], [15], 18, [30], 45, 90
Factors of 60: [1], [2], [3], 4, [5], [6], [10],12, [15], 20, [30], 60

Step 2
Identify the greatest common factor.

The greatest common factor of 90 and 60 is 30.

Therefore, the greatest number of vases the florist can use is 30 vases.

Each one of the 30 vases will hold 3 red roses ($90 \div 30 = 3$) and 2 white roses ($60 \div 30 = 2$).

Time to Try 6

Brenna is making loot bags for a party. She has 32 candies and 40 pencils to distribute equally among the loot bags. Use the greatest common factor to determine the greatest number of loot bags Brenna can make.

PRACTICE EXERCISES

1. Draw or model arrays to determine all the factors of the number 16.

2. Use multiplication to determine all the factors of the number 48.

3. Is 4 a factor of 140? Use division to explain your answer.

4. Show the factors of 18 and 24 in a Venn diagram. What are the common factors?

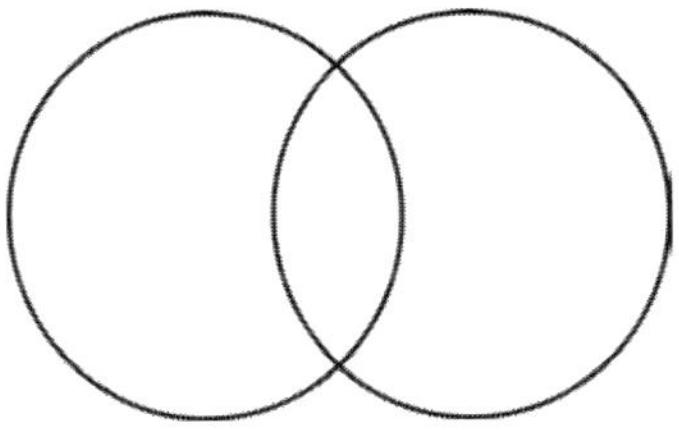

5. What is the greatest common factor for the numbers 20 and 35?

Use the following information to answer the next two questions.

Devon has 25 quarters and 35 dimes. He plans to make groups of quarters and dimes, where each group has an equal number of quarters, and each group also has an equal number of dimes.

6. Use the greatest common multiple to determine the greatest number of groups Devon can make.

7. How many quarters and dimes will be in each group?

Lesson 4 PRIME FACTORIZATION

NOTES

Prime factorization is the process of breaking down a number into its prime factors. Any composite number can be factored into prime factors.

No matter which method is used to find the prime factorization of a number, the prime factors will be the same. However, the prime factors may appear in a different order.

USING FACTOR TREES

One way to determine the prime factors of a number is to draw a **factor tree**. Each branch of a factor tree leads to a factor of the given number. Decompose each factor in the tree until all the bottom branches show a prime number.

Using a visual representation like a factor tree sometimes makes it easier to determine prime factors of a given number.

A composite number can have more than one form of factor tree. The form of the tree is different depending on which factors are used and the order in which they are used. However, the end result of prime factorization is always identical for the same number.

Example

Use a factor tree to help identify the prime factors of the number 18.

Solution

Step 1
Start drawing the factor tree.

Begin with the number 18 at the top of the tree. Draw two branches that will each lead to a factor of the number 18.

Two factors of 18 are 9 and 2 ($9 \times 2 = 18$). Place these factors in the factor tree.

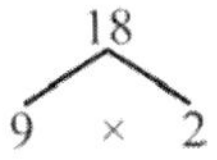

NOTES

Step 2
Change the composite number into prime factors.

Two factors of the composite number 9 are 3 and 3 $(3\times3=9)$.

Draw two branches out of the number 9 and place the factors 3 and 3 in the factor tree.

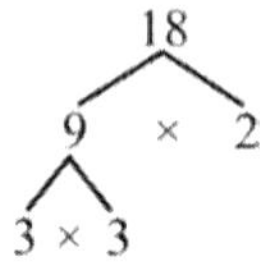

The numbers at the bottom of the branches are all prime numbers, so the prime factorization is complete.

The prime factorization of 18 can be shown as the expression $3\times3\times2$.

Step 3
Use a different factor tree.

Another factor tree can look like this.

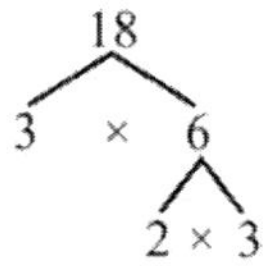

Even though different factors are used, the prime factorization is the same, just expressed in a different order.
$3\times2\times3$

Time to Try 1

Use a factor tree to help identify the prime factors of the number 36.

NOTES

USING REPEATED DIVISION BY PRIME NUMBERS

Another method used to find the prime factors of a given number is to repeatedly divide the given number by prime numbers.

Example

Use repeated division to help identify the prime factors of the number 60.

Solution

Step 1
Divide the number 60 by the smallest prime number (2) that divides evenly.

$$2\overline{)60}^{\,30}$$

Step 2
Divide 30 by the smallest prime number (2) that divides evenly.

$$\begin{array}{l} \quad 15 \\ 2\overline{)30} \\ 2\overline{)60} \end{array}$$

Step 3
Divide 15 by the smallest prime number (3) that divides evenly.

$$\begin{array}{l} \quad 5 \\ 3\overline{)15} \\ 2\overline{)30} \\ 2\overline{)60} \end{array}$$

Since 5 is a prime number, the factorization is complete.

The prime factorization of 60 is $2 \times 2 \times 3 \times 5$.

Time to Try 2

Use repeated division to help identify the prime factors of the number 45.

PRACTICE EXERCISES

1. Use a factor tree to help identify the prime factors of the number 42.

2. Use repeated division to help identify the prime factors of the number 100.

3. Identify the prime factors of 81. Explain the strategy you used.

Use the following information to answer the next question.

Evelyn decomposed the number 45 into prime factors using repeated division as follows:

$$9\overline{)45}\quad\text{quotient } 5$$

5. Is Evelyn's solution correct? Explain your answer.

REVIEW SUMMARY

- A multiple is the product of any given number and another whole number except zero.
  - Some strategies for determining the multiples of a given number are repeated addition, skip counting, and multiplication.
  - A common multiple is a number that is a multiple of two or more numbers.
  - The least or lowest common multiple (LCM) of two or more numbers is the common multiple that has the least or lowest value.
- A prime number has only two factors: 1 and itself.
- A composite number has more than two factors.
- The number 0 is neither prime nor composite because it has no factors.
- The number 1 is neither prime nor composite because it has only one factor: 1.
- A factor is a number that is multiplied by another number to result in a product.
  - Some strategies for determining all the factors of a given number are multiplication and using visual representations like arrays.
  - Common factors are factors that are the same for two or more numbers.
  - The greatest common factor (GCF) is the greatest factor that two numbers [illegible].
- Prime factorization is the process of breaking down a product into its prime [illegible]
- The prime factors of a composite number can be determined by using a visual [illegible] entation like a factor tree. A composite number can have more than one form of factor [illegible] ecause it has multiple factors.
- The prime factors of a composite number can also be determined by using repeated [illegible] ivision by the smallest prime numbers.

PRACTICE TEST

1. Use the strategy of repeated addition to determine the first five multiples of the number 13.

2. Use the strategy of skip counting to determine the first four multiples of the number 4.

3. Use the strategy of multiplication to determine the first five multiples of the number 15.

4. List the first three common multiples 3 and 4.

5. Identify the lowest common multiple of the numbers 3 and 6.

Use the following information to answer the next question.

Jenna is decorating cupcakes for a bake sale. She decides to put a red candy on every 4th cupcake, and a yellow candy on every 5th cupcake.

6. What cupcake will be the first to get both a red and yellow candy?

7. Explain why the number 7 is a prime number.

8. Use the definitions of prime numbers and composite numbers to explain why 1 is neither prime nor composite.

Use the following information to answer the next question.

3, 31, 12, 28, 39, 46, 5, 22, 17

9. Sort the given set of numbers as prime or composite.

Prime Numbers	Composite Numbers

10. Use arrays to determine all the factors of the number 20.

11. Use multiplication to determine all the factors of the number 36.

12. Is 3 a factor of 98? Use division to explain your answer.

13. Determine the common factors of 30 and 36.

14. What is the greatest common factor for the numbers 12 and 15? Show your work.

Use the following information to answer the next question.

Mrs. Palmer is dividing up her art supplies. She has 48 crayons and 42 markers. She wants to divide the crayons and markers into groups so that each group has the same number of crayons and the same number of markers.

15. What is the greatest number of groups Mrs. Palmer can make?

16. Use a factor tree to determine the prime factors of the number 24.

17. Use repeated division to determine the prime factors of the number 66.

IMPROPER FRACTIONS AND MIXED NUMBERS

When you are finished this unit, you should be able to…

- Explain, using a model, that an improper fraction represents a number greater than 1
- Express improper fractions as mixed numbers
- Express mixed numbers as improper fractions
- Translate a given improper fraction between concrete (models), pictorial, and symbolic forms
- Translate a given mixed number between concrete (models), pictorial, and symbolic forms
- Place a set of fractions, including mixed numbers and improper fractions, on a number line

Lesson	Page	Completed on
1. Relating Improper Fractions and Mixed Numbers	80	
2. Expressing Improper Fractions and Mixed Numbers	87	
3. Ordering Fractions on a Number Line	92	
Review Summary	96	
Practice Test	97	
Answers and Solutions	at the back of the book	

PREREQUISITE SKILLS AND KNOWLEDGE

Prior to beginning this unit, you should be able to…

- Create sets of equivalent fractions
- Identify equivalent fractions for given fractions
- Compare fractions with like denominators
- Compare two fractions with unlike denominators by creating equivalent fractions
- Place a set of fractions with like and unlike denominators on a number line

Lesson 1 RELATING IMPROPER FRACTIONS AND MIXED NUMBERS

NOTES

Some fractions have a numerator that is greater than the denominator. These fractions are named improper fractions. Some fractions include whole numbers. These fractions are named mixed numbers. Improper fractions are related to mixed numbers.

MODELLING IMPROPER FRACTIONS

An improper fraction has a numerator that is greater than or equal to the denominator. An improper fraction always has a value greater than or equal to 1.

The following model demonstrates how the improper fraction of $\frac{5}{4}$ represents a number greater than 1.

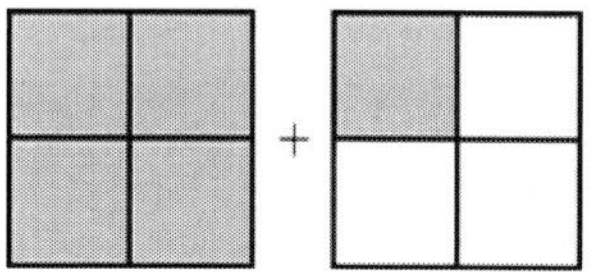

The improper fraction of $\frac{5}{4}$ represents four parts out of four parts shaded (one whole) and one part shaded out of four parts (4 shaded + 1 shaded = 5 shaded).

When modelling a given improper fraction, the denominator of the fraction tells how many equal parts each shape should have. Divide the shapes into that number of equal parts. The numerator tells how many parts to shade.

Example

Model or draw the improper fraction of $\frac{7}{4}$.

Solution

The denominator (4) shows that each shape used needs to be divided into four equal parts.

The numerator (7) shows that seven of the parts must be shaded (one whole and three parts of the other shape).

The model can look like this:

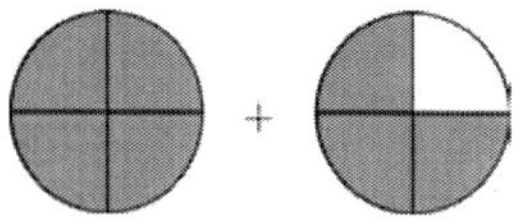

NOTES

Time to Try 1

Model or draw the improper fraction of $\frac{9}{8}$.

TRANSLATING BETWEEN CONCRETE, PICTORIAL, AND SYMBOLIC FORMS

To write the symbolic form of a concrete model or pictorial diagram, count the total number of parts that are shaded. That number becomes the numerator of the improper fraction. Count the number of equal parts in the model or diagram. The number of equal parts is the denominator.

Example

Translate the given model of an improper fraction into its symbolic form.

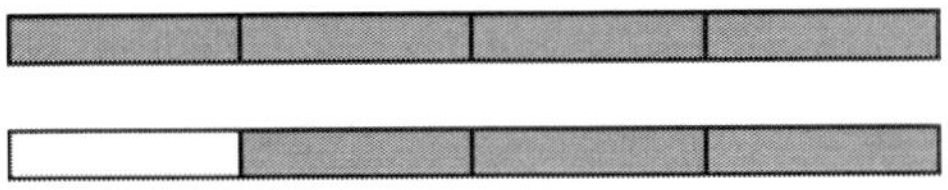

Solution

Step 1
Determine the numerator.

Count all the parts that are shaded (from both strips). There are 7 parts that are shaded. This number will be the numerator.

Step 2
Determine the denominator.

The strips are divided into four equal parts, so the denominator of the improper fraction is 4.

The symbolic form of the given model is $\frac{7}{4}$.

Time to Try 2

Translate the given model of an improper fraction into its symbolic form.

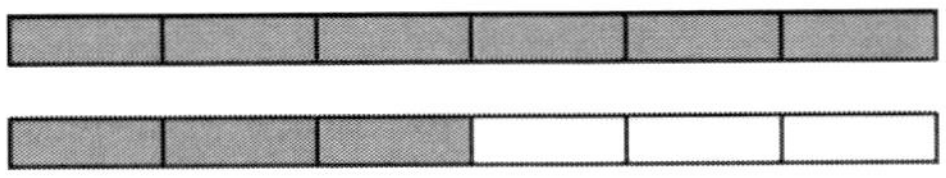

NOTES

MODELLING MIXED NUMBERS

A mixed number consists of a whole number and a proper fraction. Recall that a proper fraction has a numerator that is less than the denominator and represents a part of a whole.

The following model demonstrates how the mixed number of $1\frac{1}{4}$ is made up of the whole $\left(\frac{4}{4}=1\right)$ and the proper fraction of $\frac{1}{4}$.

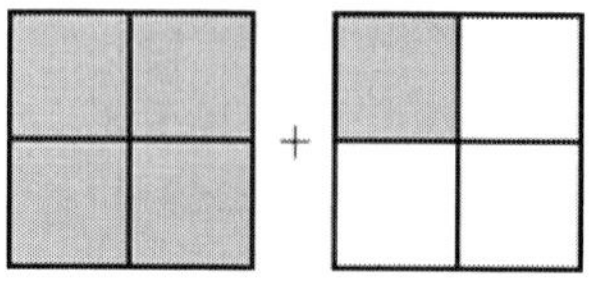

When modelling a given mixed number, the denominator of the fraction shows how many equal parts each shape should have. Divide the shapes into that number of equal parts. The whole number tells to shade all the parts of the whole. The numerator shows how many parts to shade in the partially shaded shape.

Example

Model or draw the mixed number of $3\frac{7}{10}$.

Solution

The denominator (10) shows that each shape used needs to be divided into 10 equal parts.

The whole number (3) shows that there are three whole shapes that are all shaded.

The numerator (7) shows that seven of the ten parts of the fourth shape must be shaded.

The model can look like this:

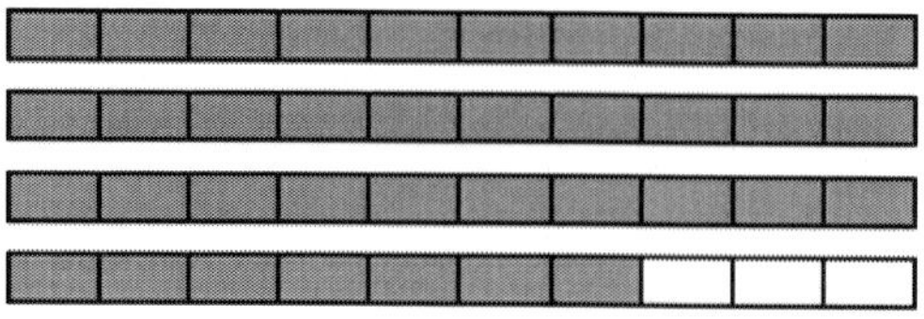

NOTES

Time to Try 3

Model or draw the mixed number of $4\frac{2}{3}$.

TRANSLATING BETWEEN CONCRETE, PICTORIAL, AND SYMBOLIC FORMS

To write the symbolic form of a concrete model or pictorial diagram, count the total number of whole shapes that are shaded. That number becomes the whole number of the mixed number.

Count the number of shaded parts in the partially shaded shape. That number becomes the numerator of the proper fraction. Count the number of equal parts in the model or diagram. The number of equal parts is the denominator of the proper fraction.

Example

Translate the given pictorial representation of a mixed number into its symbolic form.

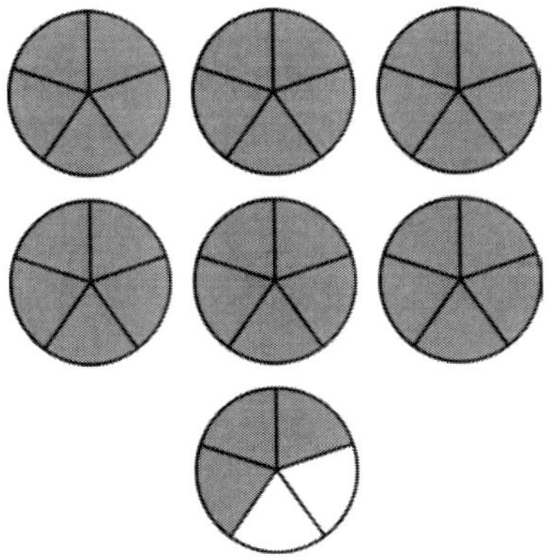

Solution

Step 1

Determine the whole number.

Count the number of shapes that are completely shaded. There are 6 completely shaded shapes, so 6 is the whole number part of the mixed number.

NOTES

Step 2
Determine the numerator.

Count the number of shaded parts in the partially shaded shape. Since three parts are shaded, the numerator of the proper fraction is 3.

Step 3
Determine the denominator.

Count the number of equal parts of the shapes. There are five equal parts in each shape, so the denominator is 5.

The symbolic form of the pictorial mixed number is $6\frac{3}{5}$.

Time to Try 4

Translate the given model of a mixed number into its symbolic form.

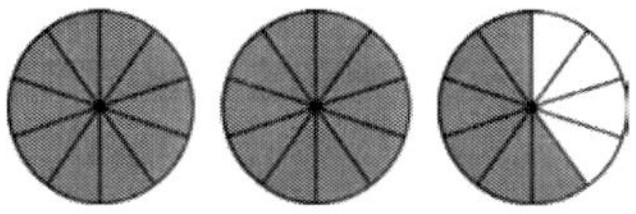

PRACTICE EXERCISES

1. Model or draw a pictorial form of the improper fraction $\frac{8}{5}$.

Use the following information to answer the next question.

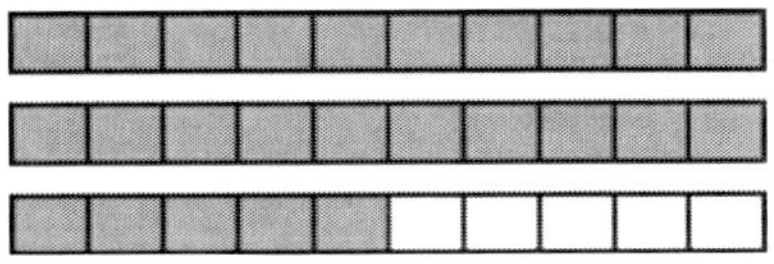

2. Translate the given model of an improper fraction into its symbolic form.

Use the following information to answer the next question.

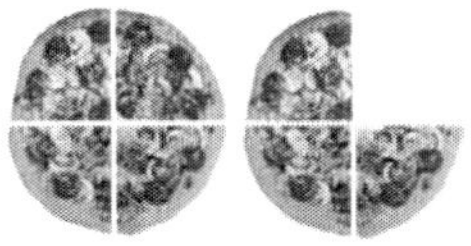

3. Translate the given pictorial representation of an improper fraction into a model form.

4. Model or draw a pictorial form of the mixed number $1\frac{3}{4}$.

5. Model or draw a pictorial form of the mixed number $2\frac{2}{6}$.

Use the following information to answer the next question.

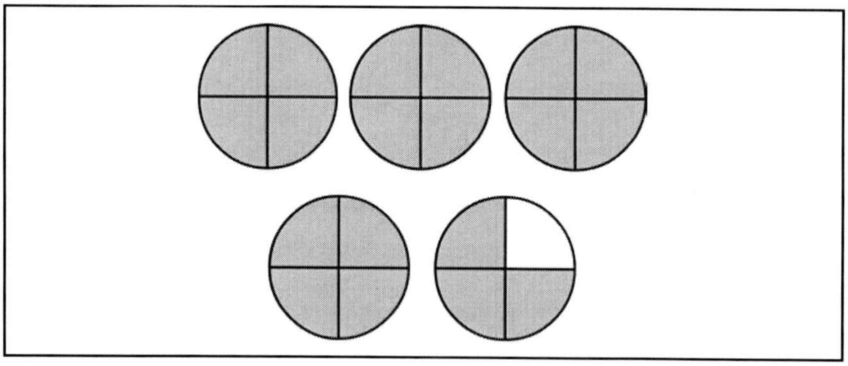

6. Translate the given model of a mixed number into its symbolic form.

Lesson 2 EXPRESSING IMPROPER FRACTIONS AND MIXED NUMBERS

NOTES

A mixed number can be expressed as a related improper fraction and an improper fraction can be expressed as a related mixed number.

This can be shown by using a model like the one below.

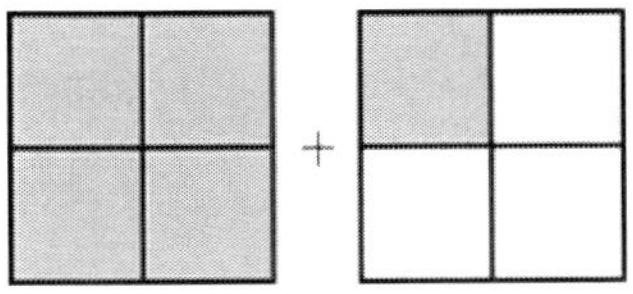

- The improper fraction that represents the given model is $\frac{5}{4}$.
- The mixed number that represents the same model is $1\frac{1}{4}$.
- $\frac{5}{4}$ can be expressed as $1\frac{1}{4}$.
- $1\frac{1}{4}$ can be expressed as $\frac{5}{4}$.

CONVERTING IMPROPER FRACTIONS TO MIXED NUMBERS

To convert an improper fraction to a mixed number, follow these steps:

1. Divide the numerator by the denominator to produce a whole number and a remainder.
2. Write the whole number to the left of the fraction part of the mixed number.
3. Use the remainder as the numerator of the fraction part of the mixed number.
4. Use the same denominator in the mixed number that was used in the improper fraction.

NOTES

Example

Express the improper fraction $\frac{19}{5}$ as a mixed number.

Solution

Step 1

Divide the numerator by the denominator to get a whole number and a remainder.

$$\begin{array}{r} 3\text{R}4 \\ 5\overline{)19} \\ \underline{15} \\ 4 \end{array}$$

Step 2

Write the mixed number.

- Write the whole number 3 on the left.
- Write the remainder 4 as the numerator of the fraction.
- Keep the same denominator (5).

The improper fraction $\frac{19}{5}$ is expressed as the mixed number $3\frac{4}{5}$.

Time to Try 1

Express the improper fraction $\frac{17}{6}$ as a mixed number.

CONVERTING MIXED NUMBERS TO IMPROPER FRACTIONS

To convert a mixed number to an improper fraction, follow these steps:

1. Multiply the whole number by the denominator.
2. Add the numerator to the product. This new number will be the numerator of the improper fraction.
3. The denominator will stay the same in both the mixed number and related improper fraction.

Mixed number $\rightarrow 2\frac{6}{10}(2\times10+6=26)=\frac{26}{10}\leftarrow$ Improper fraction

NOTES

Example

Express $2\frac{1}{3}$ as an improper fraction.

Solution

Step 1
Determine the numerator.

Multiply the whole number (2) by the denominator (3).
$2 \times 3 = 6$.

Add the numerator (1) to the product.
$6 + 1 = 7$

The number 7 becomes the numerator of the improper fraction.

Step 2
Write the improper fraction.

Recall that the denominator (3) stays the same in both the mixed number and the related improper fraction.

The mixed number $2\frac{1}{3}$ is expressed as the improper fraction $\frac{7}{3}$.

Time to Try 2

Express the mixed number $3\frac{2}{4}$ as an improper fraction.

PRACTICE EXERCISES

1. Express the mixed number $1\frac{1}{2}$ as an improper fraction. Show your work.

2. Express the improper fraction $\frac{5}{2}$ as a mixed number. Show your work.

3. Express the mixed number $12\frac{9}{10}$ as an improper fraction. Show your work.

4. Express the improper fraction $\frac{57}{9}$ as a mixed number. Show your work.

Use the following information to answer the next question.

Matt and Sophie both express the improper fraction $\frac{9}{5}$ as a mixed number. Following are their calculations.

Matt	**Sophie**
$5\overline{)9}$ = 1R4 $\underline{5}$ 4 $\frac{9}{5} = 1\frac{4}{5}$	$5\overline{)9}$ = 1R4 $\underline{5}$ 4 $\frac{9}{5} = 4\frac{1}{5}$

5. Who is correct? Justify your answer.

Use the following information to answer the next question.

Marla explains how to convert the mixed number $8\frac{7}{9}$ to an improper fraction.

Multiply $7 \times 9 = 63$.
Add $63 + 8 = 71$.
Keep the same denominator of 9.
$8\frac{7}{9} = \frac{71}{9}$

6. Is Marla correct? Justify your answer.

Lesson 3 ORDERING FRACTIONS ON A NUMBER LINE

NOTES

SAME DENOMINATORS

If the denominators of proper and improper fractions are the same, order the fractions on the number line by the numerators.

- The smaller the numerator, the smaller the fraction.
- The greater the numerator, the greater the fraction.

If an improper fraction has the same denominator as a mixed number, the improper fraction should first be converted to a mixed number. Then, the numerators can be compared more easily.

Example

Order the fractions $\frac{8}{5}$ and $1\frac{4}{5}$ on a number line marked in fifths from 1 to 2.

Solution

Step 1

Since the number line goes from 1 to 2, express the improper fraction as a mixed number.

$\frac{8}{5} = 8 \div 5 = 1\frac{3}{5}$

Both numbers now have the same whole number, as well as the same denominator.

Step 2

Order the numerators of the mixed numbers from least to greatest.

Since $3 < 4$, then $1\frac{3}{5} < 1\frac{4}{5}$.

Step 3

Place the fractions on the number line.

- Start at the 1 and count to the third tick to place $1\frac{3}{5}\left(\frac{8}{5}\right)$.
- Count to the fourth tick to place $1\frac{4}{5}$.

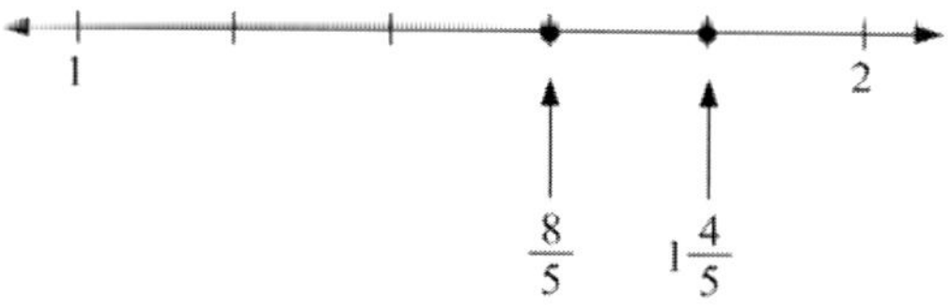

NOTES

Time to Try 1

Order the fractions $\frac{5}{10}$, $\frac{10}{10}$, and $\frac{3}{10}$ on a number line marked in tenths from 0 to 1.

DIFFERENT DENOMINATORS

When ordering a set of fractions and mixed numbers that have different denominators, the different types of fractions should first be converted into the same type of fraction.

Then, in order to place them correctly on a number line, the fractions should be changed into equivalent fractions with the same denominators.

Once the denominators are the same, order the fractions by their numerators.

Example

Order the fractions $3\frac{1}{2}$, $\frac{11}{3}$, and $3\frac{5}{6}$ on a number line marked in sixths from 3 to 4.

Solution

Step 1

Convert the improper fraction to a mixed number.

Divide 11 by 3 to make a related mixed number.

$\frac{11}{3}$ expressed as a mixed number is $3\frac{2}{3}$.

All three mixed numbers start with the whole number 3.

$3\frac{1}{2}, 3\frac{2}{3}, 3\frac{5}{6}$

It will now be easier to compare the fraction parts.

NOTES

Step 2
Make equivalent fractions.

Since the number line is marked in sixths, and 2 and 3 are factors of 6, use 6 as a common denominator for the fraction parts of the mixed numbers.

$$3\frac{5}{6} = 3\frac{5}{6}$$

$$3\frac{1}{2} = \frac{1\times 3}{2\times 3} = 3\frac{3}{6}$$

$$3\frac{2}{3} = \frac{2\times 2}{3\times 2} = 3\frac{4}{6}$$

Step 3
Order the fractions by their numerators.
$3 < 4 < 5$

- Start at the 3 and count to the third tick to place $3\frac{3}{6}$ $\left(3\frac{1}{2}\right)$.
- Count to the fourth tick to place $3\frac{4}{6}$ $\left(\frac{11}{3}\right)$.
- Count to the fifth tick to place $3\frac{5}{6}$.

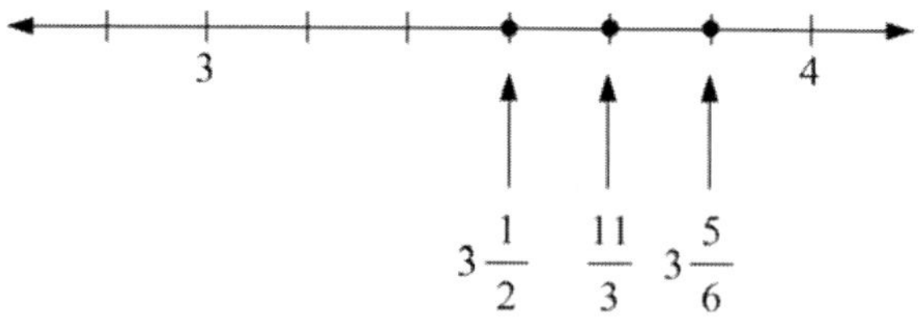

Time to Try 2

Order the fractions $\frac{12}{8}$ and $2\frac{1}{4}$ on a number line marked in fourths from 1 to 3.

PRACTICE EXERCISES

1. Order the fractions $\frac{3}{5}$ and $\frac{8}{10}$ on a number line marked in tenths from 0 to 1.

2. Order the fractions $\frac{2}{5}$, $\frac{1}{2}$, $\frac{9}{10}$, and $\frac{3}{5}$ on a number line marked in tenths from 0 to 1.

3. Order the fractions $2\frac{2}{5}$, $\frac{11}{5}$, and $2\frac{8}{10}$ on a number line marked in tenths from 2 to 3.

Use the following information to answer the next question.

Between 0 and 1, the number line is marked in tenths. From 1 to 2, the number line is marked in fourths.

4. Place the fractions $1\frac{3}{4}$, $\frac{9}{10}$, $\frac{3}{2}$, and $\frac{2}{5}$ in their correct locations on the given number line.

REVIEW SUMMARY

- An improper fraction is a fraction in which the numerator is greater than or equal to the denominator.
- A mixed number consists of a whole number and a proper fraction.
- Improper fractions and mixed numbers can be used to show the same amount.
- To convert a mixed number to an improper fraction, multiply the whole number by the denominator, and add the numerator to the product. This number becomes the numerator of the improper fraction. The denominator stays the same for both the mixed number and the improper fraction.
- To convert an improper fraction to a mixed number, divide the numerator by the denominator. The quotient becomes the whole number. The remainder becomes the numerator of the fraction. The denominator stays the same for both the improper fraction and the mixed number.
- It is easier to order a set of fractions if they are converted into the same form; i.e., either all improper fractions or all mixed numbers.
- It is easier to order a set of fractions if they have a common denominator. To change fractions so they have the same denominator, you need to be able to make equivalent fractions.

PRACTICE TEST

1. Explain what an improper fraction is. Use a model or diagram to help.

2. Model or draw the improper fraction $\frac{11}{8}$.

3. Model or draw the mixed number $3\frac{1}{3}$.

Use the following information to answer the next question.

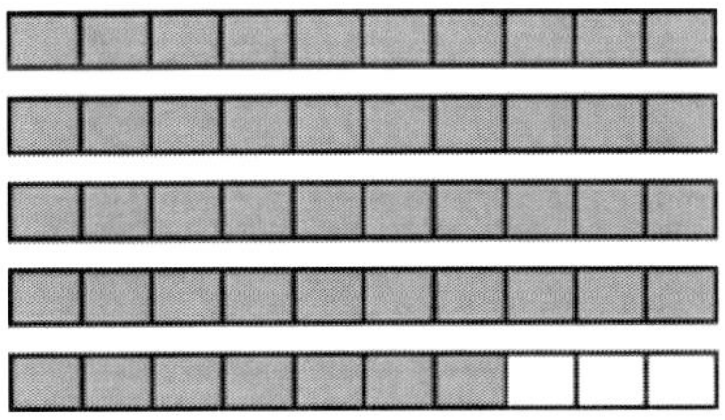

4. Translate the given model of an improper fraction into its symbolic form.

Use the following information to answer the next question.

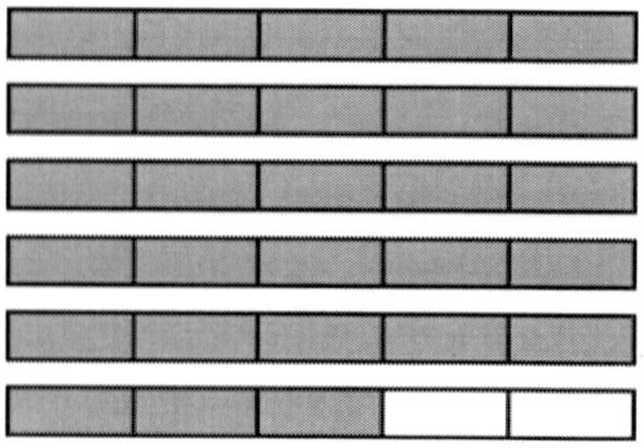

5. Translate the given model of a mixed number into its symbolic form.

6. Express the improper fraction $\frac{85}{3}$ as a mixed number. Show your work.

7. Express the mixed number $17\frac{1}{4}$ as an improper fraction. Show your work.

8. Order the fractions $\frac{18}{5}$ and $3\frac{1}{2}$ on a number line marked in tenths from 3 to 4.

9. Order the fractions $\frac{9}{10}, \frac{1}{5}, \frac{1}{2}$, and $\frac{6}{15}$ on a number line marked in tenths from 0 to 1.

RATIOS

When you are finished this unit, you will be able to…

- Model or draw a given ratio
- Write a ratio for a given concrete or pictorial representation
- Express a ratio in multiple forms
- Identify ratios in real-life contexts and record them symbolically
- Explain part/part ratios
- Explain part/whole (and whole/part) ratios
- Solve a given problem involving ratios

Lesson	Page	Completion
1. Using Ratio to Compare Two Quantities	100	
2. Part/Part and Part/Whole Ratios	107	
3. Solving Problems Involving Ratios	111	
Review Summary	113	
Practice Test	114	
Answers and Solutions	at the back of the book	

PREREQUISITE SKILLS AND KNOWLEDGE

Prior to beginning this unit, you should be able to…

- Compare objects in a group with other sets of objects in the group and with the entire group

Lesson 1 USING RATIO TO COMPARE TWO QUANTITIES

NOTES

A ratio can be used to compare two quantities. A ratio involves two numbers. The first number of the ratio is the first part or quantity of the comparison; the second number in the ratio is the second part or quantity of the comparison. In a ratio, the first quantity (first number) is compared with the second quantity (second number).

EXPRESSING RATIO IN MULTIPLE FORMS

The following diagram shows two quantities, or two parts. The first part is three squares and the second part is one heart.

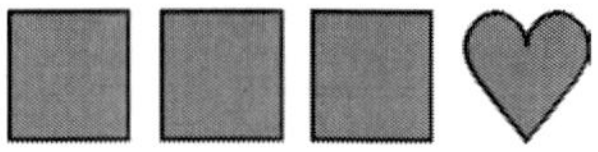

Following are three ways to express the comparison of squares to hearts as a ratio:

- Place a colon between the number of squares and the number of hearts.
  3 : 1
- Write a fraction. The number of squares becomes the numerator. The number of hearts becomes the denominator.
  $\frac{3}{1}$
- Use the word *to* between the number of squares and number of hearts.
  3 to 1

Example

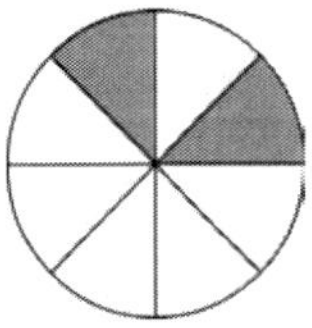

In this model, the ratio of shaded parts to white parts is 2 to 6. Express this ratio as a fraction.

Solution

The first part of the comparison is the number of shaded parts, so 2 is the numerator of the fraction.

The second part of the ratio is the number of white parts, so 6 is the denominator of the fraction.

The ratio fraction $\frac{2}{6}$ expresses the comparison of shaded parts to white parts.

NOTES

Time to Try 1

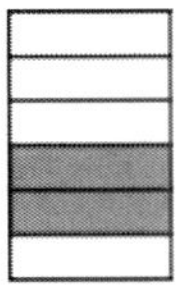

In this model, the ratio of white parts to shaded parts is $\frac{4}{2}$.

Express this ratio using the word *to*.

MODELLING OR DRAWING A GIVEN RATIO

In order to model or draw a representation of a given ratio, start with the first number. Draw or model that number of items.

Next, look at the second number. Draw or model that number of items.

Example

A rectangle is divided into eight equal squares. The ratio of white squares to shaded squares is 3 to 5. Model or draw the given ratio.

Solution

Start by drawing a rectangle divided into eight equal squares.

Since the first part of the comparison represents the number of white squares, leave three squares white. Since the number to the right of the word *to* represents the number of shaded squares, shade the remaining five squares.

The pictorial or concrete model could look like this.

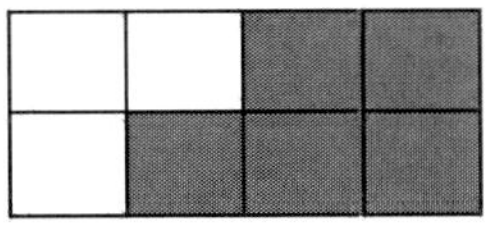

Time to Try 2

A 4 by 4 square grid is divided into 16 equal squares. The ratio of shaded squares to white squares is $4:12$. Model or draw the given ratio.

NOTES

WRITING A RATIO TO REPRESENT A MODEL OR DIAGRAM

To write a ratio from a given picture, first choose the method you plan to use—colon, fraction, or the word *to*—if you are given the choice.

1. Count the number of items in the first group being compared and write that number down.
2. Count the number of items in the second group being compared and write that number down.

Example

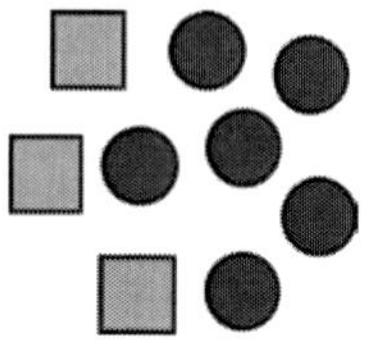

Write a ratio that compares the number of squares to the number of circles. Write the ratio in three ways: using a colon, using a fraction, and using the word to.

Solution

Step 1
Determine the first number in the ratio.

Since the first quantity being compared is the number of squares, the first number in the ratio is 3.

Step 2
Determine the second number.

Since the second quantity being compared is the number of circles, the second number in the ratio is 6.

Step 3
Write the ratio in three ways.

The ratio comparing squares to circles is expressed as $3:6$, $\frac{3}{6}$, or 3 to 6.

NOTES

Time to Try 3

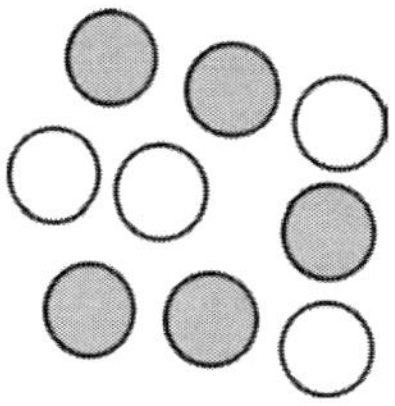

Write a ratio that compares the number of white circles with the number of shaded circles. Write the ratio in three ways: using a colon, using a fraction, and using the word *to*.

WRITING RATIOS FROM REAL-LIFE CONTEXTS

Ratios are used in everyday life for such things as cooking, medication dosages, wins and losses in sports, and scales on a map.

Example

One way to make a tasty salad dressing is to mix 2 parts olive oil with 1 part balsamic vinegar. Write a ratio using a colon to compare the amount of olive oil with the amount of balsamic vinegar. Explain what the ratio means.

Solution

Step 1
Write the first number.

The first part of the comparison is the amount of olive oil, so 2 is the first number of the ratio. Place a colon after the 2.

Step 2
Complete the ratio.

The second part of the ratio is the amount of balsamic vinegar, so 1 is the number to the right of the colon.

Using a colon, the ratio of oil to vinegar is $2:1$.

NOTES

Step 3
Explain the ratio.

In a recipe like the one given, if the ratio is $2:1$, that means that for every two parts of oil used, 1 part vinegar must be used.

If the recipe is doubled and 4 tablespoons of oil are used, then 2 tablespoons of vinegar must be used.

Time to Try 4

Robert's soccer team won 7 games and lost 3 games. Write a ratio expressed as a fraction to compare the wins with the losses.

PRACTICE EXERCISES

Use the following diagram to answer the next question.

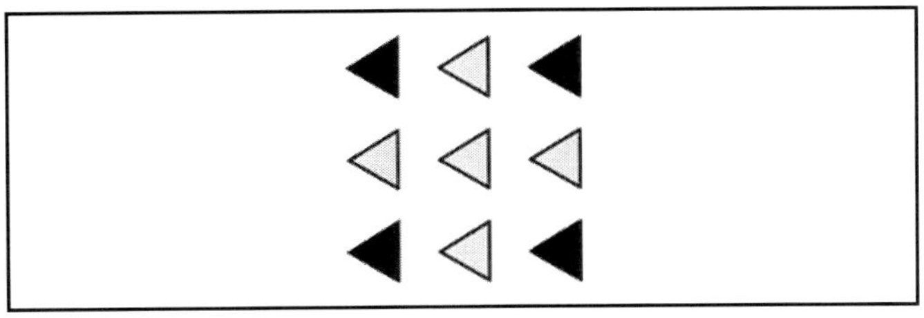

1. The ratio of grey to black triangles is 5 to 4. Express this ratio as a fraction.

2. A rectangle is divided into eight equal parts. The ratio of white parts to shaded parts is $6:2$. Model or draw the given ratio.

Use the following information to answer the next question.

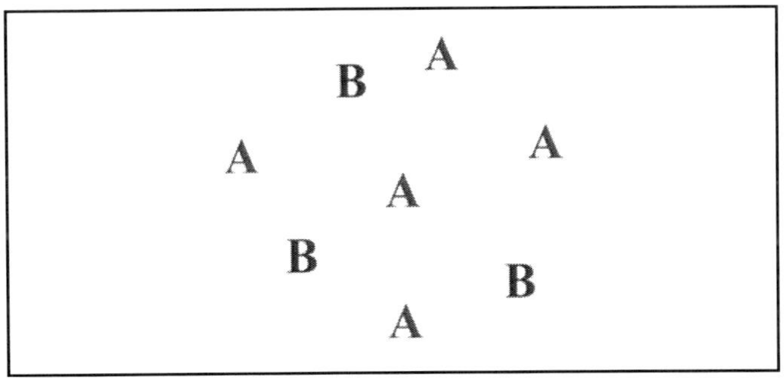

3. Write a ratio that compares the number of As with the number of Bs. Write the ratio in three ways: using a colon, using a fraction, and using the word *to*.

4. In one Olympic hockey game, it was reported that Canada beat Russia by a score of 7 to 3. Write the ratio of goals scored by the two teams as a fraction. Explain what the ratio means.

Use the following information to answer the next question.

On a particular map Ben drew for a social studies assignment, he made the following legend. 1 cm = 10 km

5. Express the comparison as a ratio using a colon. Explain what the ratio means.

Use the following information to answer the next question.

Mrs. Jenkins makes a fruit punch for a party. The recipe she uses requires two cans of frozen orange juice for every litre of ginger ale used.

6. Express the comparison of orange juice to ginger ale as a ratio using the word *to*. Explain what the ratio means.

Lesson 2 PART/PART AND PART/WHOLE RATIOS

NOTES

A ratio can compare one part of a set with another part of a set. A ratio can also compare one part of the set with the whole set, or the whole set with one part of the set.

PART/PART RATIOS

Part/part ratios show the relationship between two parts of a set. The first part of the set can be compared with the second part of the set, or the second part of the set can be compared with the first part of the set.

Example

The following diagram shows a set of five cubes: 2 grey cubes and 3 white cubes.

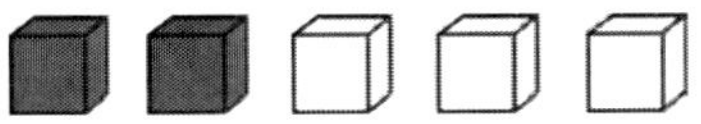

- Write a ratio comparing the grey cubes with the white cubes.
- Explain what the ratio $3:2$ means.

Solution

Step 1
Compare grey cubes to white cubes.

The ratio that expresses the number of grey cubes compared with the number of white cubes is 2 to 3, $\frac{2}{3}$, or $2:3$.

Step 2
Explain what the ratio $3:2$ means.

The ratio $3:2$ is a part/part ratio. The number of white cubes is being compared with the number of grey cubes.

Time to Try 1

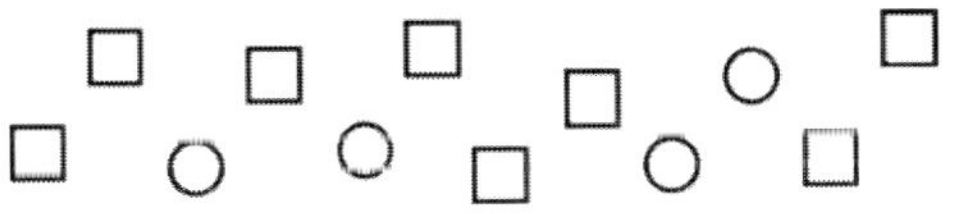

Explain what the ratio fraction $\frac{8}{4}$ means.

NOTES

PART/WHOLE RATIOS

Part/whole ratios show the relationship between a part of a set and the whole set. Either part of the set can be compared with the whole set. The whole set can also be compared with either part of the set.

Example

The following diagram shows a set of five cubes: 2 grey cubes and 3 white cubes.

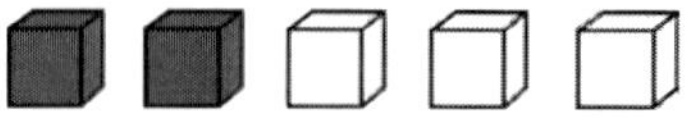

- Write a ratio comparing the grey cubes with the whole set.
- Explain what the ratio $5:3$ means.

Solution

Step 1
Compare grey cubes with the whole set.

The ratio that expresses the number of grey cubes compared with the number of cubes in the whole set is 2 to 5, $\frac{2}{5}$, or $2:5$.

Step 2
Explain what the ratio $5:3$ means.

The ratio $5:3$ is a whole/part ratio. The number of cubes in the whole set is being compared with the number of white cubes.

Time to Try 2

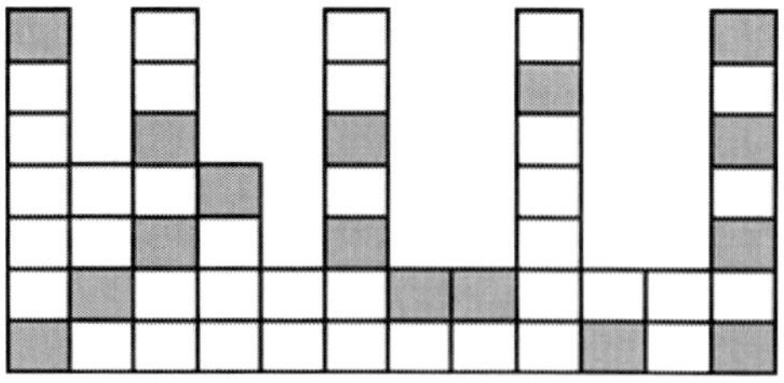

Explain what the ratio 16 to 53 means.

PRACTICE EXERCISES

Use the following information to answer the next three questions.

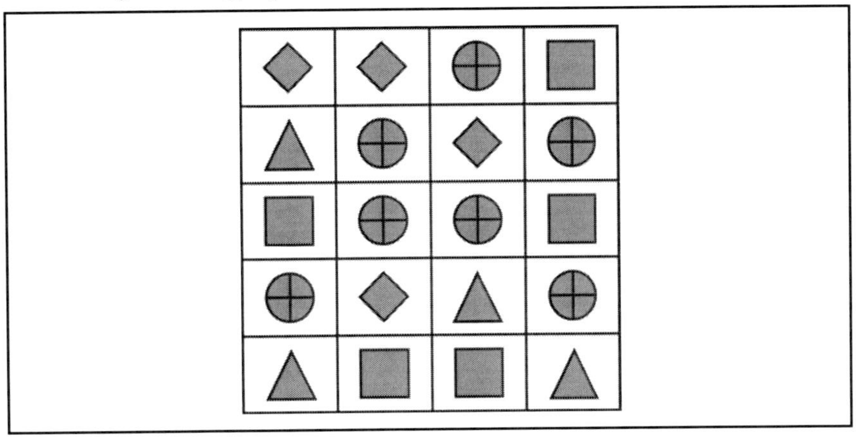

1. Explain the ratio $5:20$.

2. Explain the ratio 7 to 5.

3. Write a fraction ratio comparing the number of triangles with the total number of shapes. Explain your answer.

Use the following information to answer the next four questions.

There are 5 crows and 7 sparrows sitting on a wire.

4. Explain the ratio 7 to 5.

5. Explain the ratio $5:12$.

6. Explain the ratio $\frac{5}{7}$.

7. Write a ratio comparing the number of birds with the number of sparrows. Explain your answer.

Lesson 3 SOLVING PROBLEMS INVOLVING RATIOS

NOTES

When solving problems with ratios, read the question carefully in order to identify what is being compared.

Determine if you are looking for a part/part ratio or a part/whole ratio.

Sometimes, some of the information may be missing and you will need to calculate before you can write a ratio.

Example

A hockey team won 30 out of the 50 games played. Write a ratio comparing the team's losses with their wins.

Solution

Step 1
Identify what is being compared.

The number of games lost is being compared with the number of games won (30).

Step 2
Calculate the number of losses.

If the team won 30 games out of 50, it means they lost 20 games.
$50 - 30 = 20$

Step 3
Write the part/part ratio.

The ratio of games lost to games won can be expressed as

20 to 30, $\frac{20}{30}$, or $20:30$.

Time to Try 1

A Grade 6 class consists of 15 girls and 10 boys. What is the ratio of boys to the whole class?

PRACTICE EXERCISES

1. In Matt's class, there are a total of 30 students, of which 18 are boys. Write a ratio that expresses the total number of students compared with the number of girls in the class.

2. If there are 39 girls and 52 boys in Nadia's swim club, what is the ratio of girls to the total number of swimmers in the club?

3. The ratio of boys to girls on the soccer team is $3:1$. If there are 5 girls on the team, how many boys are on the team?

Use the following information to answer the next question.

Mrs. Jenkins makes a grape drink using the following ratio of cans of grape juice to cups of water.

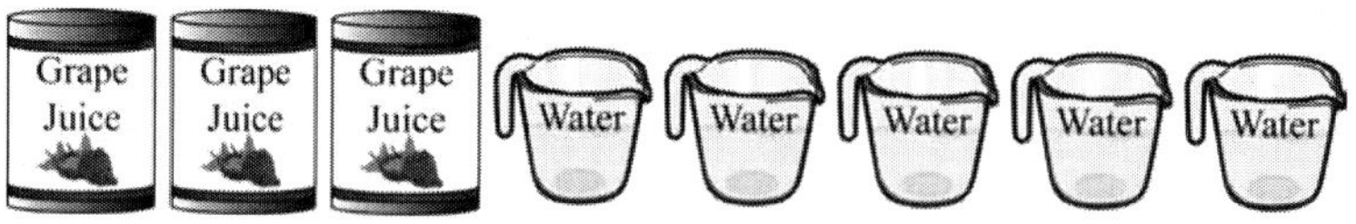

4. If Mrs. Jenkins used 12 cans of grape juice, how many cups of water does she need to use?

REVIEW SUMMARY

- A two part ratio is a comparison of two quantities.
- The first number of a ratio is the first part of the comparison; the second number of a ratio is the second part of the comparison.
- A ratio can be expressed in multiple ways: using a colon $(3:1)$, using the word *to* (3 to 1), or as a fraction $\left(\frac{3}{1}\right)$.
- A part/part ratio compares one part with another part.
- A part/whole ratio compares one part with the whole, or the whole with one part (whole/part).
- When explaining or describing a ratio, identify whether it is a part/part ratio or a part/whole ratio. Explain what each number in the ratio represents.
- When solving problems involving ratios, read the question carefully to see if a part/part ratio or a part/whole ratio is being used.
- When solving problems involving ratios, you may need to perform a calculation before you can determine the ratio.

PRACTICE TEST

Use the following information to answer the next two questions.

Jacob has a collection of 5 hats and 3 caps.

1. Write a ratio to compare the number of hats with the number caps. Express the ratio using a colon.

2. Write a ratio to compare the number of caps with the number hats. Express the ratio using the word *to*.

3. The ratio of rectangles to triangles is $\frac{3}{2}$. Model or draw the given ratio.

Use the following information to answer the next four questions.

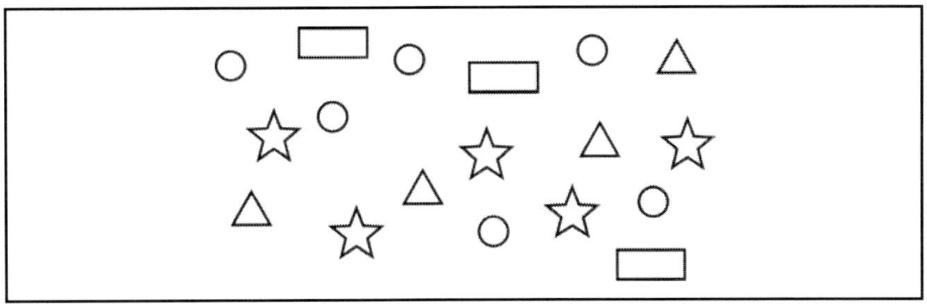

4. Explain the ratio 18 : 5.

5. Explain the ratio $\frac{4}{3}$.

6. Explain the ratio 6 to 5.

7. Write the ratio that compares the number of triangles with the total number of shapes. Express the ratio in three ways: using a colon, using a fraction, and using the word *to*.

8. Janine's hockey team lost 7 games and won 9 games in the first half of the hockey season. What was the team's win/loss ratio? Explain your answer.

9. In Karina's dance class, there are a total of 32 students, of which 19 are girls. Write a ratio that expresses the number of boys in the class compared with the total number of students.

10. The ratio of girls to boys on the track team is $2:5$. If there are six girls on the team, how many boys are on the team?

PERCENTAGES

When you are finished this unit, you should be able to…
- Understand that percentage is a ratio out of 100
- Use concrete materials and pictorial representations to illustrate a given percentage
- Record the percentage shown in a given concrete or pictorial representation
- Express a given percentage as a fraction and a decimal
- Record a percentage symbolically from a real-life context
- Solve problems involving percentage

Lesson	Page	Completed on
1. Understanding Percentage	118	
2. Expressing a Percentage as a Fraction and a Decimal	124	
3. Solving Problems Involving Percentage	127	
Review Summary	130	
Practice Test	131	
Answers and Solutions	at the back of the book	

PREREQUISITE SKILLS AND KNOWLEDGE

Prior to beginning this unit, you should be able to…
- Understand that a ratio is a comparison of two quantities
- Explain the part/part ratio and the part/whole ratio of a set
- Express a hundredth fraction with a denominator of 100
- Express a decimal in the hundredths with two digits to the right of the decimal point

Lesson 1 UNDERSTANDING PERCENTAGE

NOTES

The word *percentage* means out of 100. A percentage always represents a quantity out of 100. The symbol that is used to represent percentage is %.

EXPLAINING PERCENT AS A RATIO

Percentage can be explained in terms of a ratio. For example, 50% means a ratio of $50:100$ or $\frac{50}{100}$, where 50 is the part and 100 is the whole. The second number in a percent ratio is always 100.

USING GRIDS TO ILLUSTRATE PERCENTAGE

A pictorial representation of a given percentage can be made by shading squares on a hundredths grid.

Example

Use a hundredths grid to illustrate 40%.

Solution

Since a hundredths grid is made up of 100 squares and percentage is a ratio out of 100, shade 40 of the 100 squares to show that 40% of the grid is shaded.

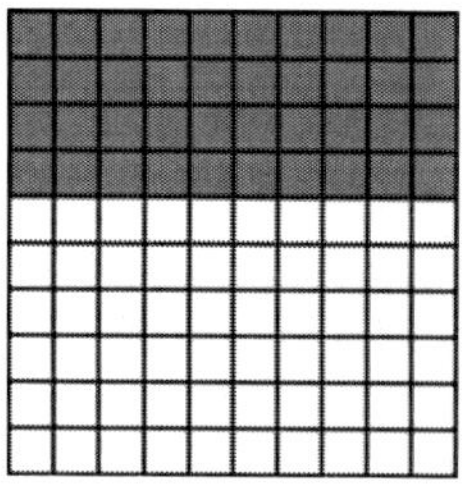

Time to Try 1

Use a hundredths grid (a 10 × 10 grid) to illustrate 9%.

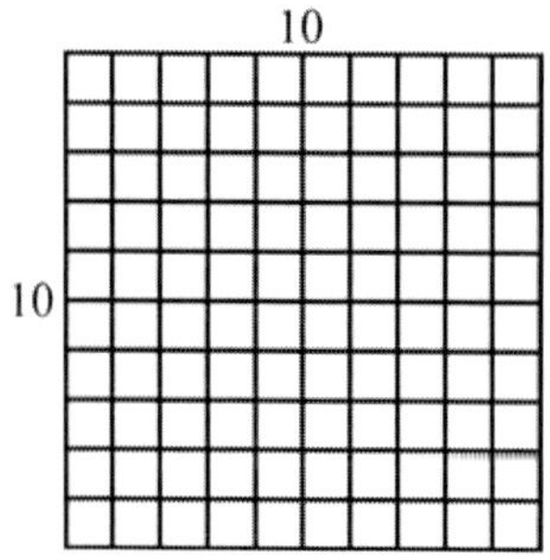

NOTES

USING CONCRETE MATERIALS TO ILLUSTRATE PERCENTAGE

Concrete materials like base ten blocks can be used to build a model that illustrates a given percentage.

Example

Use the fewest number of ten rods and units needed to build a model that illustrates 26%.

Solution

In the number 26, there are 2 tens and 6 ones. The model that illustrates 26% should be built with 2 ten rods and 6 units.

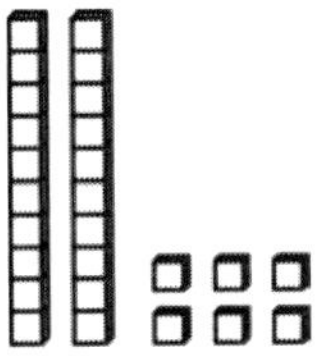

Time to Try 2

Use the fewest number of ten rods and units needed to build a model that illustrates 65%.

RECORDING PERCENTAGE FROM A GIVEN REPRESENTATION

When hundredths grids are used, count the number of squares to determine the percentage. Be sure to read the question carefully to see if the shaded parts or the white parts should be counted.

If base ten blocks are used to illustrate the percentage, count the number of ten rods by 10s and the units by 1s.

NOTES

Example

What percentage of squares in the given grid is shaded?

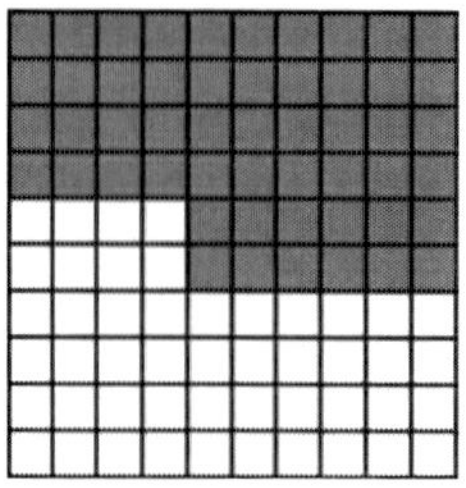

Solution

Count the number of squares that are shaded.

Since four complete rows are shaded, count the four rows by 10s.
10, 20, 30, 40

Continue counting the squares in the partially shaded rows.
…41, 41, 43,…50, 51, 52

Since 52 of the 100 squares are shaded, 52% of the grid is shaded.

Time to Try 3

What percentage does this model illustrate?

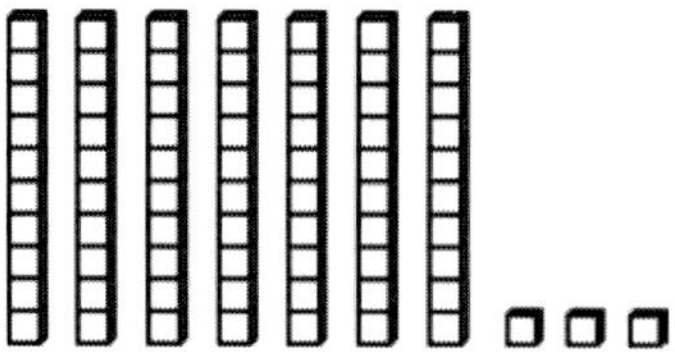

RECORDING PERCENTAGES FROM REAL-LIFE CONTEXTS

In everyday life, percentages are encountered in many situations. When shopping, a sign might state, "Sale: 10% off!" When a store advertises a 10% sale, 10% is subtracted off the original price, so you pay a lower price.

In Canada, GST (Goods and Services Tax) is added to the price of certain products. This tax is 5% of the original price and is added onto the original price.

NOTES

Students' grades and scores are often reported in percentages. If a student got 95% on a test, that student got almost all the questions correct.

Often, commonly used percentages from real-life contexts like 10%, 25%, 50%, and 75% are related to commonly used fractions like $\frac{1}{10}$, $\frac{1}{4}$, $\frac{1}{2}$, and $\frac{3}{4}$.

Example

The Canadian flag is red and white. About three-quarters of the flag is red.

About what percentage of the flag is red?

Solution

Three-quarters is the same as $\frac{75}{100}$. About 75% of the Canadian flag is red.

Time to Try 4

Jason had one dollar in his pocket. He spent about half of the money on a candy bar. What percentage of the dollar did Jason spend?

PRACTICE EXERCISES

1. Use a ratio to explain what 25% means.

2. Use a hundredths grid to illustrate 45%.

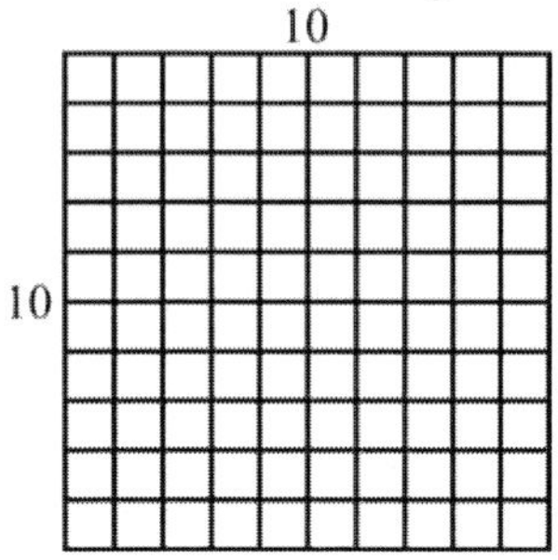

3. Use the fewest number of ten rods and units needed to build a model that illustrates 38%.

Use the following information to answer the next two questions.

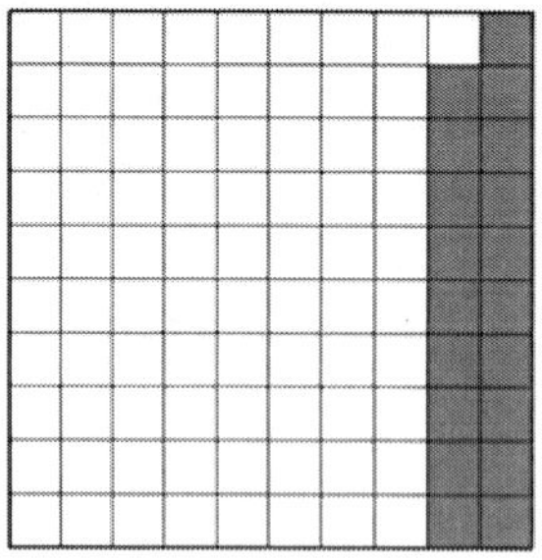

4. What percentage of the given grid is shaded?

5. What percentage of the given grid is not shaded?

Use the following information to answer the next que

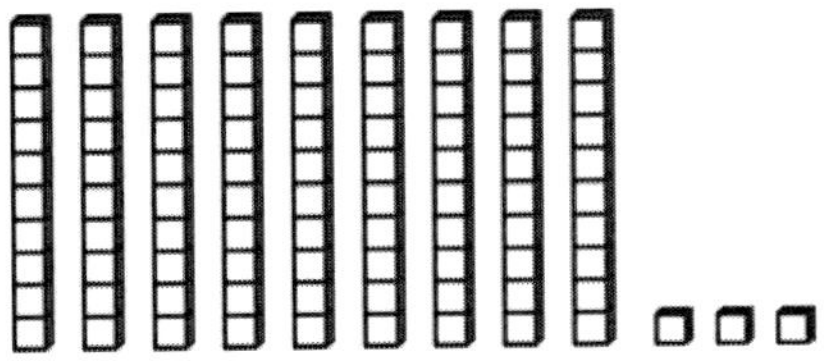

6. What percentage does the given model illustrate?

7. There are 100 students in the school choir, of which 60 are boys. What percent students are boys? Explain your answer.

8. About half of the floor in the living room in Jenna's house is covered with furniture. About what percentage of the floor is covered with furniture? Explain your answer.

Lesson 2 EXPRESSING A PERCENTAGE AS A FRACTION AND A DECIMAL

NOTES

Percentages can be expressed as factions and decimals because percentages (including ratios), fractions, and decimals are all related.

AS A FRACTION

Since percentage is a ratio out of 100, the denominator of the related fraction will be 100. The digits in the percentage will be the numerator.

Example

Express 8% as a fraction.

Solution

8% means 8 parts out of 100 parts. The denominator of the related fraction is 100. The digit in the percentage (8) becomes the numerator.

$$8\% = \frac{8}{100}$$

Time to Try 1

Express 33% as a fraction.

AS A DECIMAL

Since percentage is a ratio out of 100, the related decimal will be written in the hundredths. That means that there must be two digits to the right of the decimal point. To change a percentage into a decimal, move the decimal point in the percentage two places to the left and remove the percentage symbol.

- If there are two digits in the percentage, those two digits will be the two digits written to the right of the decimal point.
- If there is only one digit in the percentage, place a 0 in the tenths place as a place holder when writing the decimal number. The digit in the percentage is written in the hundredths place.

Remember to place a 0 to the left of the decimal point.

NOTES

Example

Express 4% as a decimal.

Solution

Recall that 4% means four parts out of 100 parts, as shown on this hundredth grid.

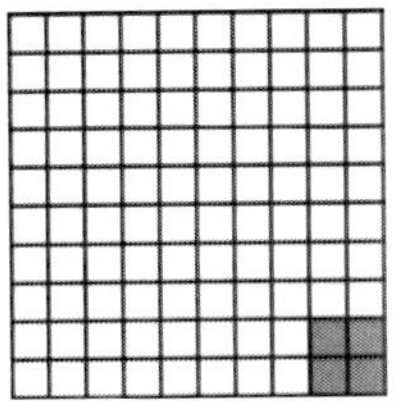

Since the related decimal number must have two digits to the right of the decimal point, place a 0 to the left of the 4 when writing the decimal number. The 0 will serve as a place holder for the tenths position. Remember to place a 0 to the left of the decimal point.

4% = 0.04

Time to Try 2

Express 89% as a decimal.

PRACTICE EXERCISES

1. Express 30% as a fraction.

2. Express 5% as a fraction.

3. Express 9% as a decimal.

4. Express 52% as a decimal.

5. Express 1% as a fraction and as a decimal.

Lesson 3 SOLVING PROBLEMS INVOLVING PERCENTAGE

NOTES

When solving problems involving percentage, remember that percentage is a ratio out of 100. In order to solve a given problem, you may first need to perform calculations using the ratio out of 100. A hundredths grid can be used to help understand or solve the problem.

Example

In a bag of one hundred candies, 42% of the candies are red in colour. How many candies are not red in colour? Use this grid to help solve the problem.

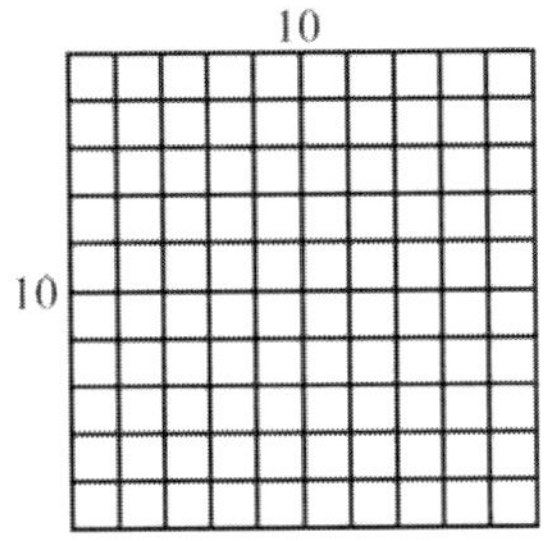

Solution

Step 1
Determine the percentage that is not red.

Since there are 100 candies represented, the percentages need to add up to 100.

Subtract 42% from 100% to determine the percentage of candies that are not red.
$100 - 42 = 58$

If you are using the grid, shade 42 squares to represent 42%. Count the number of squares that are not shaded.

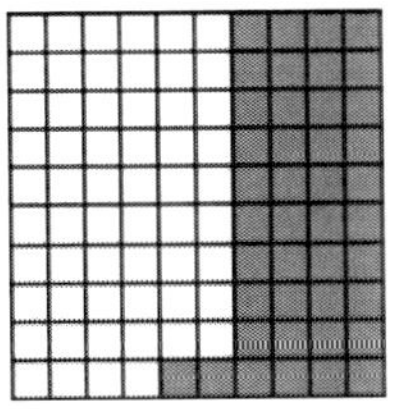

58% of the candies are not red in colour.

Step 2
Determine the number of candies that are not red.

Since percentage is a ratio out of 100, 58 out of 100 candies are not red in colour.

NOTES

Time to Try 1

Out of a group of one hundred students at Prairie School, 23% have blond hair, 12% have black hair, 54% have brown hair, and the rest have red hair.

What percentage of the students have red hair?

PRACTICE EXERCISES

1. At a grocery store, a clerk placed 34 green apples, 42 red apples, and 24 yellow apples in a basket. What percentage of the apples were either red or yellow?

2. One garden centre has one hundred fruit trees, of which 12% are plum trees, 18% are cherry trees, and the rest are apple trees. How many apple trees are in this garden centre?

Use the following information to answer the next question.

Jason earned $100 mowing lawns in his neighbourhood. Just before school started, he spent $55 buying school supplies. He put the rest of the money into his bank account.

3. What percentage of the money earned was put into the bank account?

4. On January 20, one school had a fun day to celebrate one hundred days of school. If Allie was at school for 92% of these days, how many school days was she absent from school?

REVIEW SUMMARY

- *Percentage* means *out of 100.*
- Percentage can be explained in terms of ratio: an amount out of 100.
- The percentage symbol is %.
- Percentage can be illustrated by shading squares on a hundredths grid or by building the percentage out of ten rods and units (base ten blocks).
- To record the percentage illustrated on a grid, count the number of shaded (or not shaded) squares, depending on what is asked.
- To record the percentage illustrated with base ten blocks, count the ten rods by 10s and the units by ones.
- Express a percentage as a fraction by writing 100 as the denominator and the digits in the percentage as the numerator.
- Express a percentage as a decimal by writing the two digits of the percentage to the right of the decimal point. If there is only one digit in the percentage, place a 0 to the left of the digit when writing the decimal number. The 0 acts as a place holder in the tenths position.
- When solving problems involving percentage, remember that percentage is a ratio out of 100. You may need to perform calculations involving the ratio before you can solve the problem.

PRACTICE TEST

1. Use a hundredths grid to illustrate 30%.

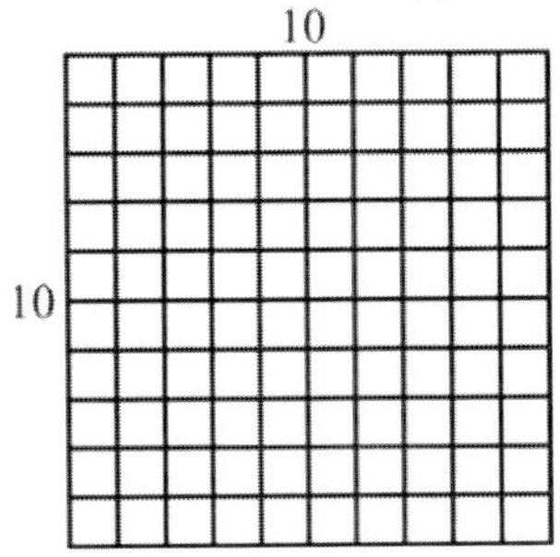

2. Use the fewest number of ten rods and units needed to build a model that illustrates 14%.

Use the following information to answer the next two questions.

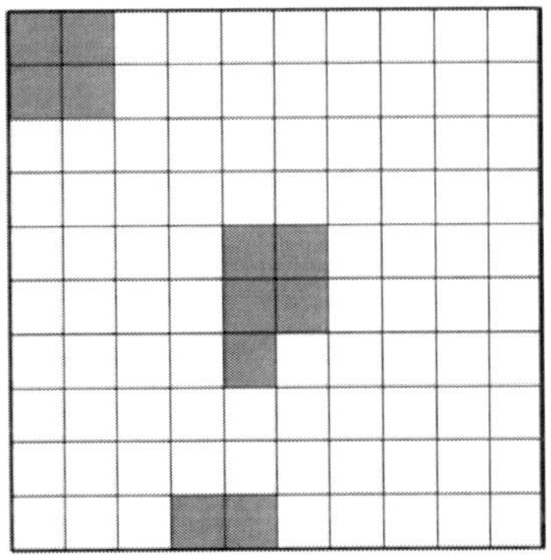

3. What percentage of the squares in the hundredths grid is shaded?

4. What percentage of the squares in the hundredths grid is not shaded?

5. In one small theatre, there are one hundred seats. To get a seat for the play being presented, a ticket must be purchased. If 83 tickets were sold, what percentage of the seats will be filled?

6. Express 43% as a fraction and a decimal.

7. Express 5% as a fraction and a decimal.

8. Pat has a collection of one hundred sports cards, of which 20% are baseball cards and 16% are soccer cards. The rest are hockey cards. How many hockey cards does Pat have in his collection? You may use the grid below to solve the problem.

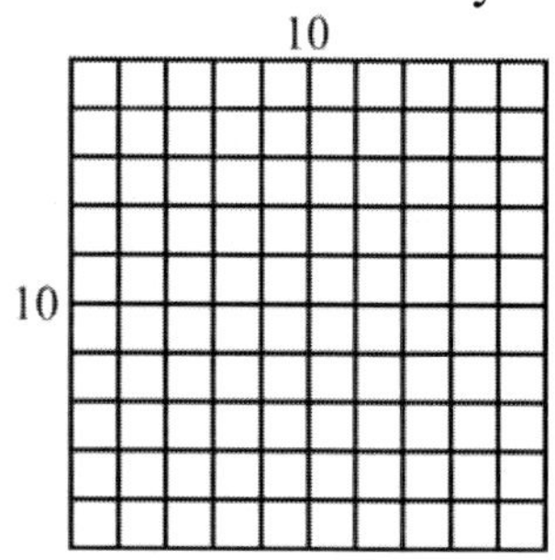

PATTERNS AND RELATIONS

When you are finished this unit, you should be able to…

- Create a table of values from a given pattern or graph
- Describe relationships shown on a graph in everyday language
- Describe patterns and relationships in tables of values using mathematical language and by formulating pattern rules
- Translate a pattern to a table of values, and then graph the table of values
- Create a concrete or pictorial representation of a relationship shown in a table of values
- Generate values in one column of a table of values when the values in the other column and the pattern rule are given
- Predict the value of an unknown term by using the relationship in a table of values, and then verify the prediction
- Identify missing elements in a given table of values
- Identify errors in a given table of values
- Create tables of values to solve problems involving patterns and relationships

Lesson	Page	Completed on
1. Using Tables and Graphs to Represent Patterns and Relationships	134	
2. Using Relationships Within Tables to Solve Problems	143	
Review Summary	158	
Practice Test	159	
Answers and Solutions	at the back of the book	

PREREQUISITE SKILLS AND KNOWLEDGE

Prior to beginning this unit, you should be able to…

- Extend given patterns and explain how each element or term differs from the previous one
- Write a mathematical expression to represent a given pattern
- Describe relationships in tables or charts using mathematical expressions
- Predict subsequent elements in a pattern
- Represent a given pattern visually to verify predictions
- Determine subsequent elements by using pattern rules

Lesson 1 USING GRAPHS AND TABLES TO REPRESENT PATTERNS AND RELATIONSHIPS

NOTES

Patterns and relationships can be represented in tables of values and graphs. A table of values is an organized way to show the term numbers (the sequence or order of the terms) and the terms (the numbers that make up the pattern). Such patterns are often more easy to see when they are presented in a graph.

CREATING A TABLE OF VALUES FROM A GIVEN PATTERN

When you make a table of values, the order of terms (term numbers) can be placed on the left side of the table or in the top row of the table. The term numbers (order of terms) can be written as cardinal numbers (1, 2, 3) or as ordinal numbers (1st, 2nd, 3rd). Sometimes, term numbers are expressed as words (first, second, third).

When given a sequence of numbers for the pattern, place them in the term column on the right side of the table (or the bottom row of the table) in the order they appear.

Example

Create a table of values from the pattern 64, 32, 16, 8, 4.

Solution

Step 1
Write the term numbers.

Since there are five terms in the pattern of numbers, write the numbers 1, 2, 3, 4, and 5 on the left side of the table of values.

Term Number	Term
1	
2	
3	
4	
5	

NOTES

Step 2
Write the terms.

Write the five terms in the term column, in the order they appear.

Term Number	Term
1	64
2	32
3	16
4	8
5	4

Time to Try 1

Create a table of values from the pattern 21, 30, 39, 48.

CREATING A TABLE OF VALUES FROM A GIVEN GRAPH

A table of values contains the term number and the term. An ordered pair is created using the term number (the x-coordinate) and its corresponding term (the y-coordinate). The x- and y-coordinates are placed in brackets and separated by a comma to create ordered pairs in the form of (x, y).

To create a table of values from a graph, first identify the coordinates.

- The first number in the ordered pair will be the term number.
- The second number in the ordered pair will be the term.

NOTES

Example

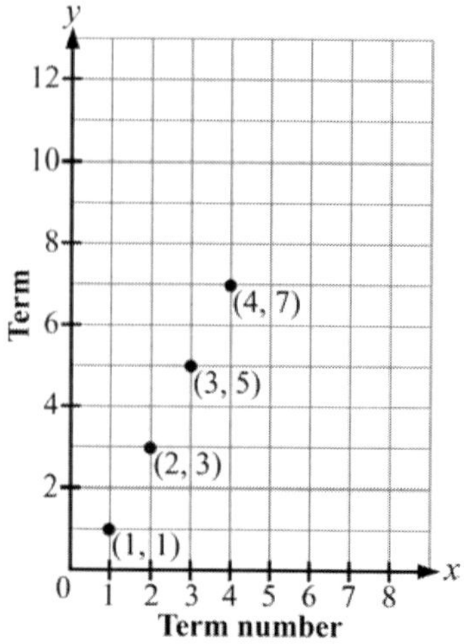

The ordered pairs (1, 1), (2, 3), (3, 5), and (4, 7) identify the locations of the four plotted points on the given graph. Create a table of values from the graph.

Solution

Step 1
Write the term numbers.

The term numbers will be the first numbers of the ordered pairs.

Term Number	Term
1	
2	
3	
4	

Step 2
Write the terms.

The terms will be the second numbers in the ordered pairs. Write them in the order they appear.

Term Number	Term
1	1
2	3
3	5
4	7

NOTES

Time to Try 2

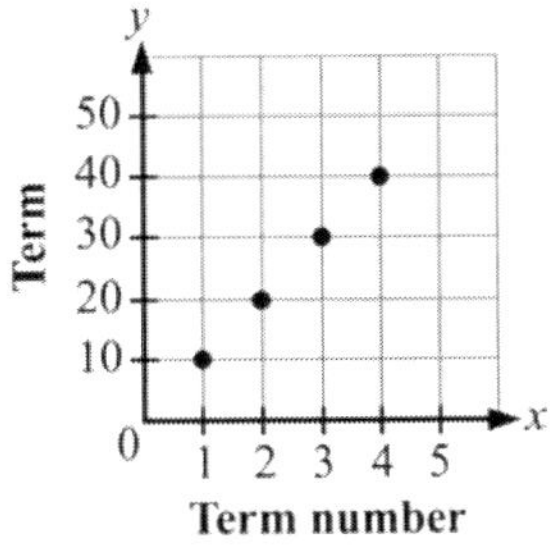

Create a table of values from the graph.

TRANSLATING A PATTERN TO A TABLE OF VALUES AND GRAPHING THE TABLE OF VALUES

A pattern can be translated to a table of values and the term numbers and terms can be written as ordered pairs. Then, the ordered pairs can be plotted on a graph.

Example

Translate the pattern 4, 5, 6, 7 into a table of values, and then graph the table of values.

Solution

Step 1
Translate the pattern to a table of values.

Term Number	Term
1	4
2	5
3	6
4	7

Step 2
Write the term numbers and terms as ordered pairs.

The ordered pairs are (1, 4), (2, 5), (3, 6), (4, 7).

NOTES

Step 3
Plot the ordered pairs on a graph.

- Start at 0, and move to the right along the *x*-axis until you reach the given term number.
- Move vertically until you reach the line that the given term from the *y*-axis is on.
- Make a dot where the two lines intersect.

The graph could look like this.

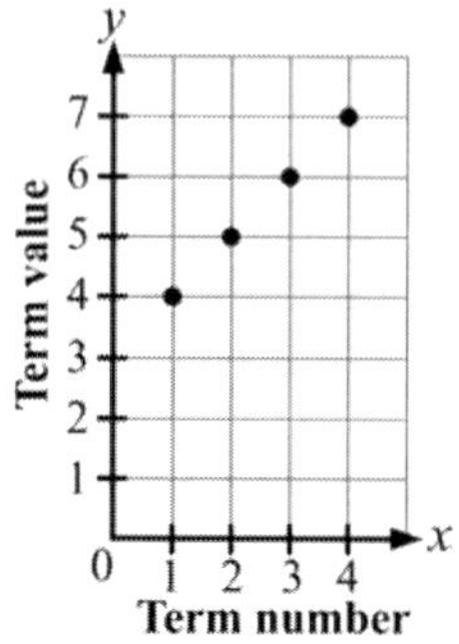

Time to Try 3

Translate the pattern 7, 5, 3, 1 into a table of values, and then graph the table of values.

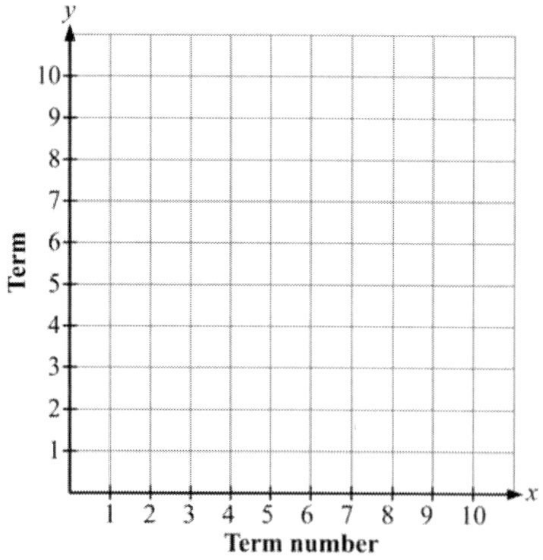

DESCRIBING RELATIONSHIPS SHOWN ON GRAPHS

The main relationship that is usually described in a graph is the relationship between the term numbers and their corresponding terms. Such patterns are often more easily seen when presented on a graph.

NOTES

A relationship between term numbers and terms describes the changes in a pattern from the term number to the term value. It describes the relationship between the numbers on the x-axis and the y-axis on a graph.

Example

On the given graph, the term numbers are the times on the x-axis. The terms are the temperatures on the y-axis. Dotted lines connect the term numbers (times) and their corresponding terms (temperatures) to help you see the relationship between each term number and term.

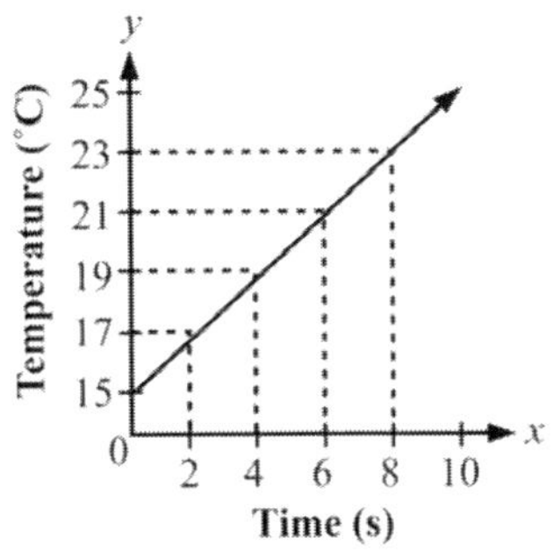

What is the relationship between the times and the corresponding temperatures?

Solution

Step 1
Determine how the term numbers change.

The term numbers (time in seconds) start at 2 and increase by 2 each time.

Step 2
Determine how the terms change.

The corresponding terms (temperatures) start at 17 and increase by 2 each time.

Step 3
Determine the relationship between term numbers and corresponding terms.

The relationship between the term numbers and their corresponding terms (time and its corresponding temperature) is that every time the time increases by 2 seconds, the temperature increases by 2 degrees.

NOTES

Time to Try 4

On the given graph, the minutes a tortoise travels is shown on the horizontal x-axis. These numbers represent the term numbers.

The distance the tortoise travels is shown on the vertical y-axis. These numbers represent the corresponding terms.

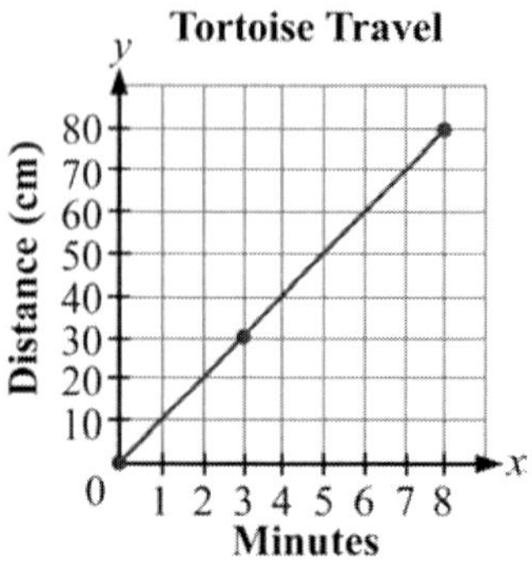

What is the relationship between the times the tortoise travels and the corresponding distances the tortoise travels?

PRACTICE EXERCISES

1. Create a table of values from the pattern 2, 3, 5, 8, 12.

Use the following graph to answer the next question.

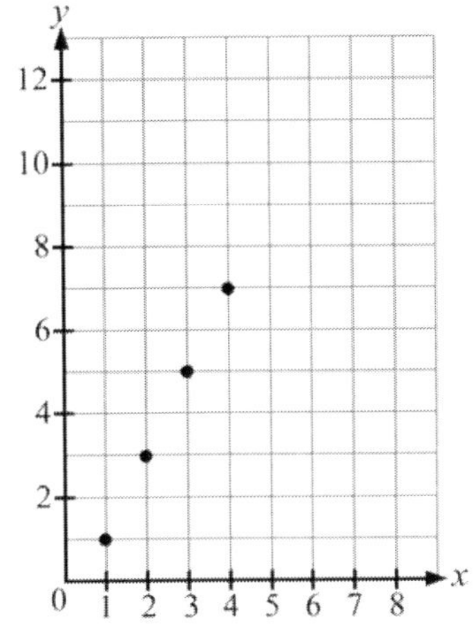

2. Create a table of values from the graph.

3. Translate the pattern 2, 7, 12 into a table of values and then graph the table of values.

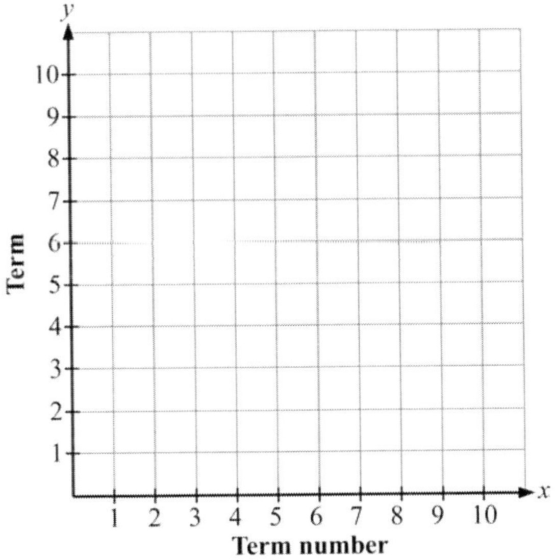

Use the following graph to answer the next question.

On the given graph, the number of tables used is shown on the horizontal x-axis. These numbers represent the term numbers.

The number of people who sit at each table is shown on the vertical y-axis. These numbers represent the corresponding terms.

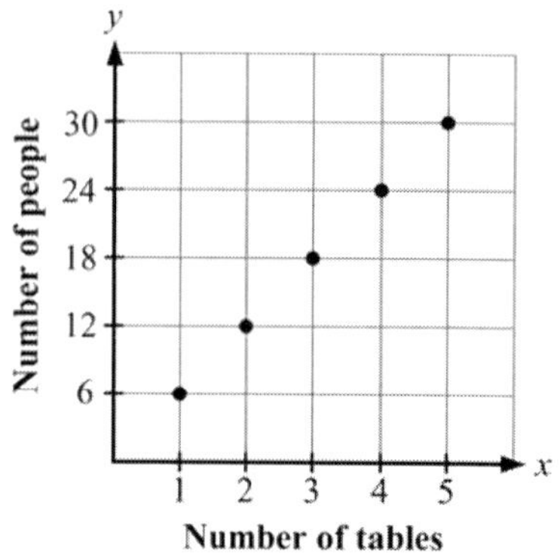

4. Describe the relationship between tables and people seated at tables.

Use the following graph to answer the next question.

On the given graph, the number of pens purchased is shown on the horizontal x-axis. These numbers represent the term numbers.

The number of pencils purchased for the same price is shown on the vertical y-axis. These numbers represent the corresponding terms.

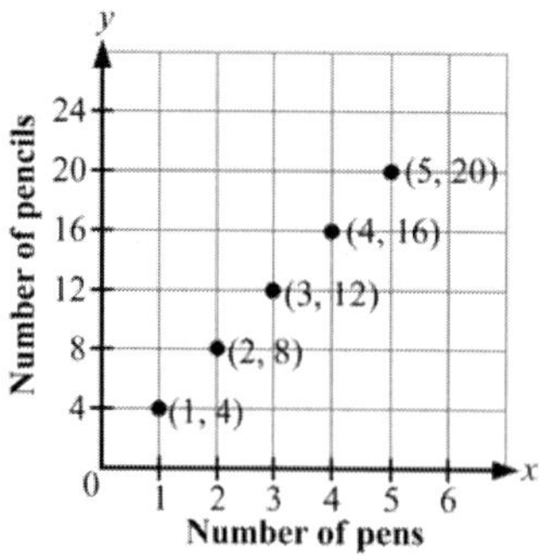

5. Describe the relationship between pens and pencils.

Lesson 2 USING RELATIONSHIPS WITHIN TABLES TO SOLVE PROBLEMS

NOTES

DESCRIBING PATTERNS WITHIN EACH COLUMN AND FORMULATING RULES

There are two main types of relationships that can be represented in a table of values. There are relationships among terms and relationships between term numbers and terms.

Relationships Among Terms

A relationship between terms describes the change that occurs from one term to the next term in a given pattern of numbers. The relationship among the terms is the same as the pattern rule used to create the numeric pattern.

Example

Describe the relationship among the terms in this table of values. Using mathematical language, express the relationship as a pattern rule.

Term Number	Term
1	8
2	11
3	14
4	17

Solution

Step 1
Determine how the numbers change from term to term.

The pattern starts with 8 and adds 3 to each term to get the next term.

$8+3=11$
$11+3=14$
$14+3=17$

Step 2
Describe the relationship.

The relationship between any two consecutive terms in the pattern is that each successive term is 3 greater than the previous term. The relationship can be expressed as the rule +3 or $n+3$, where n represents any given term.

NOTES

Time to Try 1

Describe the relationship among the terms in this table of values. Using mathematical language, express the relationship as a pattern rule.

Term Number	Term
1	3
2	6
3	12
4	24

Relationships Among Term Numbers and Terms

A relationship between term numbers and terms describes the changes in a pattern from the term number to its term value. It describes the relationship between the two columns in a table of values.

Example

Describe the relationship between each term number and its corresponding term. Using mathematical language, express the relationship as a pattern rule.

Term Number	Term
1	4
2	8
3	12
4	16

Solution

Step 1

Determine how each term number changes to its corresponding term.

$1 \times 4 = 4$

$2 \times 4 = 8$

$3 \times 4 = 12$

$4 \times 4 = 16$

NOTES

Step 2
Describe the relationship.

The relationship between any term number and its corresponding term is that the term number is multiplied by 4. The relationship can be expressed as the rule $\times 4$ or $k \times 4$, where *k* represents any given term number.

Time to Try 2

Describe the relationship between each term number and its corresponding term. Using mathematical language, express the relationship as a pattern rule.

Term Number	Term
1	8
2	9
3	10
4	11

NOTES

PREDICTING THE VALUE OF AN UNKNOWN TERM

When a pattern is presented in a table of values, the value of an unknown term can be predicted by using the relationship between the terms. First, determine the pattern rule used to create the pattern. Then, apply the pattern rule to make a prediction and verify the prediction made.

Example

Term Number	Term
1	4
2	7
3	10
4	13

Predict the value of the 7th term of the pattern. Verify your prediction.

Solution

Step 1

Determine the relationship.

$4+3=7$

$7+3=10$

$10+3=13$

Each number is 3 more than the previous number.

Step 2

Make a prediction.

To get from term number 4 to 7, add 3 for three more counts. A good prediction is 22.

Step 3

Verify the prediction.

Term Number	Term
5	$13+3=16$
6	$16+3=19$
7	$19+3=22$

NOTES

Time to Try 3

Term Number	Term
1	87
2	82
3	77
4	72

Predict the value of the 7th term of the pattern. Verify your prediction.

GENERATING VALUES IN ONE COLUMN GIVEN A PATTERN RULE AND VALUES IN THE OTHER COLUMN

Values in one column of a table of values can be generated if the values in the other column and a pattern rule are given. A pattern rule is used to find each successive term in a pattern. By studying the pattern, the term for a specific term number can be determined. The term number of a specific term can also be determined.

Example

The pattern rule is to start with 1 and add 2 to each term to get the next term. The first three terms of the pattern are shown in the table of values.

Term Number	Term
	1
	3
	5

Complete the table of values to determine which term number will have a value of 13.

Solution

Step 1
Continue applying the pattern rule.

Fill in the terms up to the number 13.

Term Number	Term
	7
	9
	11
	13

NOTES

Step 2
Number the term numbers, starting with 1.

Term Number	Term
1	1
2	3
3	5
4	7
5	9
6	11
7	13

The 7th term will have a value of 13.

Time to Try 4

The pattern rule is to start with 2 and add 6 to each term to get the next term. The first six term numbers of the pattern are shown in the table of values.

Term Number	Term
1	
2	
3	
4	
5	
6	

Complete the table of values to determine which term number will have a value of 26.

IDENTIFYING MISSING ELEMENTS IN A TABLE OF VALUES

In some problems, the missing terms or missing term numbers may need to be identified. To do this, the numbers should be examined to see how they change from term to term. Once the relationship between the numbers is determined, the missing terms or term numbers can then be identified.

NOTES

Example

The first, fifth, and sixth terms of a sequence are 1, 16, and 32.

Term Number	Term
1	1
2	
3	
4	
5	16
6	32

Fill in the missing terms.

Solution

Step 1
Determine the relationship.

Only 16 and 32 are consecutive numbers, which means that 16 is doubled (multiplied by 2) to get 32. This shows that the pattern rule is to multiply each term by 2 to get the next term. The rule can be expressed as $\times 2$ or $n \times 2$ where n represents any term.

Step 2
Start with 1 and apply the pattern rule.

Term Number	Term
1	1
2	$1 \times 2 = 2$
3	$2 \times 2 = 4$
4	$4 \times 2 = 8$

Step 3
Verify the pattern rule for terms 5 and 6.

5	$8 \times 2 = 16$
6	$16 \times 2 - 32$

The missing terms are 2, 4, and 8.

NOTES

Time to Try 5

Kurt created the following table of values showing that the first term is 96, the fourth term is 12, and the fifth term is 6.

Term Number	Term
1	96
2	
3	
4	12
5	6
6	

Identify the values of the 2nd, 3rd, and 6th terms in Kurt's pattern.

CREATING A CONCRETE OR PICTORIAL REPRESENTATION OF A RELATIONSHIP

A relationship among numbers shown in a table of values can be represented with concrete objects like blocks or tiles, or it can be represented with drawings of shapes.

Example

Use counters or draw circles to represent the relationships among the given terms.

Term Number	Term
1	1
2	2
3	4
4	7
5	11

Solution

Step 1

Determine how the numbers change from term to term.

$1+1=2$

$2+2=4$

$4+3=7$

$7+4=11$

The pattern rule can be described as $n+1, +2, +3, +4$

NOTES

Step 2
Apply the rule.

- Start with one circle to make the first figure.
- Add one more circle to make the second figure.
- Add two more circles to make the third figure.
- Add three more circles to make the fourth figure.
- Add four more circles to make the fifth figure.

The concrete or pictorial representation can look like this.

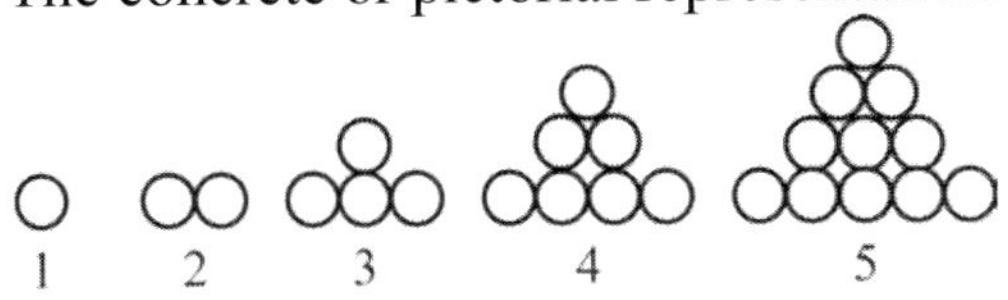

Time to Try 6

Use star tiles or draw stars to represent the relationships among the given terms.

Term Number	Term
1	16
2	14
3	12
4	10
5	8

IDENTIFYING ERRORS IN TABLES

In order to identify errors in a table of values, the relationships between the numbers of the pattern must first bc understood. Once the relationships are understood, check to see if the rule has been followed throughout.

NOTES

Example

Identify the error in this table of values.

Term Number	Term
1	8
2	16
3	32
4	64
5	96

Solution

Step 1

Determine the relationship.

Each term is multiplied by 2 to get the next term.

$8 \times 2 = 16$

$16 \times 2 = 32$

$32 \times 2 = 64$

$64 \times 2 = 128$

Step 2

Identify the error.

The fifth term is incorrect. It does not follow the pattern rule of $\times 2$.

$64 \times 2 = 128$, not 96.

Time to Try 7

Identify the error in this table of values.

Term Number	Term
1	324
2	108
3	36
4	12
5	3

NOTES

CREATING TABLES TO SOLVE PROBLEMS

When working with patterns and relationships of numbers, it can be helpful to create a table of values to organize the information, understand the relationships, and solve the problem.

Example

Jenna reads independently for 3 hours each week. If she reads for 30 minutes a day, how many days will it take her to read for a total of 3 hours? Create a table of values to help solve the problem.

Solution

Step 1
Identify the relationship.

Since Jennifer must read 30 minutes each day, the pattern starts with 30 and 30 is added to each term to get the next term.

Step 2
Create a table of values.

Since there are 180 minutes in three hours, keep adding 30 to each term up to the number 180.

$1 \text{ h} = 60 \text{ min}$

$3 \text{ hr} = 60 \times 3 = 180 \text{ min}$

Term Number (Number of days)	**Term** (Number of minutes)
1	30
2	$30 + 30 = 60$
3	$60 + 30 = 90$
4	$90 + 30 = 120$
5	$120 + 30 = 150$
6	$150 + 30 = 180$

It would take Jenna six days to read for three hours.

Time to Try 8

Mrs. Smith has to seat 112 people for the community league dinner. She uses same-sized square tables. If four people can sit at each side of a square table, how many tables does Mrs. Smith need? Create a table of values to help solve the problem.

PRACTICE EXERCISES

Use the following table of values to answer the next question.

Term Number	Term
1	2
2	8
3	32
4	128

1. Describe the relationship among the terms in the table of values. Using mathematical language, express the relationship as a pattern rule.

Use the following table of values to answer the next question.

Term Number	Term
1	13
2	14
3	15
4	16

2. Describe the relationship between each term number and its corresponding term. Using mathematical language, express the relationship as a pattern rule.

3. Predict the value of the 7th term. Verify your prediction.

Term Number	Term
1	1
2	12
3	23
4	34
5	
6	
7	

4. The pattern rule is to start with 5 and add 13 to each term to get the next te he table of values to determine which term number will have a value of 70.

Term Number	Term
	5
	18
	31

5. The first, fifth, and sixth terms of a sequence are 64, 4, and 2. Fill in the missing terms.

Term Number	Term
1	64
2	
3	
4	
5	4
6	2

Use the following table of values to answer the next question.

Term Number	Term
1	1
2	5
3	9
4	13

6. Use square tiles or draw squares to represent the relationships among the given terms.

Use the following table of values to answer the next question.

Term Number	Term
1	87
2	92
3	97
4	112
5	117

7. Identify the error in this table of values.

Use the following information to answer the next question.

> Every day for one whole week, some students collected food for the food bank.
> - On the first day, they collected 13 cans of food.
> - On the second day, they collected 28 cans of food.
> - On the third day, they collected 43 cans of food.

8. If the number of cans they collected each day follows the given pattern, how many cans will they have collected on the 7th day? Create a table of values to help solve this problem.

REVIEW SUMMARY

- In a table of values, the Term Number column shows the order in which the terms of the pattern are presented. The Term column shows the consecutive numbers of the given pattern.
- A table of values can be created from a given pattern or from a given graph.
- Relationships between terms and between term numbers and their corresponding terms can be described in everyday language, in mathematical language, and by formulating pattern rules.
- A pattern can be translated into a table of values. The term numbers and their corresponding terms can then be written as ordered pairs, which can then be plotted on a graph.
- The first number of the ordered pair will be located on the horizontal x-axis. The second number of the ordered pair will be located on the vertical y-axis.
- Values can be generated in one column of a table of values if the pattern rule and the numbers in the other column of the table are known.
- To predict the value of an unknown term, use the relationship seen between the terms. Apply the relationship (rule) to make the prediction, and then verify the prediction by extending the pattern.
- To find missing elements, first determine the relationship between the numbers. Then, apply the rule, remembering to verify the rule.
- Since it is often easier to see relationships when information is presented in tables, it may be helpful to create a table of values to help solve problems involving patterns and relationships.

PRACTICE TEST

1. Create a table of values from the pattern 1, 2, 4, 8, 16.

Use the following graph to answer the next question.

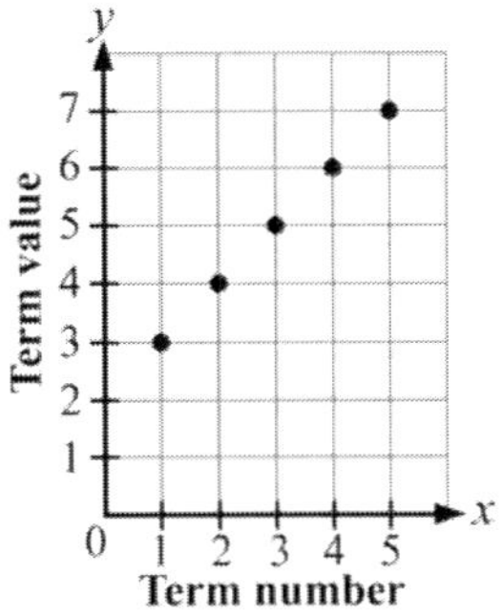

2. Create a table of values from the graph.

3. Translate the pattern 0, 1, 2, 3 into a table of values and then graph the table of values.

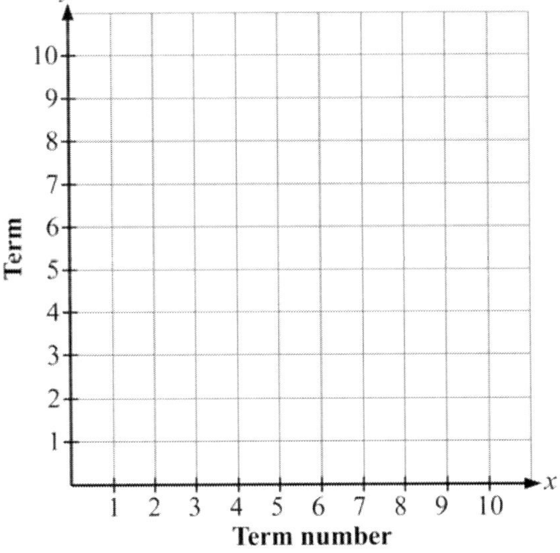

Use the following graph to answer the next question.

Jordan plotted points on a graph to show the temperatures he recorded on a particular day from 10:00 A.M. to 3:00 P.M.

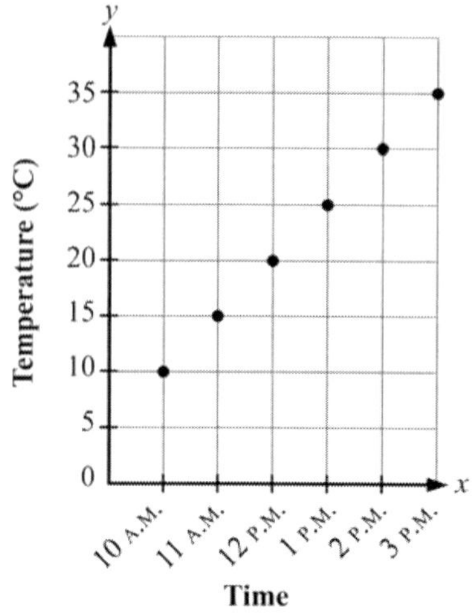

4. Describe the relationship between the times recorded (term numbers) and the corresponding temperatures (terms).

Use the following table of values to answer the next question.

Term Number	Term
1	625
2	125
3	25
4	5

5. Describe the relationship among the terms in this table of values. Using mathematical language, express the relationship as a pattern rule.

Use the following table of values to answer the next question.

Term Number	Term
2	22
4	44
6	66
8	88

6. Describe the relationship between each term number and its corresponding term. Using mathematical language, express the relationship as a pattern rule.

7. Predict the value of the 7th term of the pattern. Verify your prediction.

Term Number	Term
1	13
2	21
3	29
4	37
5	
6	
7	

Use the following information to answer the next question.

The pattern rule is to start with 3 and multiply each term by 2 to get the next term. The first three terms of the pattern are shown in the table of values.

8. Complete the table of values to determine which term number will have a value of 192.

Term Number	Term
	3
	6
	12

9. The first, second, and sixth terms of a sequence are 4, 11, and 39. Fill in the missing terms.

Term Number	Term
1	4
2	11
3	
4	
5	
6	39

Use the following table of values to answer the next question.

Term Number	Term
1	1
2	3
3	9
4	27

10. Use square tiles or draw squares to represent the relationships among the given terms.

Use the following table of values to answer the next question.

Term Number	Term
3	21
4	32
5	35
6	42
7	49

11. Identify the error in the table of values.

12. Josh rides in a bike-a-thon to raise money for the local animal shelter. He maintains a constant speed of 18 km per hour. How many hours will it take Josh to bike 108 km? Create a table of values to help solve the problem.

ANGLES

When you are finished this unit, you should be able to…

- Identify examples of angles in the environment
- Classify angles as acute, right, obtuse, straight, or reflex
- Estimate the measure of angles, using 45°, 90°, and 180° as reference angles
- Describe the relationships among the reference angles
- Sketch reference angles without the use of a protractor
- Measure angles in degrees
- Draw and label a specified angle with the use of a protractor
- Demonstrate that the sum of the interior angles of any triangle is alw[illegible]
- Demonstrate that the sum of the interior angles of any quadrilateral is al[illegible]°

Lesson	Page	Com[illegible] on
1. Identifying and Classifying Angles According to Their Measures	164	
2. Estimating and Determining Angle Measures in Degrees	168	
3. Drawing Angles With and Without a Protractor	175	
4. The Sum of Interior Angles	181	
Review Summary	186	
Practice Test	187	
Answers and Solutions	at the back of the book	

PREREQUISITE SKILLS AND KNOWLEDGE

Prior to beginning this unit, you should be able to…

- Recognize examples of 90° angles in the environment
- Sketch a 90° angle without the use of a protractor
- Label a 90° angle using a symbol

Lesson 1 IDENTIFYING AND CLASSIFYING ANGLES ACCORDING TO THEIR MEASURES

NOTES

An angle is formed by two lines that share the same endpoint. The point where the lines intersect (meet) is called the vertex of the angle. An angle is often shown with an arc or an arc with an arrow. A right angle is often shown with a square corner. The size of an angle is measured in degrees (°).

The following chart shows five classifications of angles. Each classification has an illustration of the angle and an explanation of the measure of the angle.

Name of Angle	**Example of Angle**	**Explanation of Angle's Measure**
Acute angle	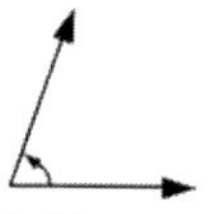	An angle that measures less than 90°.
Right angle	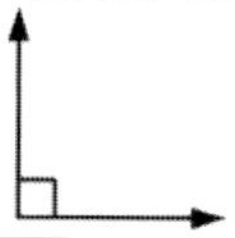	An angle that measures exactly 90°.
Obtuse angle		An angle that measures more than 90° but less than 180°.
Straight angle		An angle that measures exactly 180°. It forms a straight line.
Reflex angle		An angle that measures more than 180° but less than 360°.

Example

Name each angle according to its measure. Explain your answer.

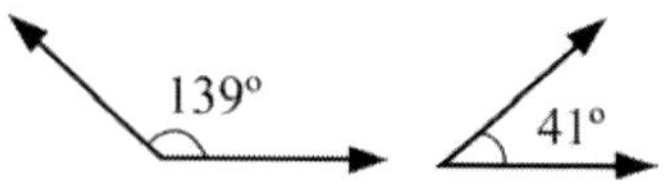

Solution

The first angle is an example of an obtuse angle because its measure is greater than 90° and less than 180°.

The second angle is an example of an acute angle because its measure is less than 90°.

NOTES

Time to Try 1

Name each angle according to its measure. Explain your answer.

IDENTIFYING ANGLES FOUND IN THE ENVIRONMENT

Angles exist everywhere. It is easy to find examples of angles in the environment because everyday objects are made up of a variety of angles. Almost everything you see has angles.

This bridge structure shows several examples of right angles and straight angles

The tips of the sails on this sailboat show examples of acute and obtuse angles.

PRACTICE EXERCISES

Use the following information to answer the next question.

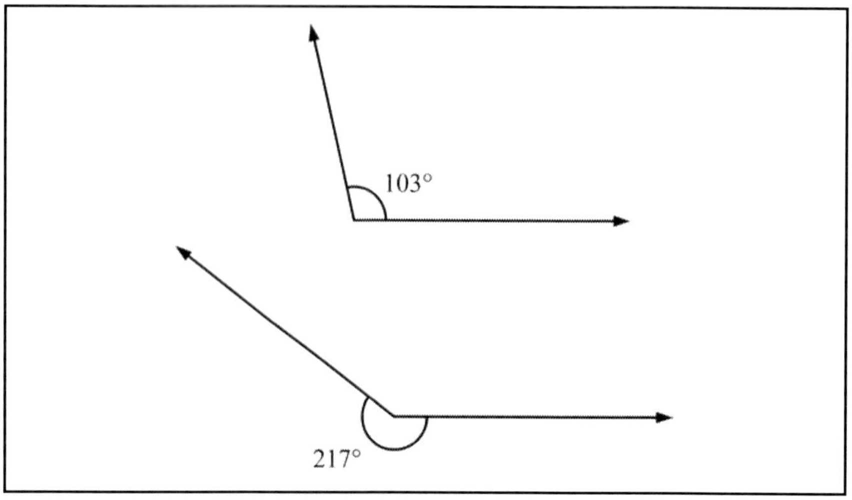

1. Classify the angles shown according to their measures. Justify your answer.

Use the following information to answer the next question.

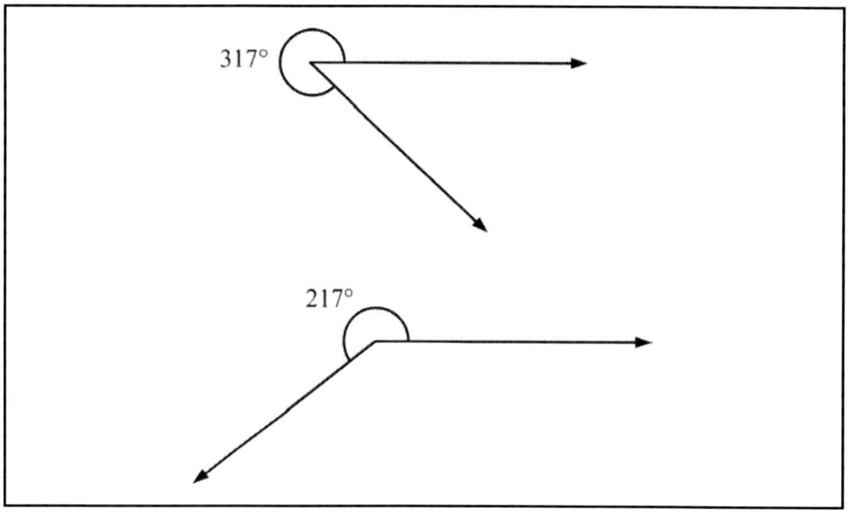

2. Ira's teacher asked him to label the two given angles. He told her that the top angle is an acute angle and the bottom angle is a reflex angle. Is Ira correct? Justify your answer.

Use the following information to answer the next question.

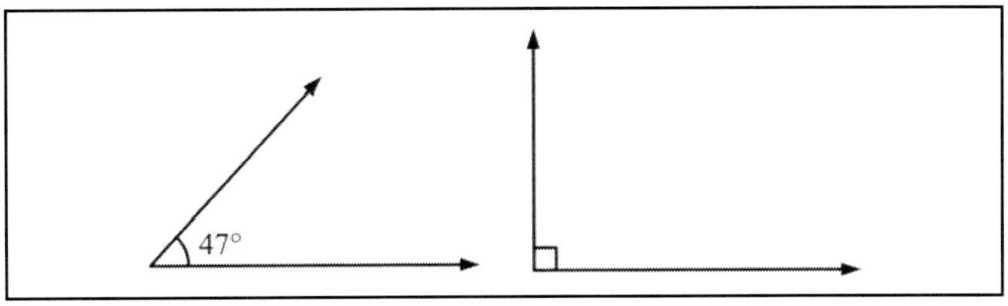

3. Classify the angles shown according to their measures. Justify your answer.

Lesson 2 ESTIMATING AND DETERMINING ANGLE MEASURES IN DEGREES

NOTES

Notice the relationship between these reference angles: 90° is twice the size of 45°, and 180° is twice the size of 90°.

You can estimate the size of angles using reference angles, and you can determine actual sizes of angles by measuring with a protractor.

USING REFERENCE ANGLES TO ESTIMATE ANGLE MEASURES

You can use your knowledge of angle size to estimate the measurement of an angle. You can do this by comparing an angle size to reference angles like 180°, 90°, or 45° to help you make a reasonable estimate.

- The halfway point between 0° and 90° is 45°.
- The halfway point between 90° and 180° is 135°.
- The halfway point between 0° and 180° is 90°.

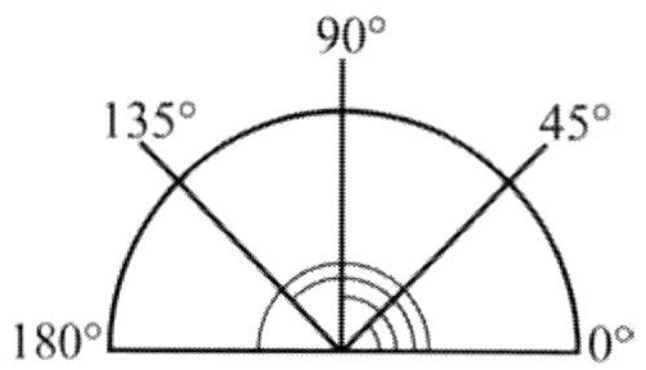

Example

Estimate the measure of the following angle. Explain your estimate.

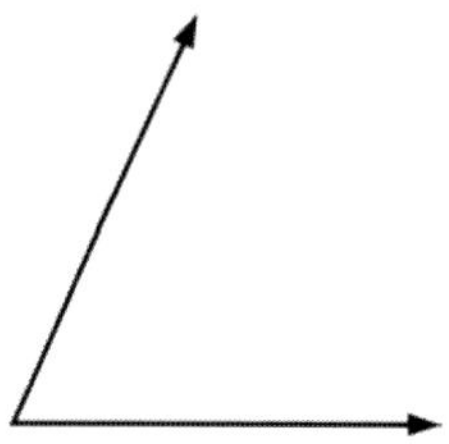

Solution

The given angle is greater than a 45° angle, but is less than a 90° angle. It appears to be closer to a 90° angle than a 45° angle.

A good estimate of the angle measure could be about 70°, which is closer to 90° than 45°.

NOTES

Time to Try 1

Estimate the measure of the following angle. Explain your estimate.

USING A PROTRACTOR TO DETERMINE ANGLE MEASURES

A protractor is an instrument used for measuring angles. It is usually in the shape of a semicircle and is marked around the edge in degrees.

A semicircle protractor goes from 0° to 180°. Most semicircle protractors can be read from both the left side (outer scale, starts with 0) and from the right side (inner scale, starts with 0), as shown in the following diagram.

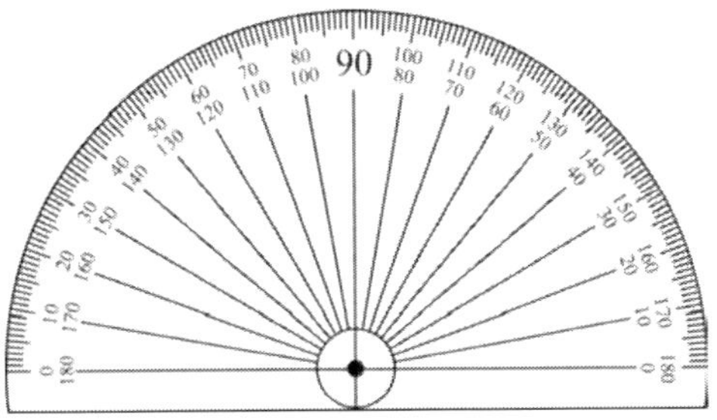

It is important to examine the direction of the numbers on the two scales. On some protractors, the inner scale starts with the 0 on the right and ends with 180 on the left, like the one shown. On other protractors, however, the outer scale starts with 0 on the right and ends with 180 on the left.

When measuring an angle, follow these steps:

1. Place the midpoint (centre) of the protractor on the vertex of the angle.
2. Line up the midpoint and the 0° line of the protractor with one ray of the angle.
3. Start from 0° and count the number of degrees to where the other ray lines up.

NOTES

The following diagram shows how to place the protractor on the given angle to measure the angle.

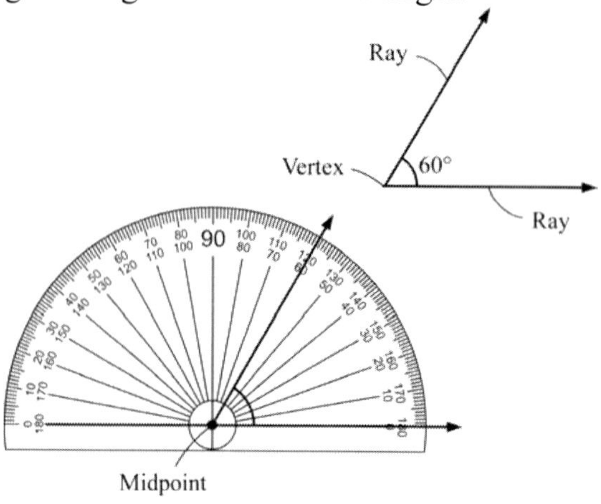

Example

Use a protractor to measure this angle.

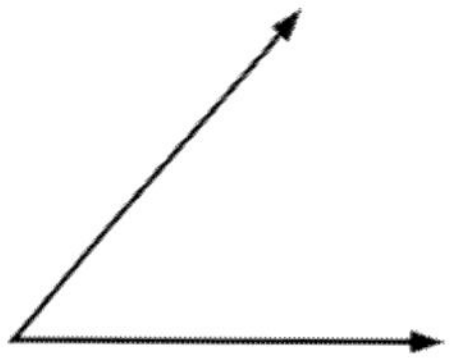

Solution

Step 1
Place the midpoint of the protractor (centre) on the vertex of the angle.

Step 2
Line up the midpoint and the 0° line of the protractor with the bottom ray of the angle.

Step 3
Start from 0° on the inner scale, and count the number of degrees to where the other ray lines up.

The measure of the given angle is 48°.

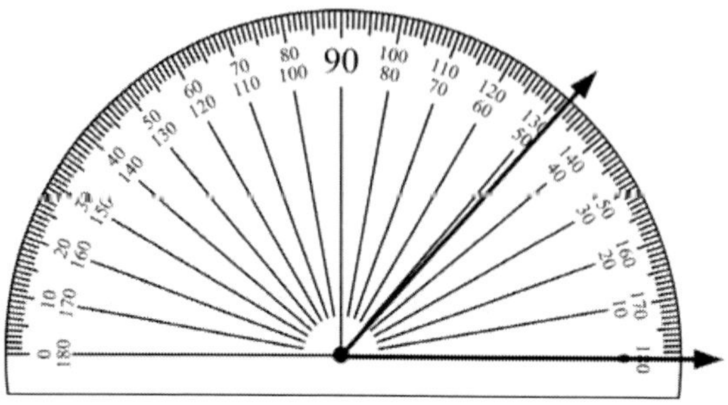

NOTES

Time to Try 2

Use a protractor to measure the given angle.

MEASURING REFLEX ANGLES

There are two methods for measuring a reflex angle.

1. Start with a straight angle of 180°. Flip the protractor vertically, lining it up with the midpoint and the 0° line. Continue counting the degrees to the outer ray. Add the two sets of degrees to get the measure of the angle.
2. Measure the smaller angle (the interior or inside angle) that is less than 180°, and subtract that measure from 360°, which is a full circular turn.

Example

Explain how to measure the following reflex angle by flipping the protractor and adding the measures.

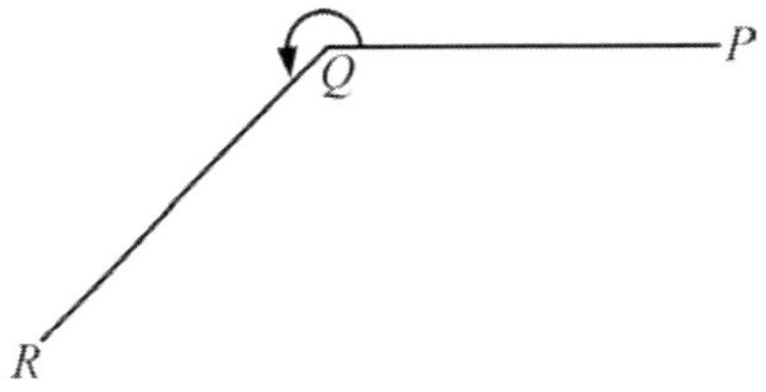

Solution

Step 1

Extend ray *PQ* to the left, creating a straight angle of 180° (angle *PQA* in the following diagram), and then flip the protractor vertically, keeping it lined up from the midpoint to the 0° line.

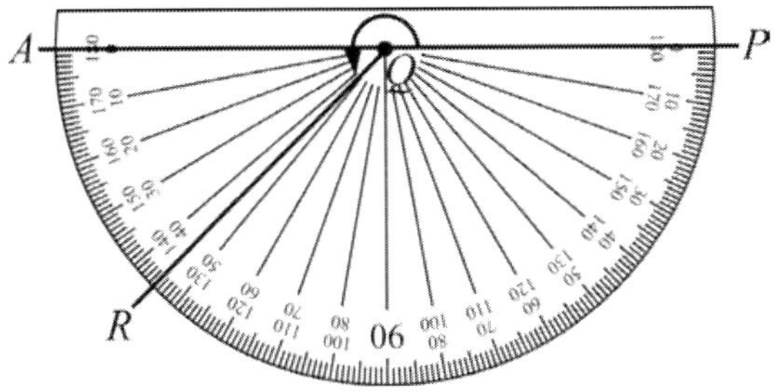

NOTES

Step 2
Start from 0° on the inner scale (left side) and count the number of degrees to where the other ray (*QR*) lines up. It lines up at 45°.

Step 3
Add the two sets of degrees.
$180° + 45° = 225°$

Reflex angle *PQR* has a measure of 225°.

Time to Try 3

Explain how to determine the measure of the given reflex angle using subtraction from 360°.

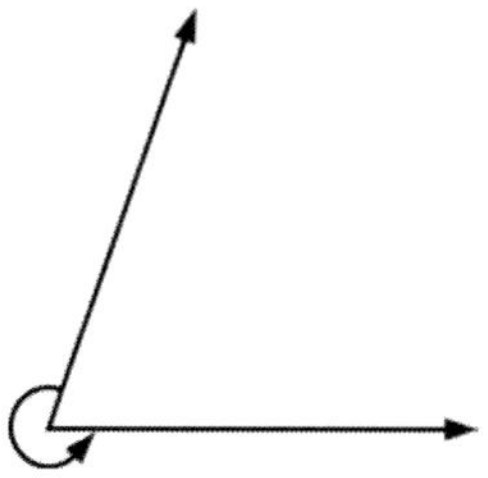

PRACTICE EXERCISES

Use the following diagram to answer the next question.

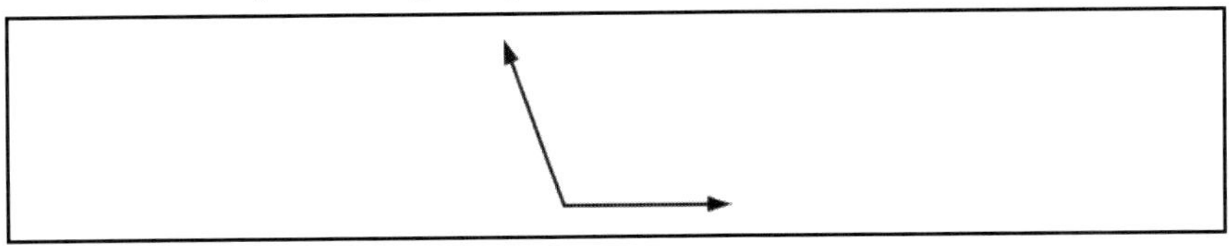

1. Estimate the measure of the given angle by comparing it to reference angles.

Use the following diagram to answer the next question.

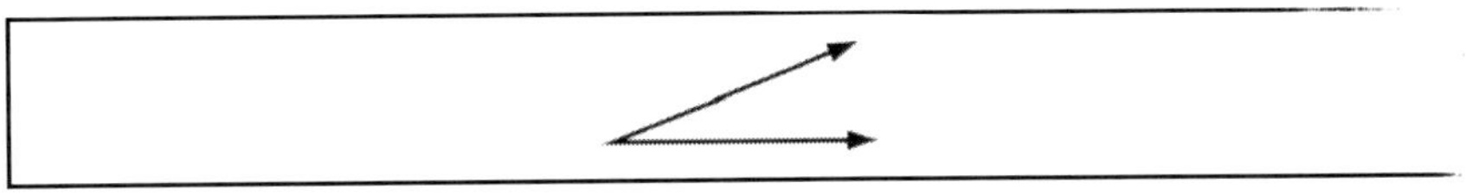

2. Estimate the measure of the given angle by comparing it to reference angles.

Use the following diagram to answer the next question.

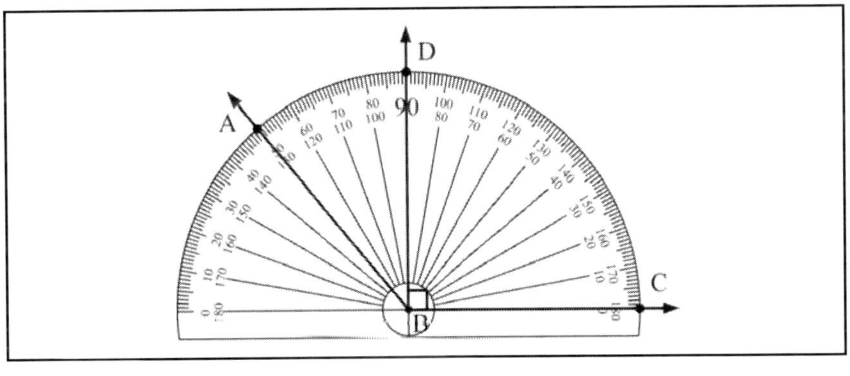

3. What is the measure of angle *ABC*?

Use a protractor to measure the following angles.

4.

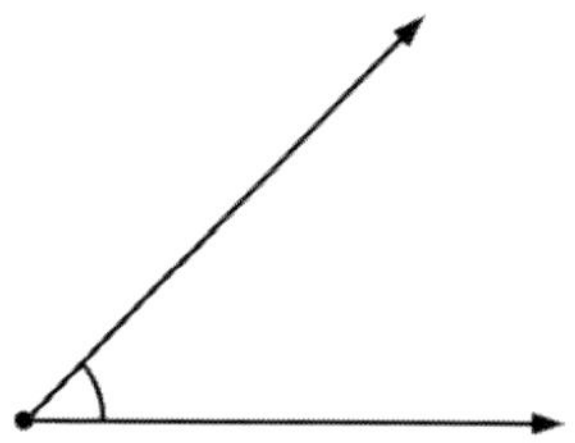

5.

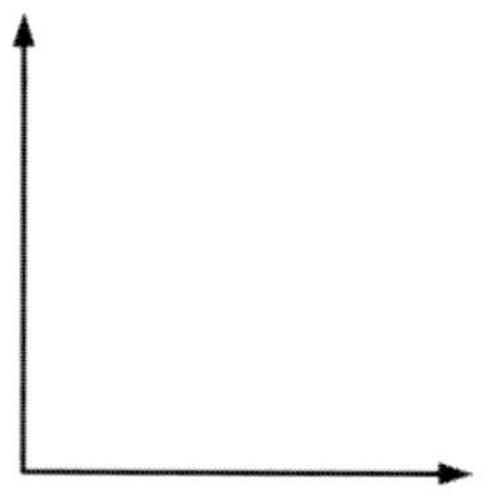

6.

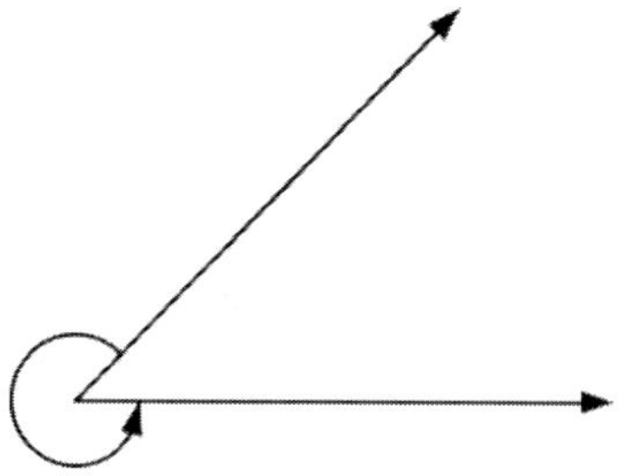

Lesson 3 DRAWING ANGLES WITH AND WITHOUT A PROTRACTOR

NOTES

When drawing angles, you can choose to use or not use a protractor.

SKETCHING REFERENCE ANGLES WITHOUT A PROTRACTOR

You can easily sketch angles like 45° angles, 90° angles, and 180° angles if you understand the relationship among the angles.

The following diagram shows the relationships among the reference angles.

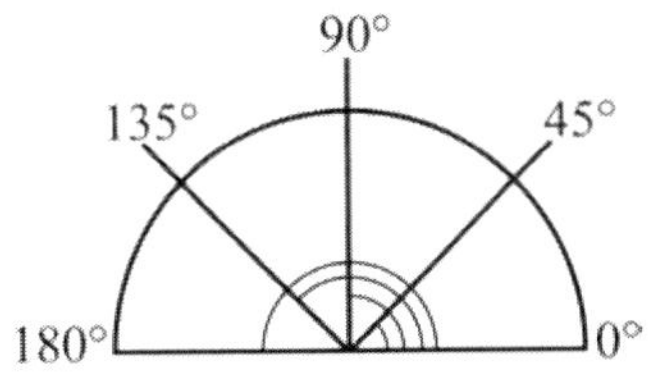

- A straight angle of 180° looks like a straight line and is twice the size of a right angle. Two right angles make a straight angle.
- A right angle of 90° looks like a squared corner with the two rays being perpendicular to each other. A right angle is twice the size of a 45° angle.
- A 45° angle is half the size of a 90° angle and one-quarter the size of a 180° angle.
- Just like the 45° angle is half the size of the right angle on the right of the diagram, the 135° angle is half the size of the right angle on the left of the diagram.

Example

Sketch and label a 45° angle.

Solution

Think of a right angle.

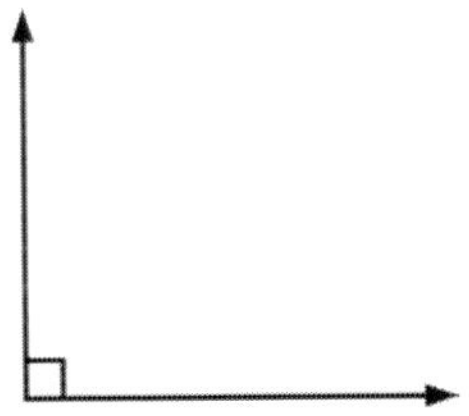

NOTES

Estimate the halfway mark between the two rays.

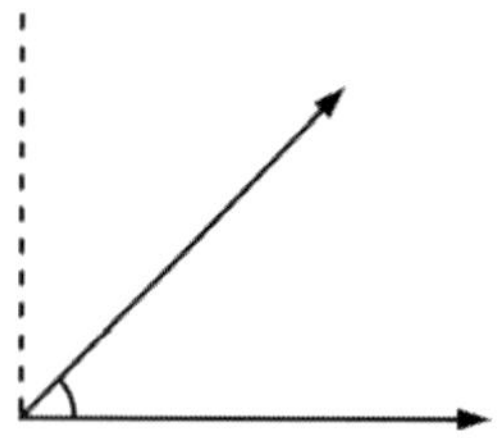

Draw and label the angle.

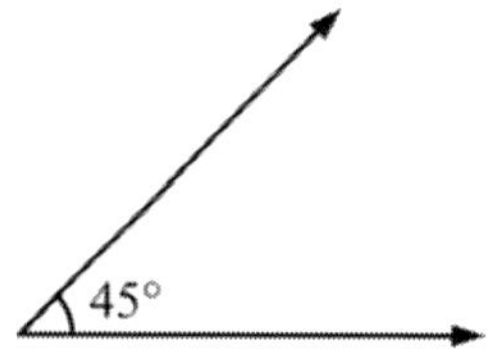

Time to Try 1

Sketch and label a 180° angle.

DRAWING ANGLES USING A PROTRACTOR

When constructing an angle using a protractor, follow these steps:
Draw one ray of your angle.

Place the centre of the protractor (midpoint) on one end of the line where you want the vertex to be.

Line up the centre and the 0° tick of the protractor with the ray you drew.

Starting from 0°, count the number of degrees on the protractor that you want the angle to be. Make a tick.

Draw a line between the tick and the end of the ray where you wanted the vertex.

NOTES

Remember, angles do not always face right. They can be drawn in different orientations. For example, these two angles are the same, but are drawn in different orientations.

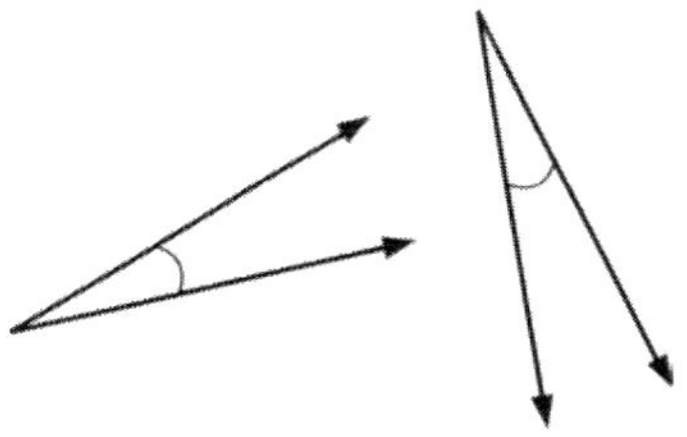

Example

Draw an angle that has a measure of 135°.

Solution

Step 1
Draw a ray.

Step 2
Place the protractor on the ray so that the midpoint of the protractor lines up with the endpoint of the ray.

Start at the 0 on the inner scale, as shown on this protractor, and count up to 135°. Make a tick at 135°.

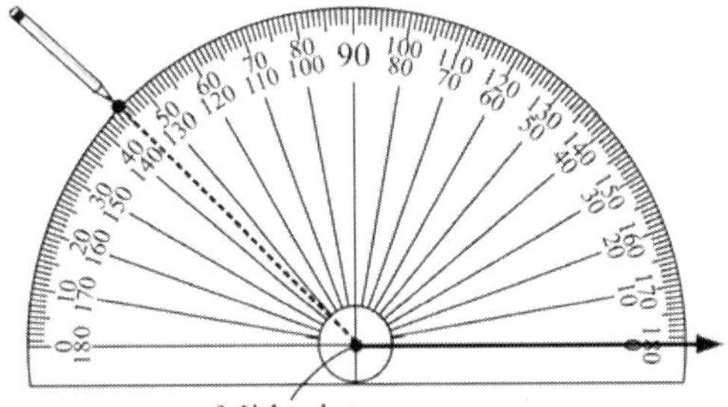

Step 3
Remove the protractor.

Using a ruler, join the tick at 135° to the endpoint of the original ray.

NOTES

Time to Try 2

Using a protractor and a ruler, construct an angle that has a measure of 77°.

CONSTRUCTING REFLEX ANGLES

To construct reflex angles, follow theses steps:

1. Calculate the size of the smaller (inner) angle by subtracting the size of the reflex angle from 360° (a full turn).
2. Draw the smaller angle.
3. Mark the outside rotation with an arc, and label the measurement. The outside rotation is the reflex angle.

Example

Draw an angle measuring 240°.

Solution

Step 1

Calculate the size of the smaller (inner) angle.

$360° - 240° = 120°$

Step 2

Draw the 120° angle.

Label the outside rotation as 240°.

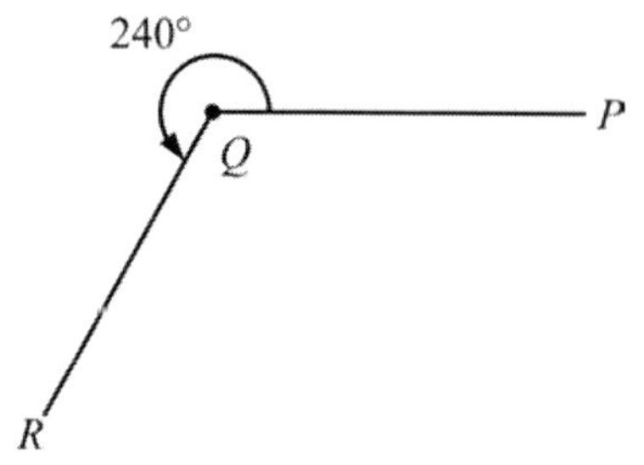

Time to Try 3

Draw an angle that has a measure of 300°.

NOTES

PRACTICE EXERCISES

1. Using a protractor, draw an angle that has a measure of 30°.

2. Using a protractor, draw an angle that has a measure of 145°.

3. Sketch a right angle without using a protractor. Draw the angle in three different orientations.

4. Draw an angle that has a measure of 265°.

5. Draw and label an angle that faces left and has a measure of 63°.

Lesson 4 THE SUM OF INTERIOR ANGLES

NOTES

The interior angles of a triangle will always add up to 180°, while the interior angles of a quadrilateral will always add up to 360°. Knowing these rules will be helpful.

INTERIOR ANGLES OF TRIANGLES

The interior angle measures of any triangle add up to 180°.

The sum of the interior angles of any triangle can be illustrated by placing all three angles next to each other, as shown in the following diagram.

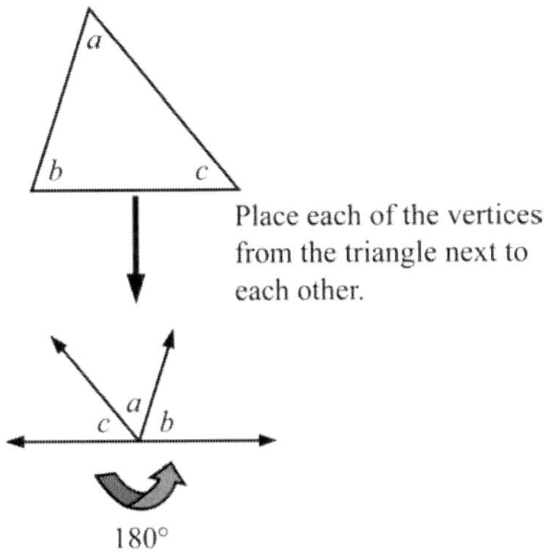

Therefore, the sum of the interior angles of a triangle can be represented as $\angle a + \angle b + \angle c = 180°$.

To find a missing angle measure in a triangle, first calculate the sum of the given angles. Then, subtract the sum from 180°.

NOTES

Example

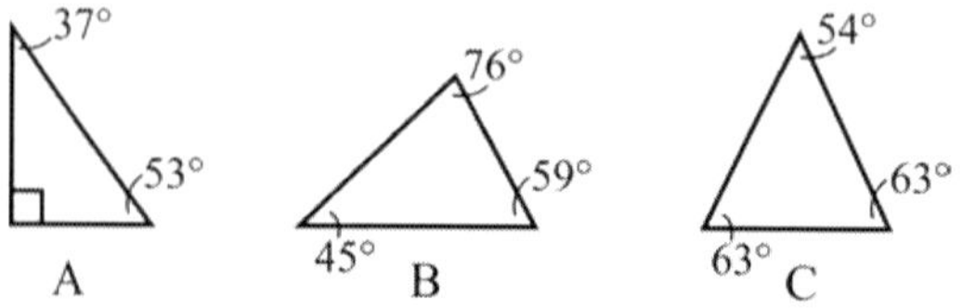

Use the given triangles to prove that the sum of the measures of all the interior angles of any triangle always equals 180°.

Solution

Calculate the sum of the angles in each triangle.

Triangle A: $90 + 37 + 53 = 180$
Triangle B: $45 + 76 + 59 = 180$
Triangle C: $63 + 54 + 64 = 180$

No matter what type of triangle is used, the sum of its angles is always 180°.

Time to Try 1

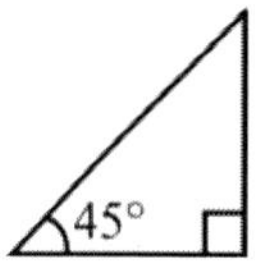

Find the missing angle measures for the given triangle.

INTERIOR ANGLES OF QUADRILATERALS

The four interior angles of any quadrilateral add up to 360°.

The sum of the interior angles can be illustrated by dividing a quadrilateral in half by drawing a line from a corner to its diagonally opposite corner.

The quadrilateral is now two triangles, as shown in the following diagram.

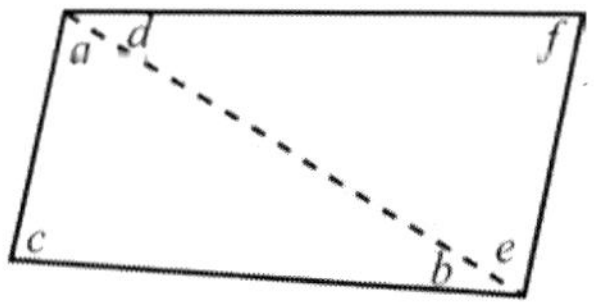

NOTES

The sum of all the angles in a triangle is 180°. It follows that the sum of two triangles together (a quadrilateral) is 360°.

To find the missing angle measure in any quadrilateral, first calculate the sum of the given angles. Then, subtract the sum from 360°.

Example

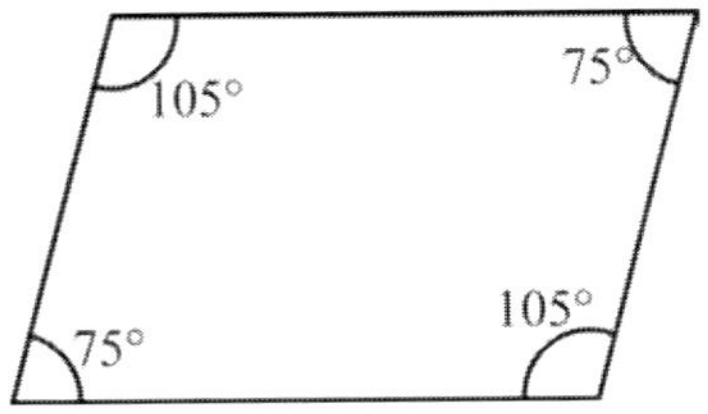

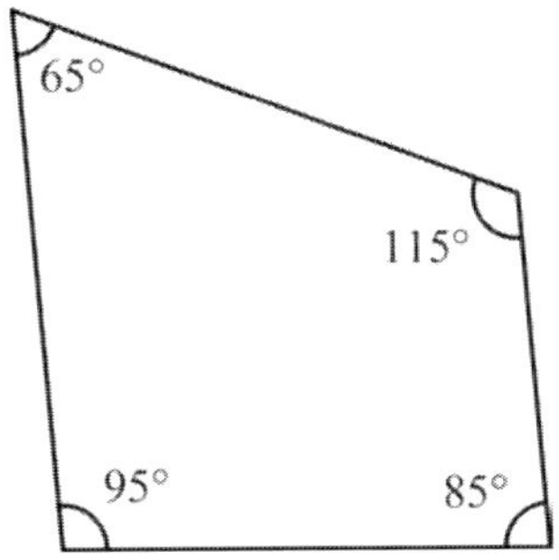

Use the given quadrilaterals to prove that the sum of all the interior angles in any quadrilateral will always be 360°.

Solution

Calculate the sum of the angles.

$105 + 105 + 75 + 75 = 360$

$65 + 115 + 95 + 85 = 360$

No matter what type of quadrilateral is used, the sum of its angles is always360°.

Time to Try 2

Find the missing angle measure of trapezoid *ABCD*.

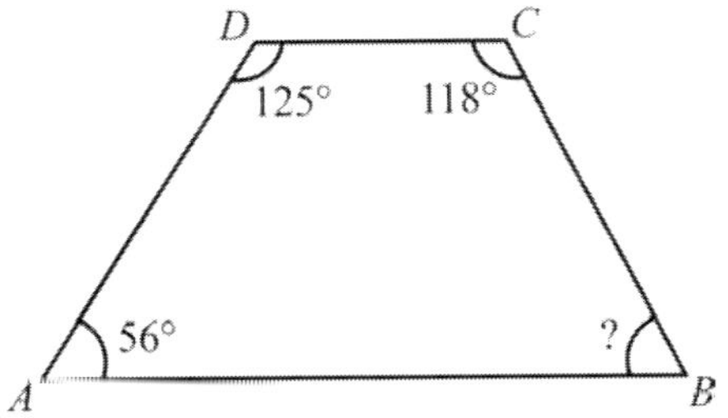

PRACTICE EXERCISES

Use the following information to answer the next question.

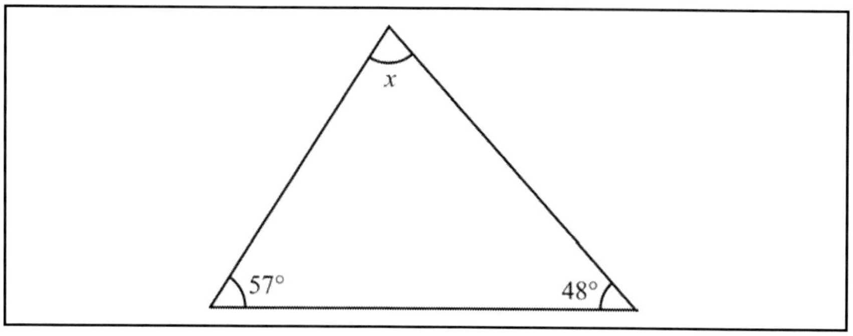

1. Explain how you can find the measure of the angle marked with an x.

Use the following information to answer the next question.

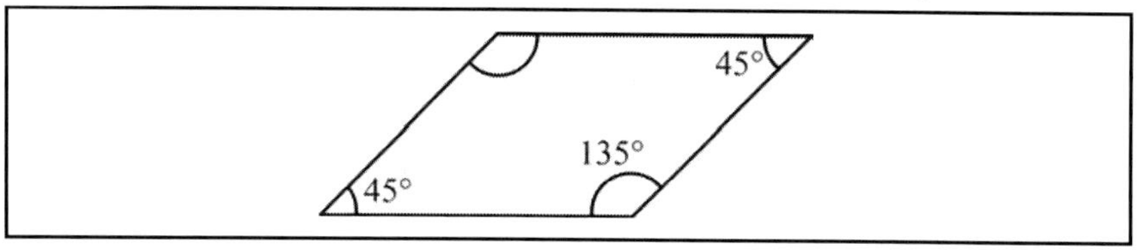

2. Explain how you can find the missing measure of the angle in the quadrilateral.

Use the following information to answer the next question.

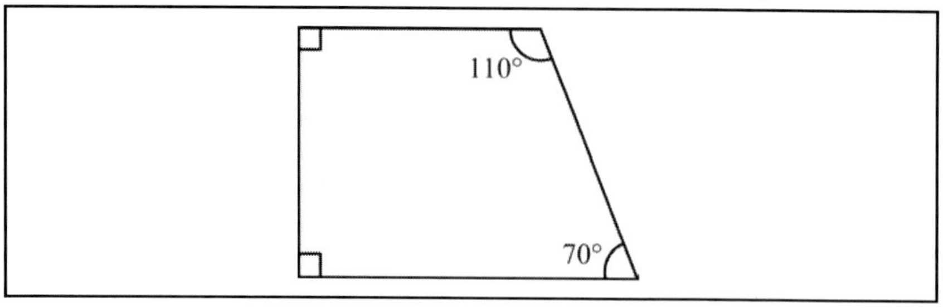

3. Explain how you can find the missing measures of the two angles. Justify your answer.

Use the following information to answer the next question.

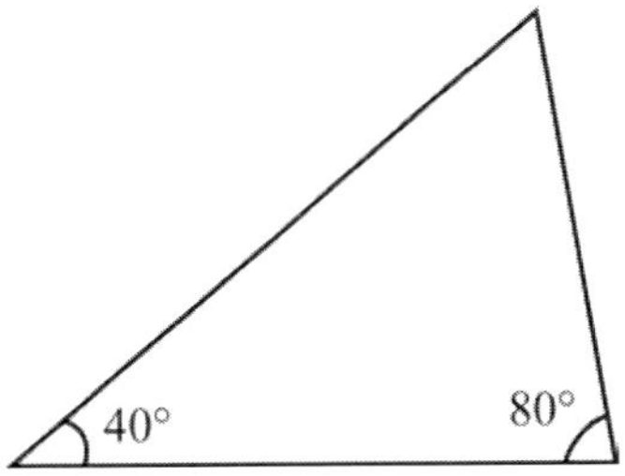

4. What is the missing measure of the unmarked angle?

Use the following information to answer the next question.

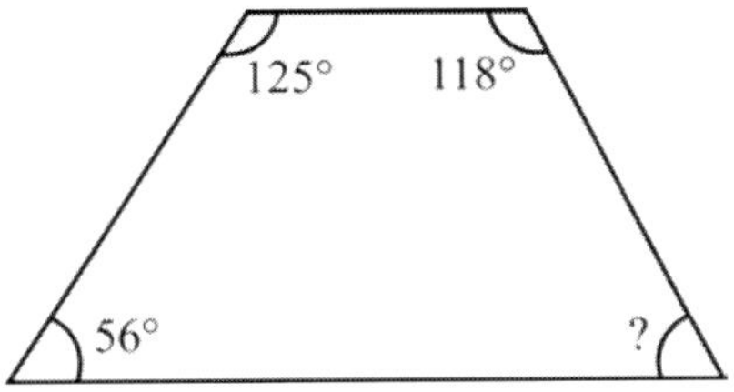

5. What is the missing measure of the unmarked angle?

REVIEW SUMMARY

- Angles are measured using degrees.
- Angles can be classified by their measure.
- An acute angle measures less than 90°; a right angle measure exactly 90°; an obtuse angle measure more than 90° but less than 180°; a straight angle measures exactly 180°; and a reflex angle measures more than 180° and less than 360°.
- You can estimate the measures of angles using 45°, 90°, and 180° as reference angles.
- The relationships among the reference angles is that a 45° angle is half the size of a 90° angle and one-quarter the size of a 180° angle. A 90° angle is half the size of a 180° angle and twice the size of a 45° angle. A 180° angle is twice the size of a 90° angle and four times the size of a 45° angle.
- A protractor is a tool used to measure angles. It can also be used to draw angles for a given measure.
- Most protractors have an inner and an outer scale. The scale you use depends on the direction the angle is facing.
- The interior angles of any triangle add up to 180°.
- To find a missing angle measure of a triangle, first calculate the sum of the given angles, then subtract the sum from 180°.
- The interior angles of any quadrilateral add up to 360°.
- To find a missing angle measure of a quadrilateral, first calculate the sum of the given angles, then subtract the sum from 360°.

PRACTICE TEST

Use the following diagrams to answer the next question.

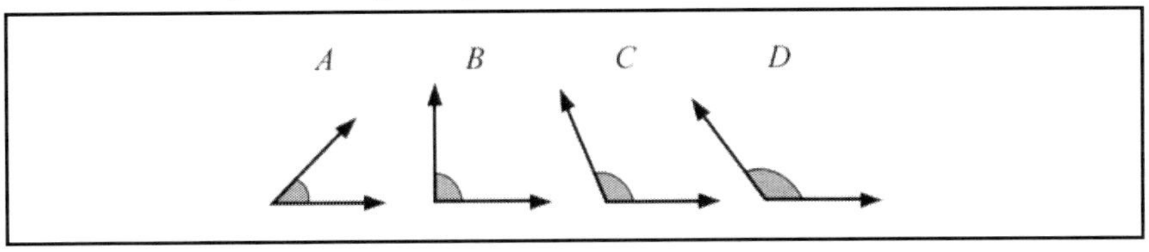

1. What is the name of each angle shown? Explain your answers.

Using reference angles, estimate the measure of the following angles. Explain your answers.

2.

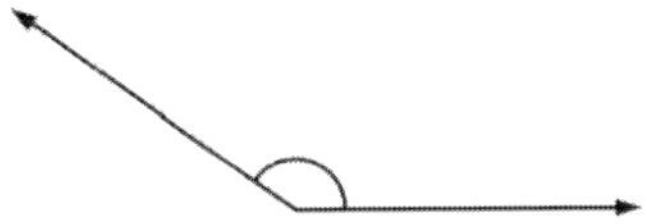

3.

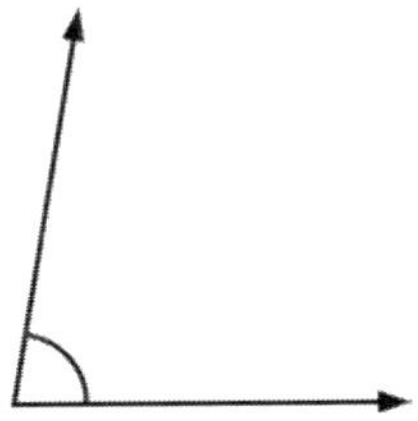

Use the following information to answer the next question.

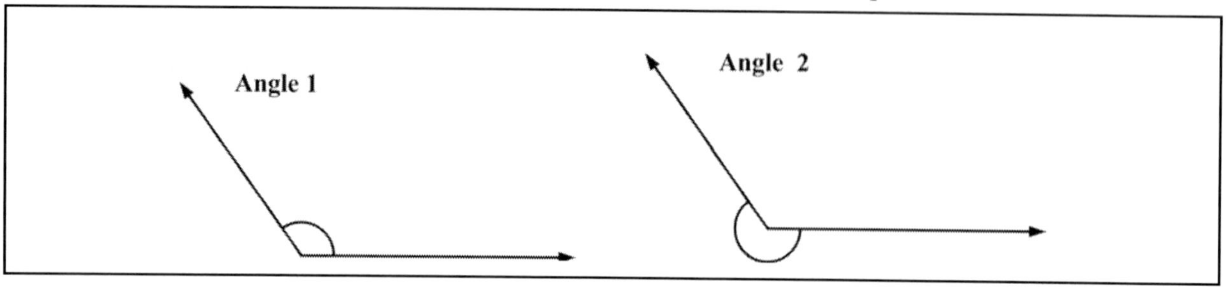

4. Classify the given set of angles according to their measures.

Use the following information to answer the next question.

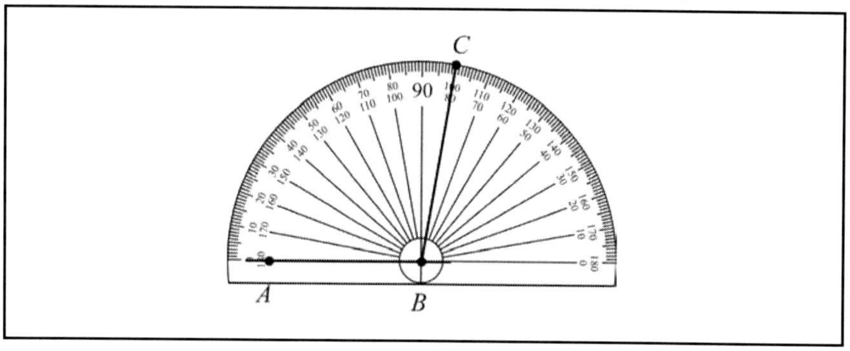

5. What is the measure of angle *ABC*? Explain how you arrived at your answer.

Use a protractor to measure the following angles.

6.

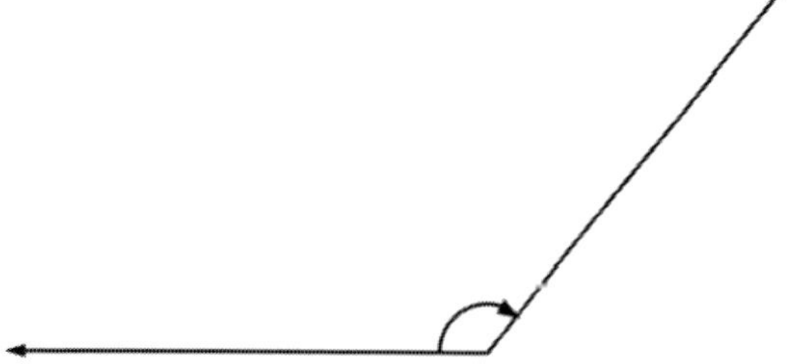

7.

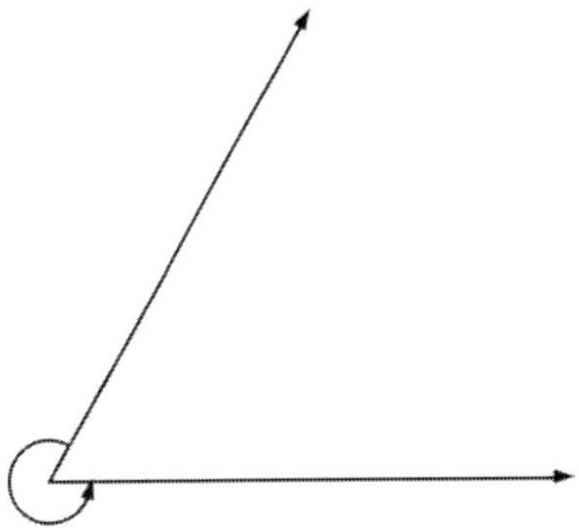

8. Without the use of a protractor, sketch a 45° angle and a 180° angle. Explain the relationship between the two angles.

Use a protractor to draw angles for the following measures.

9. An acute angle that has a measure of 55°.

10. An obtuse angle with a measure of 160°.

11. A reflex angle with a measure of 245°.

Use the following information to answer the next question.

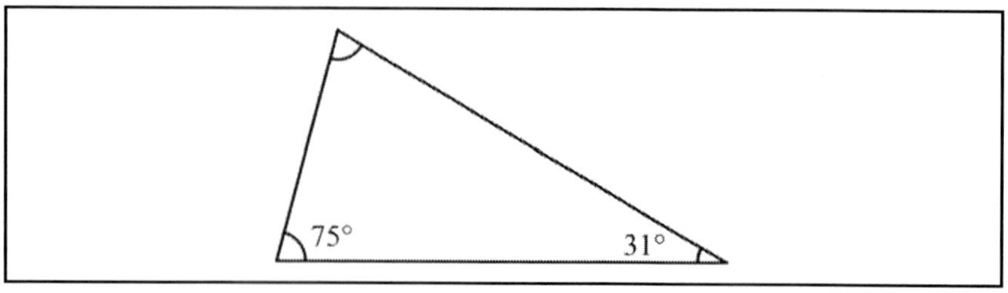

12. Explain how you can find the missing measure of the given triangle.

Use the following information to answer the next question.

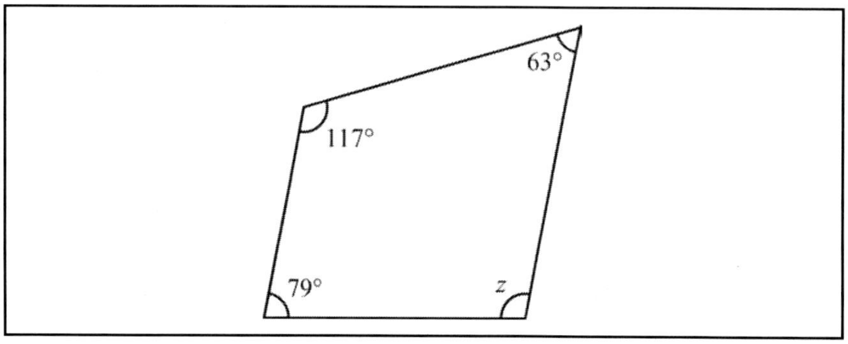

13. Explain how you can find the measure of the angle marked z.

TWO-DIMENSIONAL SHAPES

When you are finished this unit, you should be able to…

- Compare triangles according to side lengths and angle measures
- Sort a given set of triangles, and explain the sorting rule
- Draw a specified angle
- Replicate a given triangle in a different orientation, and show that the two triangles are congruent
- Sort a set of 2-D shapes into polygons and non-polygons, and explain the sorting rule
- Sort a set of polygons as regular or irregular polygons, and justify the sorting
- Demonstrate that the sides of a regular polygon have the same length and that the angles have the same measure
- Demonstrate congruency in regular polygons by superimposing or measuring

Lesson	Page	Completed on
1. Comparing Triangles	192	
2. Drawing and Replicating Triangles	200	
3. Regular and Irregular Polygons	204	
4. Congruency in Regular Polygons	209	
Review Summary	212	
Practice Test	213	
Answers and Solutions	at the back of the book	

PREREQUISITE SKILLS AND KNOWLEDGE

Prior to beginning this unit, you should be able to…

- Identify and sort quadrilaterals according to different attributes
- Identify and describe a single translation, rotation, or reflection of a 2-D shape
- Perform and draw a single translation, rotation, or reflection of a 2-D shape

Lesson 1 COMPARING TRIANGLES

NOTES

DESCRIBING TRIANGLES USING SIDE LENGTHS

An **equilateral triangle** has three sides of equal length. An equilateral triangle is unique because it has three angles of equal measure. This triangle is an example of an equilateral triangle:

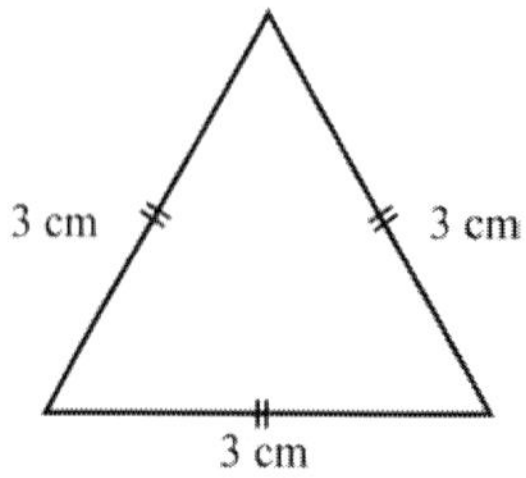

An **isosceles triangle** has two sides of equal length and one side that is of a different length. This triangle is an example of an isosceles triangle:

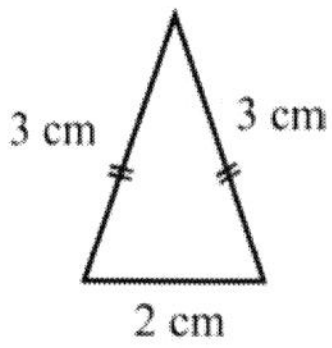

A **scalene triangle** has different side lengths for all three sides. It has no equal sides. This triangle is an example of a scalene triangle:

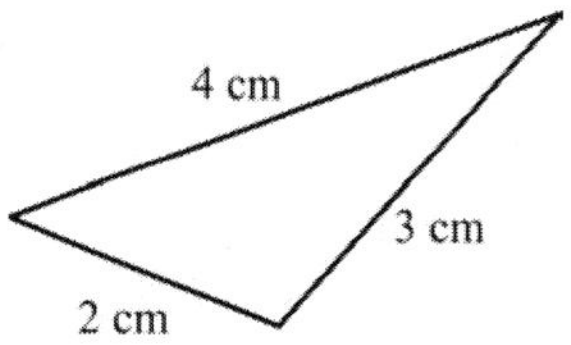

Example

Use side lengths to identify the given triangle. Explain your answer.

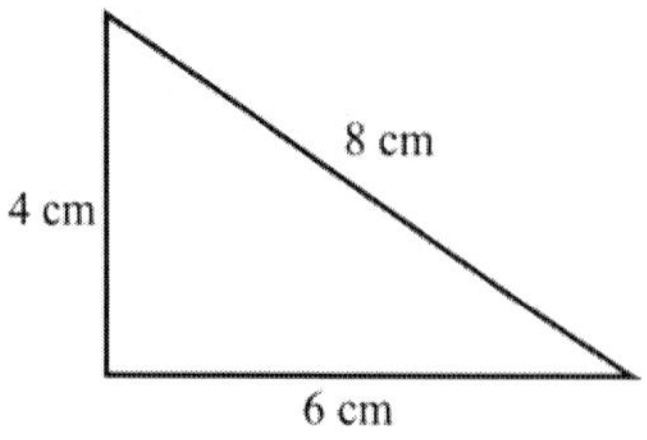

Solution

This is a scalene triangle because the three side lengths are all different.

NOTES

Time to Try 1

Use side lengths to identify the given triangle. Explain your answer.

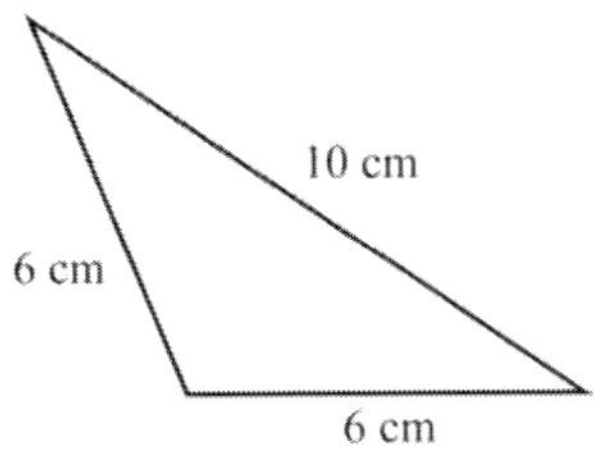

DESCRIBING TRIANGLES USING INTERIOR ANGLES

An **acute triangle** has three acute angles. Each angle must be less than 90°. This triangle is an example of an acute triangle:

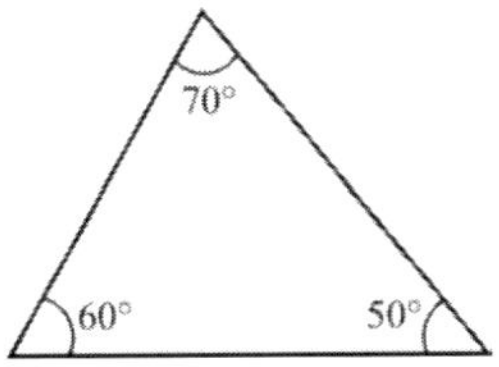

A **right triangle** has one right angle. A right angle has a measure of 90°. This triangle is an example of a right triangle:

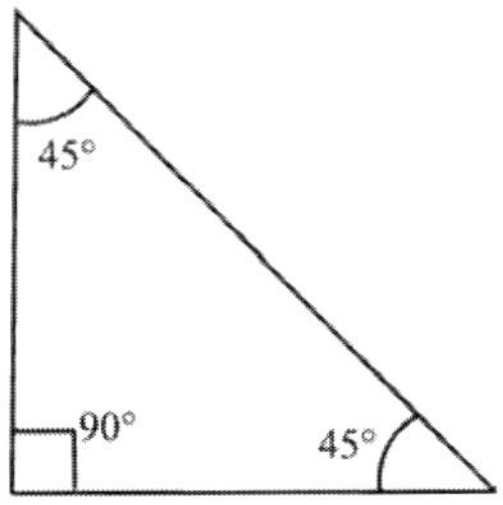

An **obtuse triangle** has one obtuse angle. An obtuse angle is greater than 90° (but less than 180°). The other two angles are acute angles. This triangle is an example of an obtuse triangle:

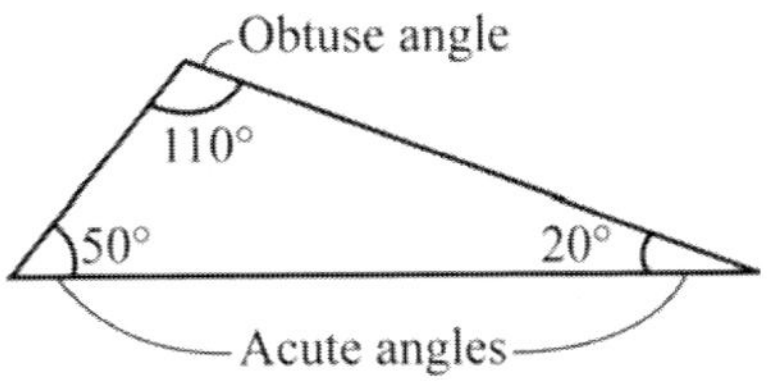

NOTES

Example

Use interior angles to identify the given triangle.
Explain your answer.

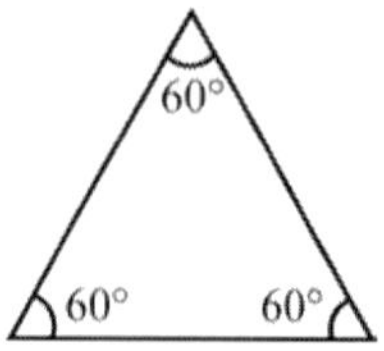

Solution

This is an acute triangle because it has three acute angles. All the angles are less than 90°.

Time to Try 2

Use interior angles to identify the given triangle. Explain your answer.

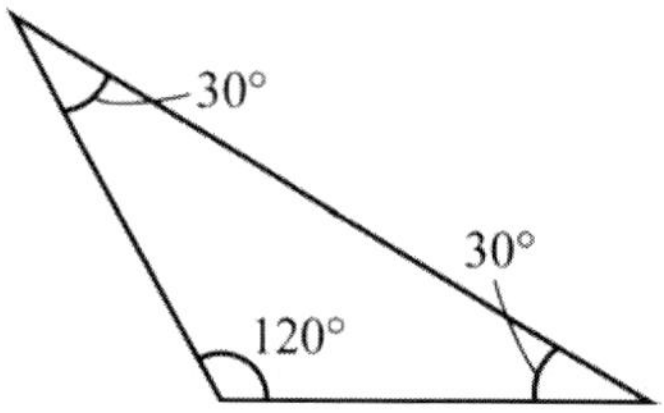

DESCRIBING TRIANGLES USING SIDE LENGTHS AND INTERIOR ANGLES

Triangles can be described by both their side lengths and their angle measures.

- Acute triangles may also be classified as scalene, isosceles, or equilateral triangles.
- Every equilateral triangle has three 60° angles. All equilateral triangles can also be classified as acute triangles.
- Obtuse triangles may also be classified as isosceles or scalene triangles.
- Right triangles may also be classified as either isosceles or scalene triangles.

NOTES

Example

This diagram shows a triangle with three different side lengths and three different angle measures.

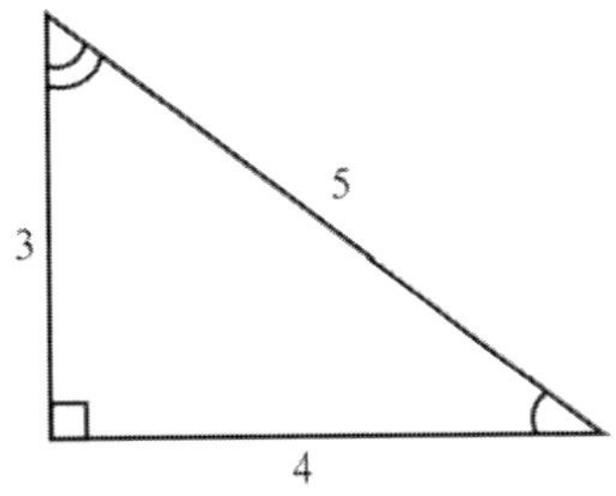

Classify the triangle according to its side lengths and angle measures. Explain your answer.

Solution

The given triangle can be classified as a scalene triangle because all three sides have different side lengths.

The triangle can also be classified as a right triangle because it has one right angle (90°), as shown by the square corner.

Time to Try 3

Classify this triangle according to its side lengths and angle measures.

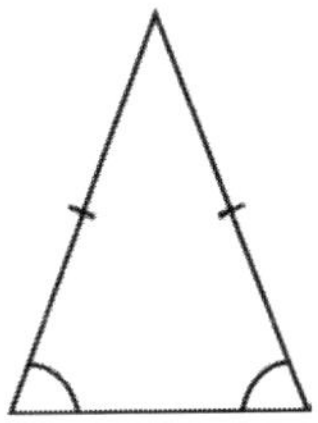

SORTING TRIANGLES

Sets of triangles can be sorted into two groups when a sorting rule is given. A sorting rule explains what characteristic belongs to each group.

NOTES

If the triangles are already sorted into groups, you need to be able to identify the characteristics of each group in order to explain the sorting rule used.

Example

Sort the given triangles into two groups. The sorting rule is *right angles or no right angles.*

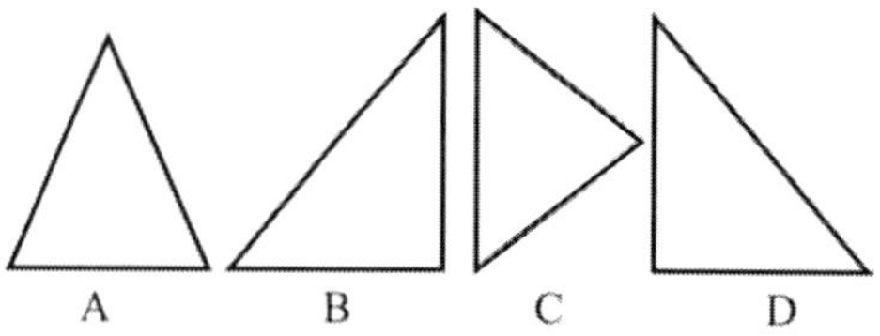

Solution

Step 1
Identify the triangles that have right angles.

Triangles B and D have right angles. They belong to the first group.

Step 2
Identify the triangles that do not have right angles.

Triangles A and C do not have right angles. They belong to the second group.

Time to Try 4

The following triangles have been sorted into two groups.
Explain the sorting rule used.

Group 1

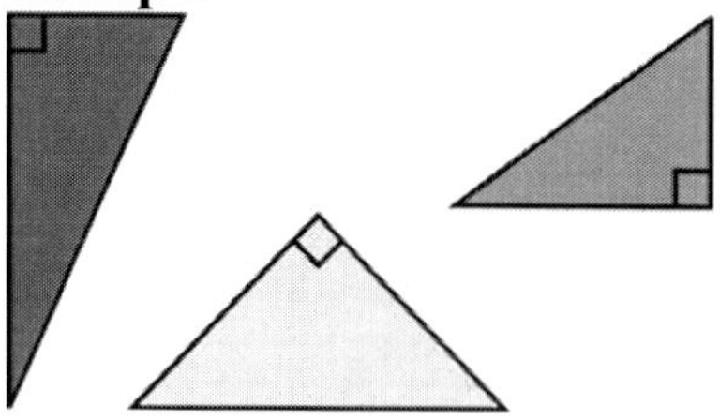

Group 2

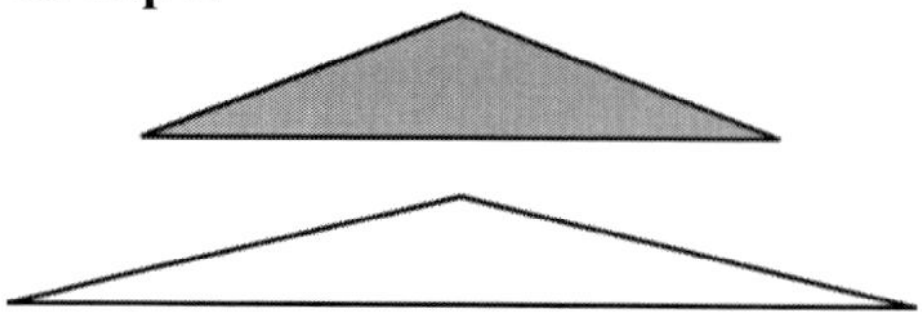

PRACTICE EXERCISES

Use the following information to answer the next question.

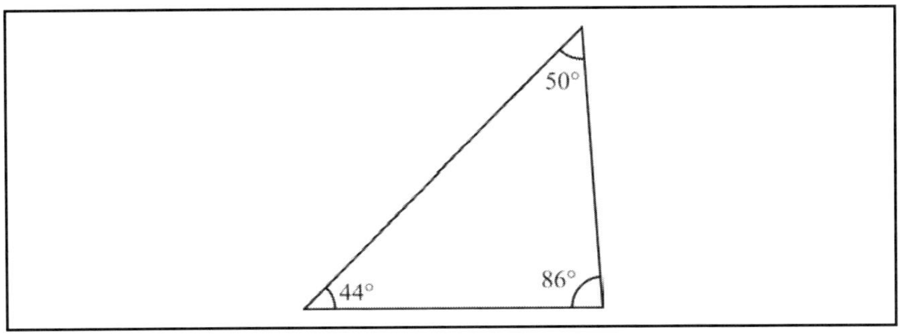

1. Classify the given triangle based on its interior angles. Explain your answer.

Use the following information to answer the next question.

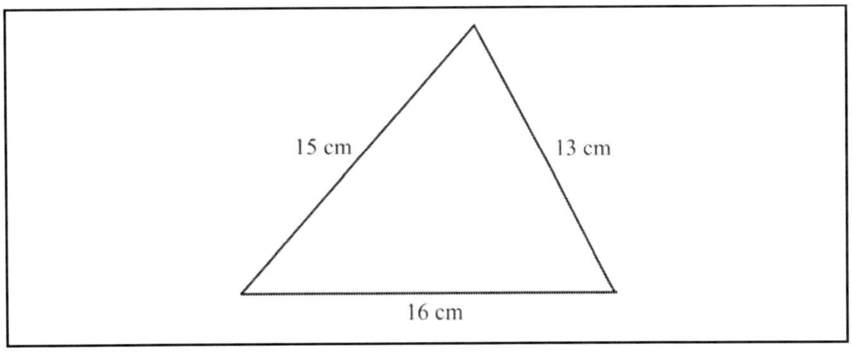

2. Classify the given triangle based on its side lengths. Explain your answer.

Use the following information to answer the next question.

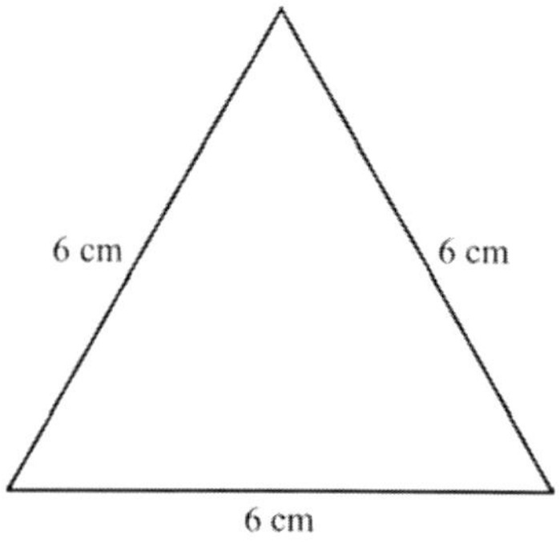

3. Classify the given triangle based on its side lengths and interior angles. Explain your answer.

Use the following information to answer the next question.

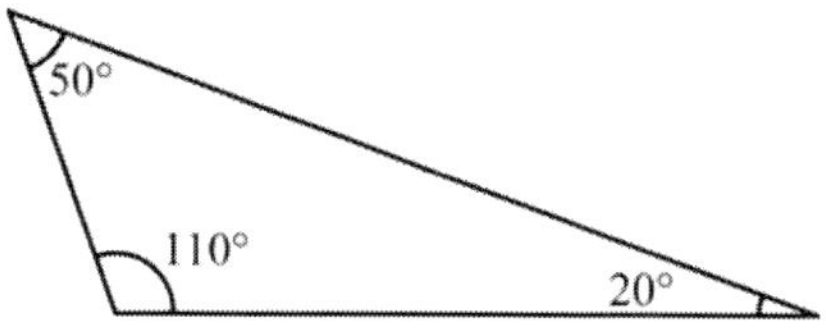

4. Classify the given triangle based on its side lengths and interior angles. Explain your answer.

Use the following diagram to answer the next question.

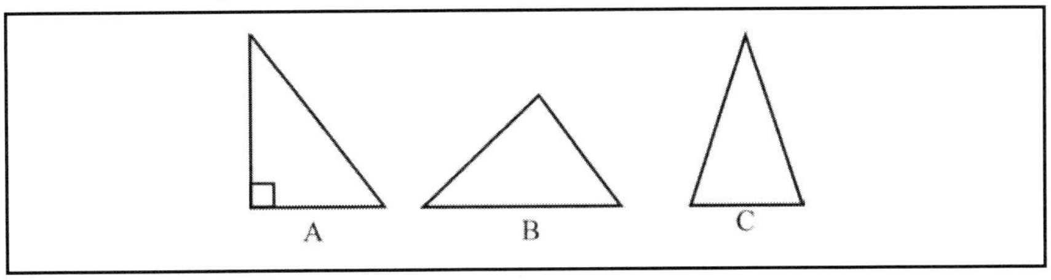

5. Sort the given triangles into two groups. Use the sorting rule *all equal side lengths or not all equal side lengths.*

Use the following information to answer the next question.

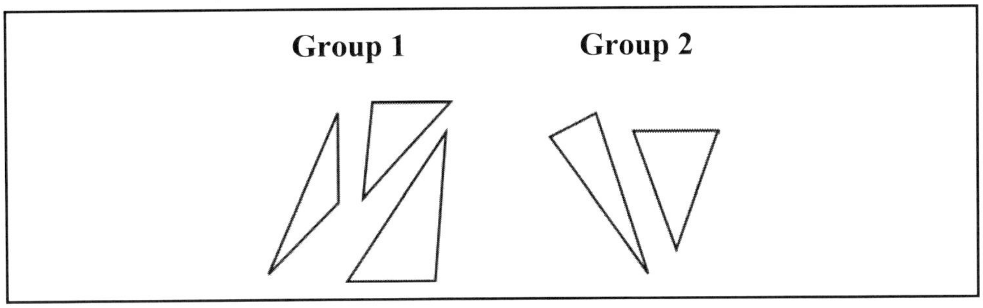

6. Explain the sorting rule used to sort the given triangles into the two groups.

Lesson 2 DRAWING AND REPLICATING TRIANGLES

NOTES

DRAWING A SPECIFIED TRIANGLE

One method for constructing a specified triangle is to use a protractor and a ruler. A protractor can be used to measure a specified angle. A ruler can be used to measure a specified side length, as well as to draw straight lines.

Example

Draw an isosceles triangle that has two 45° angles and a side length of 6 cm.

Solution

Step 1
Using a centimetre ruler, draw a line segment that is 6 cm long.

Step 2
At one endpoint of the line segment, measure a 45° angle using a protractor. At the other endpoint of the line segment, measure another 45° angle.

Extend the two rays of the drawn angles until they intersect. The intersection forms the third angle of the isosceles triangle.

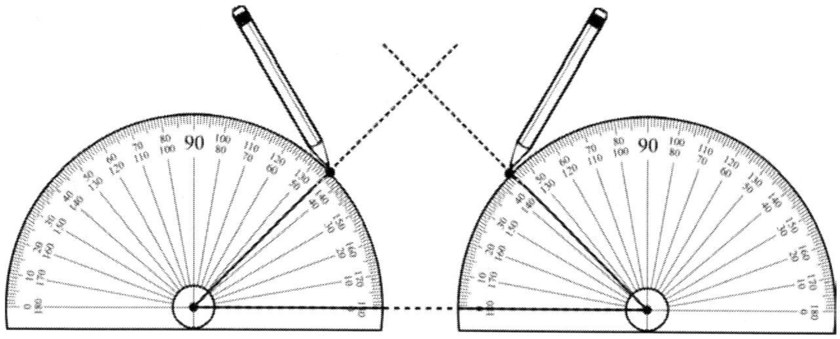

Step 3
Label the base line and the two angles you measured.

The triangle will look like this:

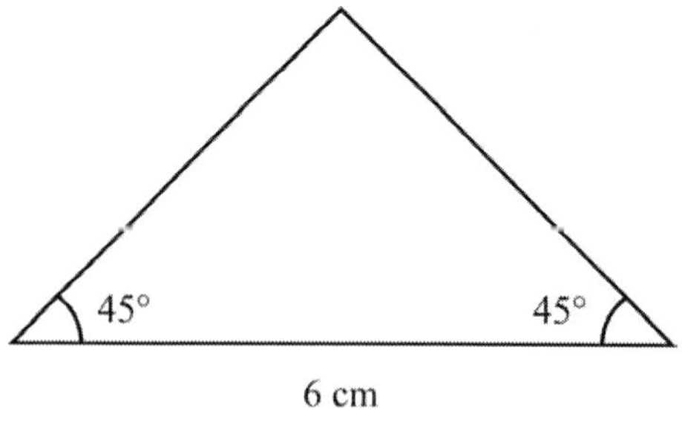

NOTES

Time to Try 1

Draw an equilateral triangle with a base of 7 cm and three congruent angles of 60°.

REPLICATING TRIANGLES IN DIFFERENT ORIENTATIONS

In order to replicate a given triangle in a different orientation, you can measure and draw or trace the given triangle in different positions on the page. When replicating triangles, all the triangles will be congruent to each other. They must all have the same side lengths and angle measures.

Replicate means to duplicate, copy, reproduce, or repeat.

Congruent means coinciding exactly when superimposed.

Example

Replicate the given triangle in three different orientations.

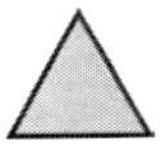

Solution

Step 1
Method 1
Measure the sides of the given triangle.

The triangle has three sides that each measure 1 cm, making it an equilateral triangle.

The triangles you draw must all have 1 cm side lengths.

Method 2
Trace the shape and carefully cut it out.

NOTES

Step 2
Replicate the shape.

Draw or trace three 1 cm equilateral triangles that are in different positions.

Here are three replicated triangles.

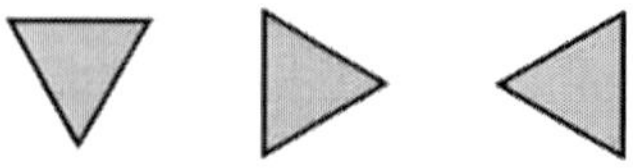

Time to Try 2

Replicate the given triangle in a different orientation, and show that the two triangles are congruent.

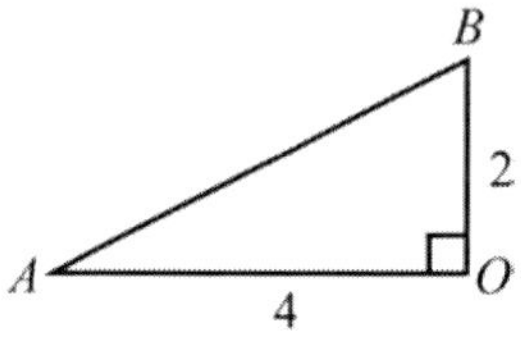

PRACTICE EXERCISES

1. Draw a right triangle with a base line of 7 cm and acute angles of 30° and 60°.

2. Draw an equilateral triangle with a base line of 4 cm.

Use the following diagram to answer the next question.

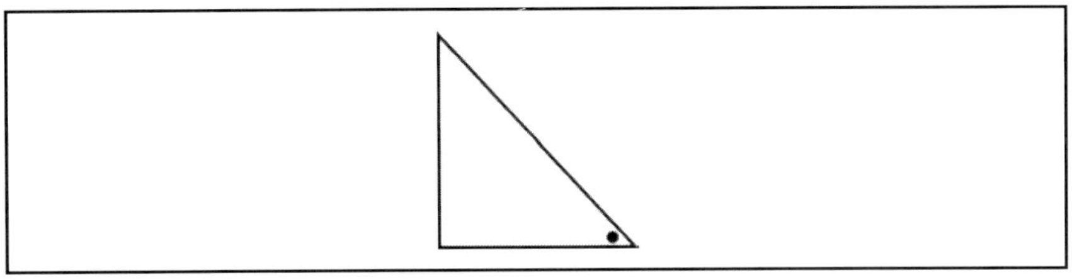

3. Replicate the given triangle in three different orientations.

Use the following diagram to answer the next question.

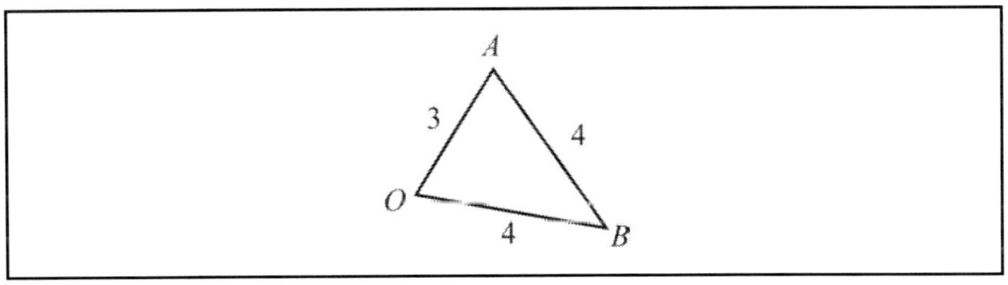

4. Replicate the given triangle in a different orientation, and show that the two triangles are congruent.

Lesson 3 REGULAR AND IRREGULAR POLYGONS

NOTES

A **polygon** is a closed two-dimensional (2-D) shape with three or more straight sides.

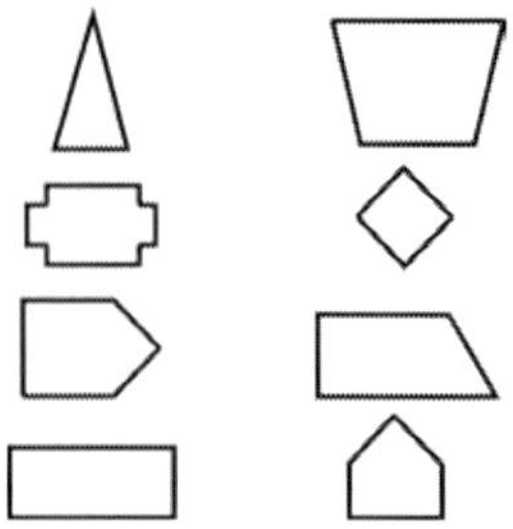

A **regular polygon** has all congruent sides and congruent angles.

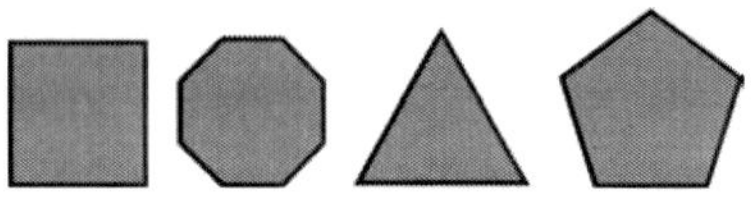

An **irregular polygon** is any other polygon whose sides and angles are not all congruent.

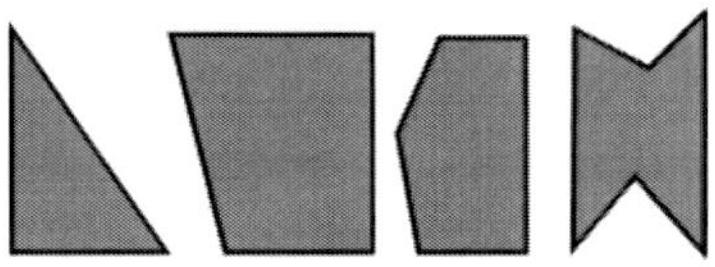

IDENTIFYING POLYGONS IN THE ENVIRONMENT

There are many polygons in the environment. This table shows some examples of patterns formed by polygons as seen in the environment.

	In the world of geology, crystals have flat faces that form polygons.
	Polygons result when cooling lava forms hexagonal columns of basalt.

NOTES

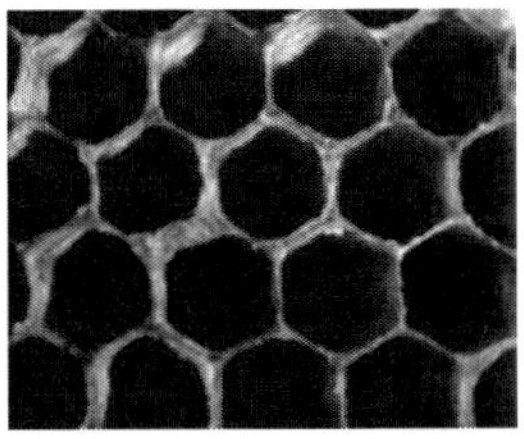	The wax honeycomb made by bees is a tessellation of regular hexagons.
	A giraffe's body is completely covered in irregular polygons.

SORTING POLYGONS

The following list shows some properties by which 2-D shapes can be sorted:

- Polygons or non-polygons
- Regular polygons or irregular polygons
- Size of interior angles
- Lengths of sides

A sorting rule tells you which characteristics are being used to describe the polygons.

Example

Use the sorting rule *polygons or non- polygons* to sort these shapes into two groups. Explain the sorting rule.

Solution

Step 1

Sort the shapes using the sorting rule.

Polygons	Non-polygons
	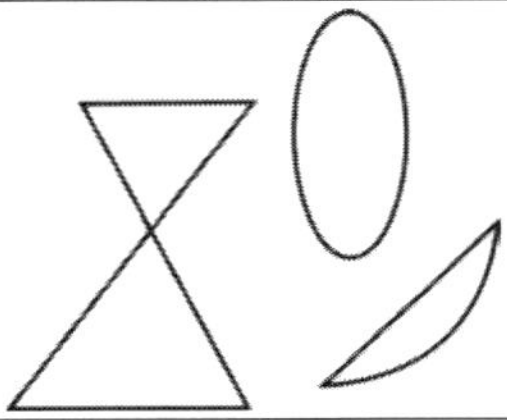

NOTES

Step 2
Explain the sorting rule.

All shapes that are closed, made of straight lines only, and do not intersect at places other than the ends of sides are to be placed in the group labelled "polygons."

All other shapes are to be placed in the group labelled "non-polygons."

Time to Try 1

Use the sorting rule *regular polygons or irregular polygons* to sort these shapes into two groups. Explain the sorting rule.

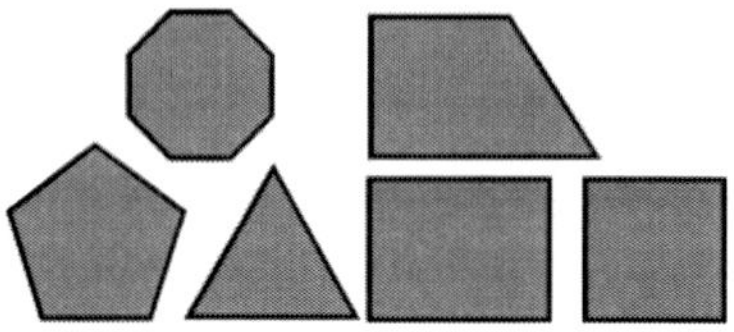

PRACTICE EXERCISES

Use the following information to answer the next question.

Carla makes this design out of polygons.

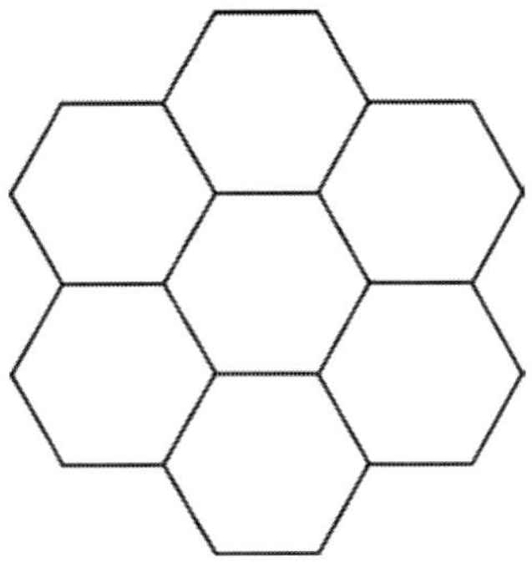

1. Did Carla use a regular polygon, an irregular polygon, or a non-polygon to create this design? Justify your answer.

Use the following information to answer the next question.

Shawn drew two three-sided polygons.

2. Label the shapes as regular or irregular polygons. Justify your answer.

Use the following information to answer the next question.

Robert made four photo frames.

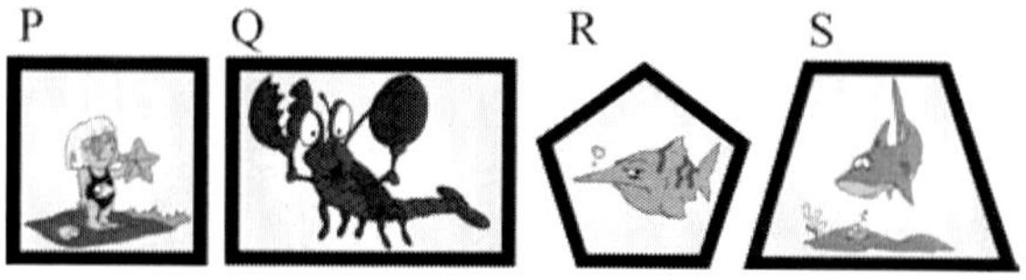

3. Sort the photo frames according to the sorting rule *all equal angles or not all equal angles.* Justify the sorting.

Use the following information to answer the next question.

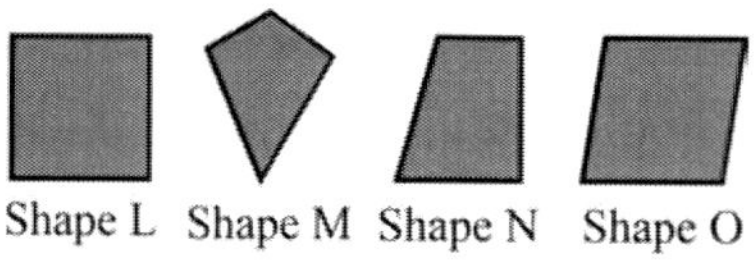

4. Sort these shapes according to the sorting rule *all equal side lengths or not all equal side lengths.* Justify the sorting.

Use the following information to answer the next question.

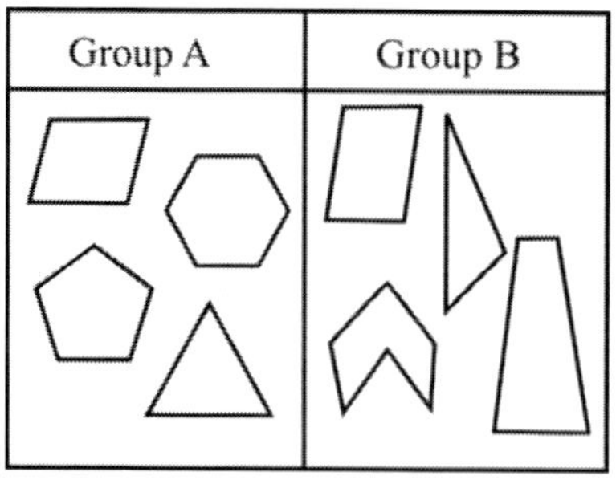

5. Identify and explain the sorting rule used to sort the shapes into group A and group B.

Lesson 4 CONGRUENCY IN REGULAR POLYGONS

NOTES

When two or more regular polygons are described as being congruent, this means that all the side lengths match and all the angles match. Polygons can be congruent, but have different orientations, as shown by these three congruent squares.

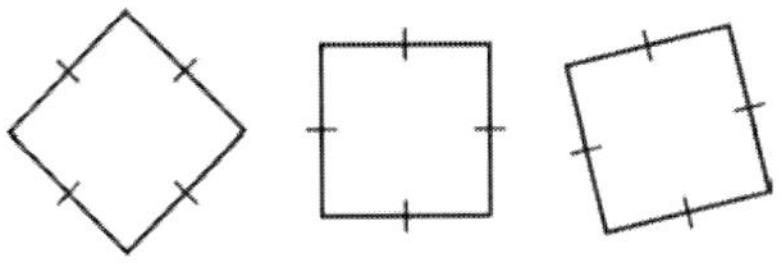

There are two methods for determining congruency:

1. **Superimposing**—Cut one shape out and place it on top of the other shape. If all the side lengths and angles fit identically, the shapes are congruent.
2. **Measuring**—Measure the side lengths or angles of the two shapes. If all the measurements are identical, the shapes are congruent.

Example

Measure the side lengths of these pentagons to determine if they are congruent.

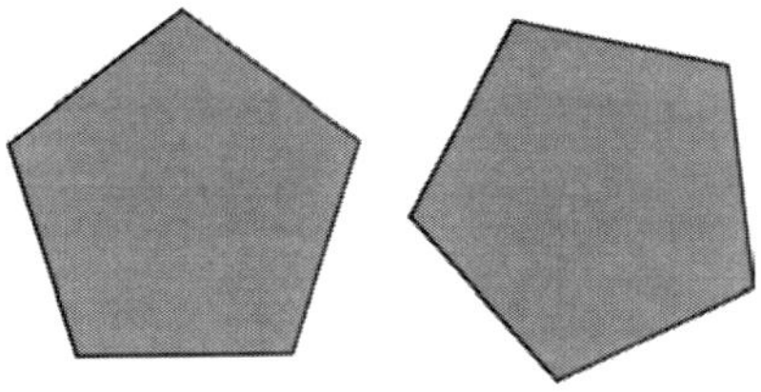

Solution

Both regular pentagons have five side lengths of 2 cm. Since all the side lengths are the same, you know that all the five angle measures will be the same as well. The two pentagons are congruent.

Time to Try 1

Of the five given triangles, identify the triangle that is congruent to triangle *A*. Explain how you can show the congruency of the two triangles.

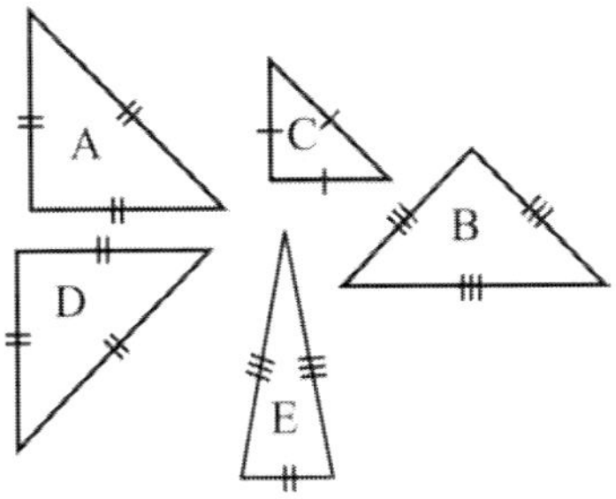

PRACTICE EXERCISES

Use the following information to answer the next question.

Gino traced one of these pattern blocks to make a set of congruent triangles.

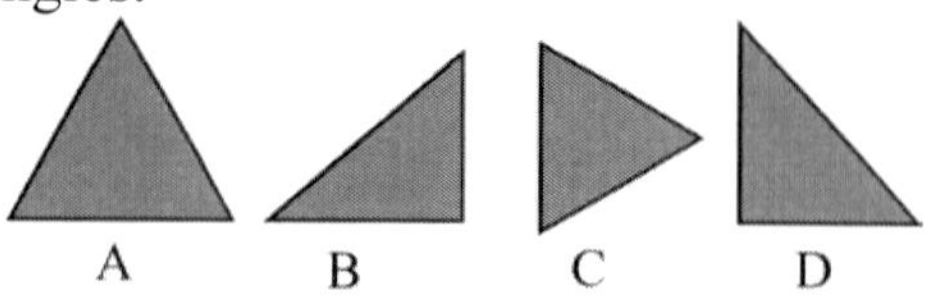

This is the set of shapes he made:

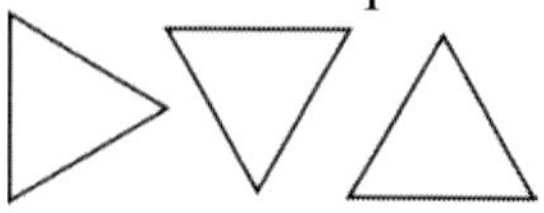

1. Which pattern block did Gino use to create the set of congruent shapes he made? Explain your answer.

Use the following information to answer the next question.

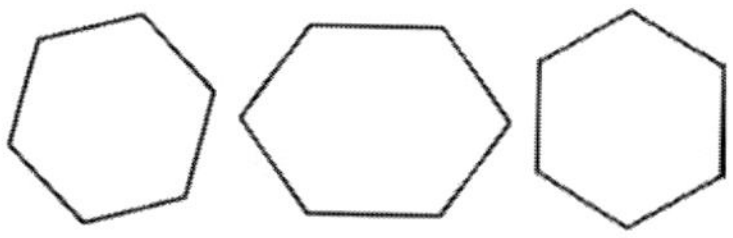

2. Are any of these hexagons congruent? Justify your answer.

Use the following information to answer the next question.

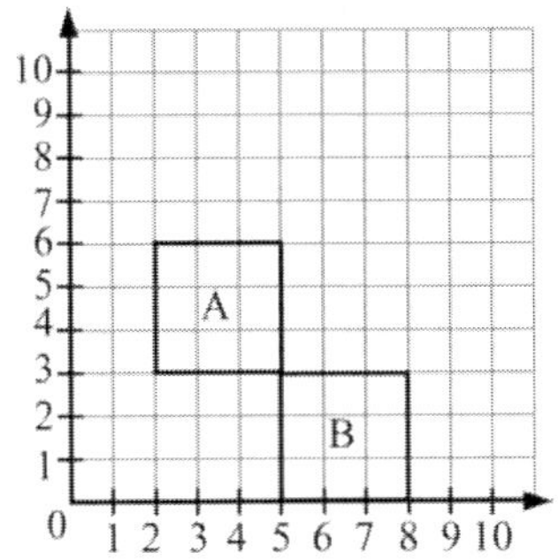

3. Are square A and square B congruent? Use measurement to justify your answer.

REVIEW SUMMARY

- Triangles can be classified according to the measurement of their interior angles and/or according to their side lengths.
- An acute triangle has three acute angles (angles that are less than 90°).
- A right triangle has one right angle (a 90° angle).
- An obtuse triangle has one obtuse angle (an angle larger than 90° and less than 180°).
- An equilateral triangle has three equal side lengths and three equal angles.
- An isosceles triangle has two equal sides and two equal angles.
- A scalene triangle has no equal sides and no equal angles.
- In order to accurately draw a triangle, you will need a protractor and a ruler. The protractor is used to measure out the exact angle(s) that you require. The ruler is used to measure out the exact side lengths that you require.
- To replicate a given triangle in a different orientation, trace the triangle in different positions, labeling the congruent sides and angles if the measures are given. Replicated shapes are always congruent.
- A polygon is a closed 2-D figure with three or more straight sides.
- 2-D shapes can be sorted into polygons and non-polygons.
- A regular polygon has all congruent sides and angles.
- An irregular polygon is any other polygon whose sides and angles are not all congruent.
- Polygons can be sorted and compared according to their angles and their sides.
- Congruent polygons have the same angle measures and side lengths.
- To check for congruency, you can superimpose the two images, putting them on top of one another, or you can measure their sides and angles.

PRACTICE TEST

Use the following information to answer the next question.

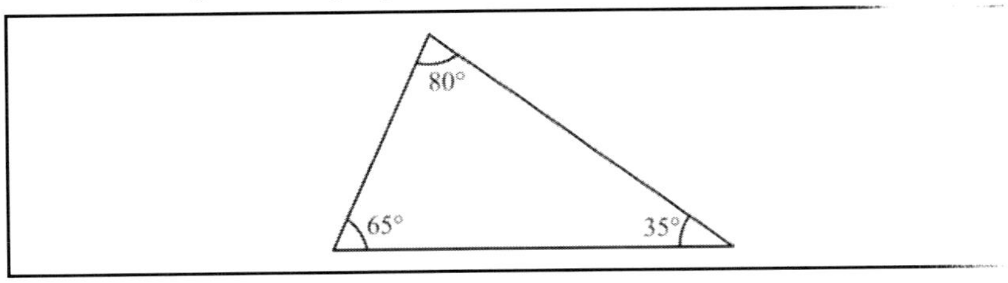

1. Classify the given triangle based on its interior angles. Justify your answer.

Use the following information to answer the next question.

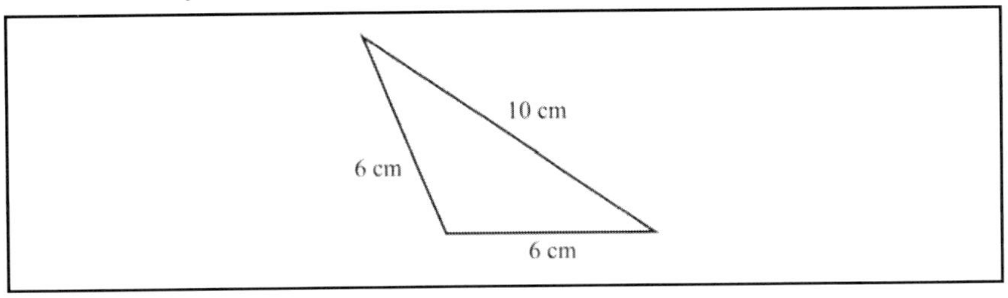

2. Classify the given triangle based on its side lengths. Justify your answer.

Use the following information to answer the next question.

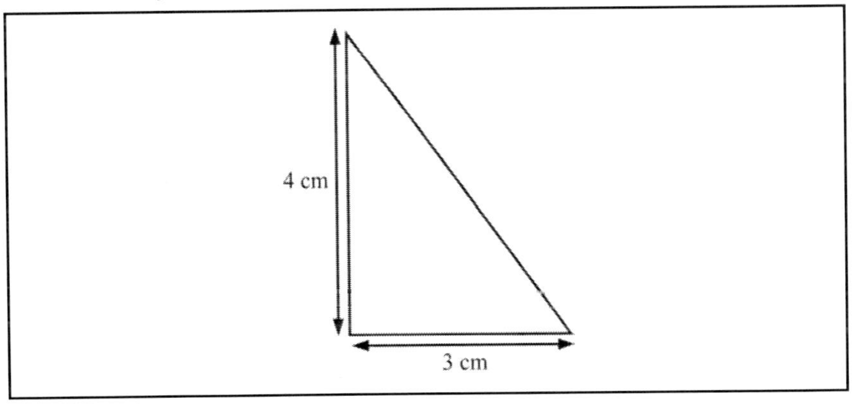

3. Classify the given triangle based on both its interior angles and its side lengths. Justify your answer.

Use the following information to answer the next question.

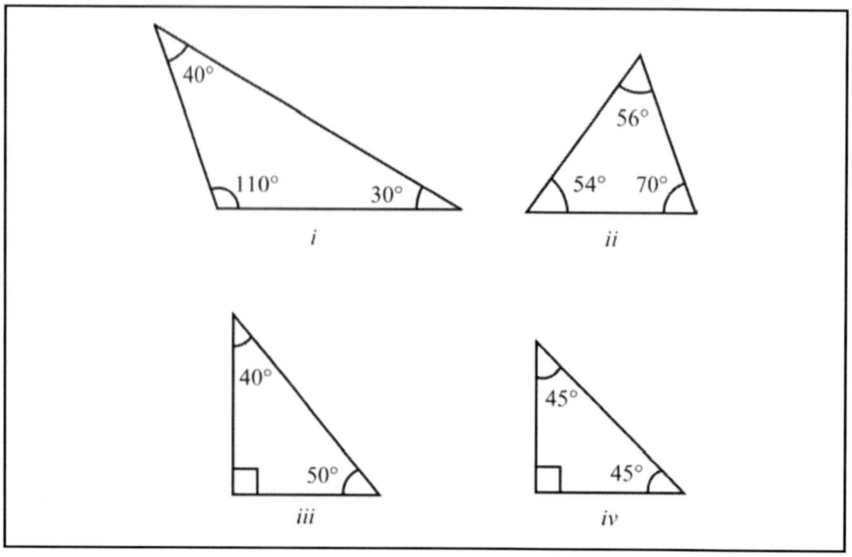

4. Sort the given triangles according to the sorting rule *acute triangles or not acute triangles*. Explain the sorting rule.

5. Draw an equilateral triangle with a base measuring 6 cm.

Use the following diagram to answer the next question.

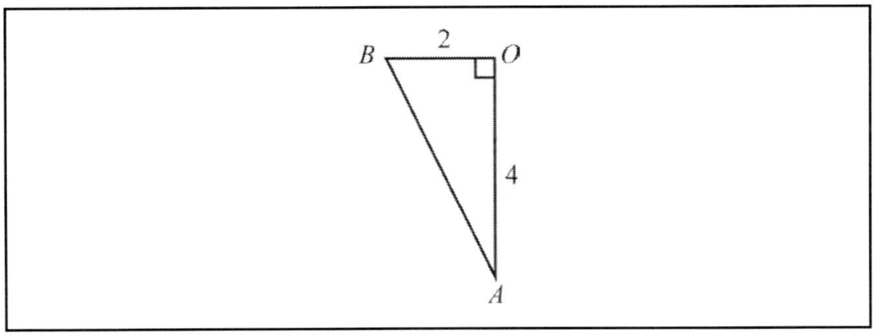

6. Replicate the given triangle in two different orientations, and show that the triangles are congruent.

Use the following diagram to answer the next question.

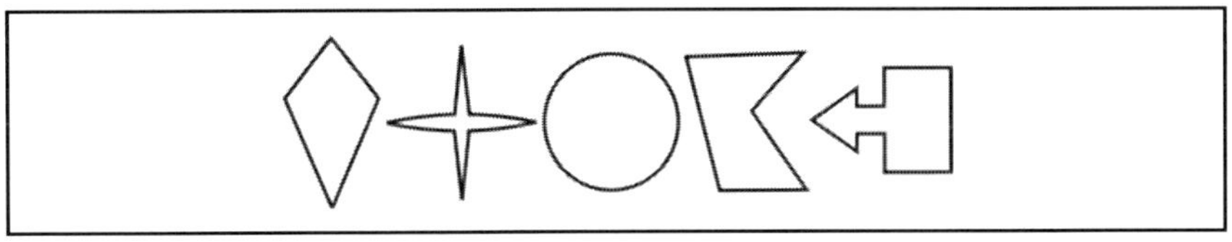

7. Sort the given set of shapes into polygons and non-polygons. Explain the sorting rule.

Use the following diagram to answer the next question.

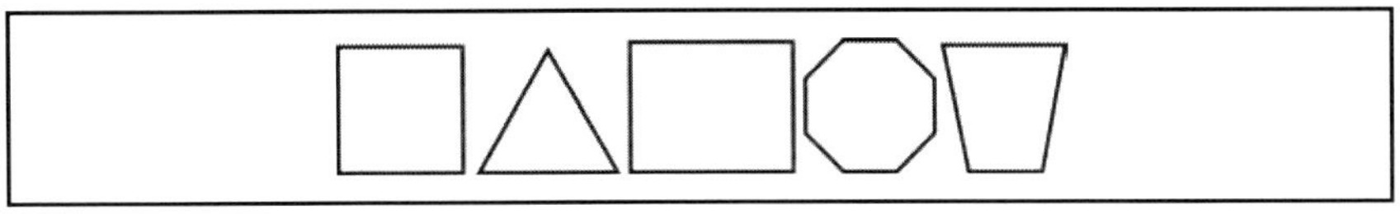

8. Sort the given shapes as regular or irregular polygons. Justify your sorting.

Use the following information to answer the next question.

Mr. Wilson asked his students to each draw four congruent shapes. These are the shapes Eric drew.

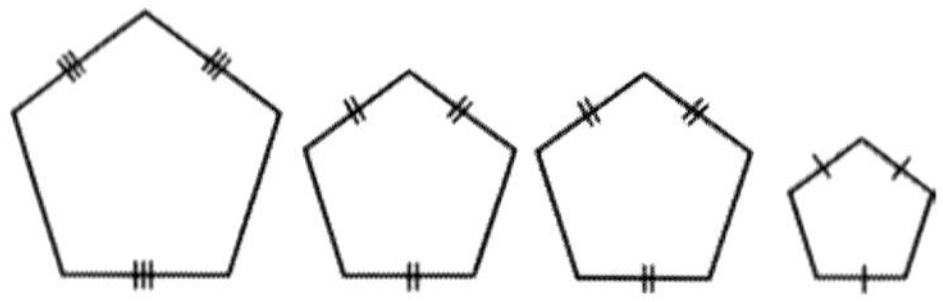

9. Are all the shapes Eric drew congruent? Justify your answer.

Use the following information to answer the next question.

Shari drew a square on 1 cm grid paper to determine congruency of the given square.

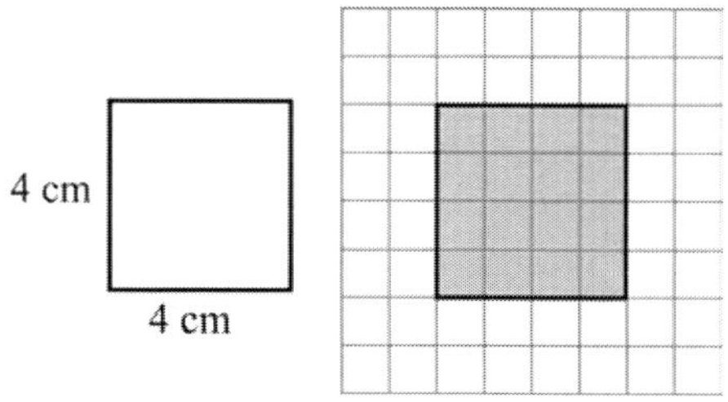

10. Are the two squares congruent? Explain how you can determine whether they are congruent or not.

TRANSFORMATIONS

When you are finished this unit, you should be able to…

- Describe successive translations, rotations, and reflections performed on a 2-D shape to produce a given image
- Identify and describe a combination of different types of transformations on a 2-D shape to produce a given image
- Draw the resulting image of a 2-D shape given a combination of transformations
- Perform a combination of successive transformations to create a design
- Analyze a given design by describing the original shape and the transformations used to create the design

Lesson	Page	Completion
1. Successive and Combination Transformations	218	
2. Designs Created from Successive Transformations	228	
Review Summary	233	
Practice Test	234	
Answers and Solutions	at the back of the book	

PREREQUISITE SKILLS AND KNOWLEDGE

Prior to beginning this unit, you should be able to…

- Identify and describe a single translation, rotation, or reflection of a 2-D shape
- Perform and draw a single translation, rotation, or reflection of a 2-D shape

Lesson 1 SUCCESSIVE AND COMBINATION TRANSFORMATIONS

NOTES

Clockwise movements are completed from left to right in the direction the hands on a clock move.

Counterclockwise movements are completed in a direction opposite to the direction the hands on a clock move.

Transformations of a two-dimensional (2-D) shape can be combined to show more than one movement. Regardless of the combinations used, the images must always be congruent to the original shape. All side lengths and angle measures in the image must be the same as in the figure; only the positions and orientations of the images can change.

- When describing the movement of a slide, give the horizontal movement first (left or right) and then the vertical movement (up or down).
- When describing the movement of a flip, describe the distance of the vertices from the flip line.
- When describing the movement of a turn, describe the point of rotation and the degree of rotation.

DESCRIBING SUCCESSIVE TRANSFORMATIONS

You can perform successive translations, reflections, or rotations on a 2-D shape. Each transformed image must be congruent to the original shape.

Example

Describe the two successive transformations that occurred to the given shape in position 1 for it to result in the image in position 2. Explain how you can determine congruency through measurement.

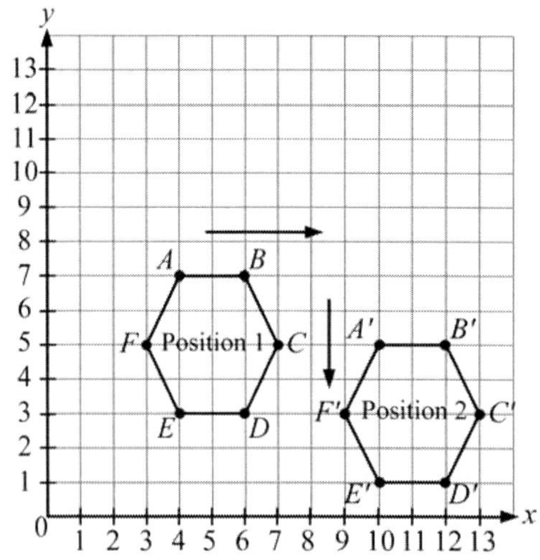

NOTES

Solution

Step 1
Identify the first transformation.

The first transformation is a slide, as shown by the horizontal arrow pointing to the right. The figure slides six units to the right. The slide can also be expressed as R6.

Step 2
Identify the second transformation.

The second transformation is also a slide, as shown by the vertical arrow pointing downward. The figure slides vertically downward for two units. The slide can also be expressed as D2.

Step 3
Describe the successive transformations.

To result in the image shown, the given shape slid to the right for six units and then slid down for two units. The transformations can also be expressed as R6, D2.

Step 4
Determine congruency through measurement.

Count the number of squares between each set of vertices in position 1 and each set of corresponding vertices in position 2.
- Two squares between AB and $A'B'$
- Four squares between FC and $F'C'$
- Two squares between ED and $E'D'$
- Four squares between AE and $A'E'$
- Four squares between BD and $B'D'$

This means that the figure and its image have the same side lengths. Since both figures have the same side lengths, they also have the same angle measures. Therefore, the image in position 2 is congruent to the figure in position 1.

NOTES

Time to Try 1

Describe the two successive transformations that occurred to the given shape in position 1 for it to result in the image in position 2. Explain how you can determine congruency through measurement.

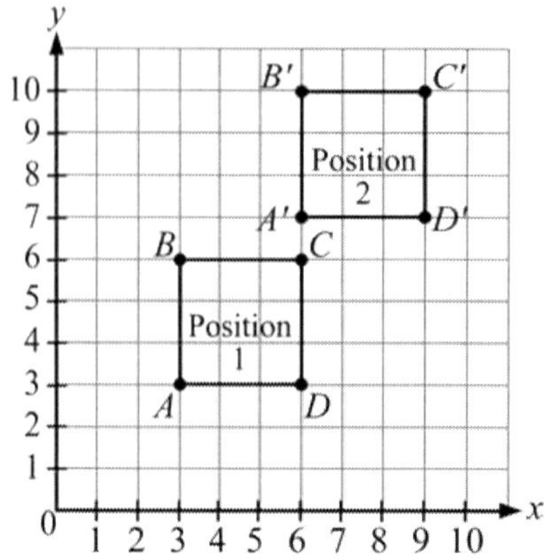

DESCRIBING COMBINATION TRANSFORMATIONS

You can also perform a combination of transformations on a given shape, such as a translation (slide) and then a reflection (flip).

Example

Describe the three transformations that occurred for the given shape in position *A* to result in the image in position *C*.

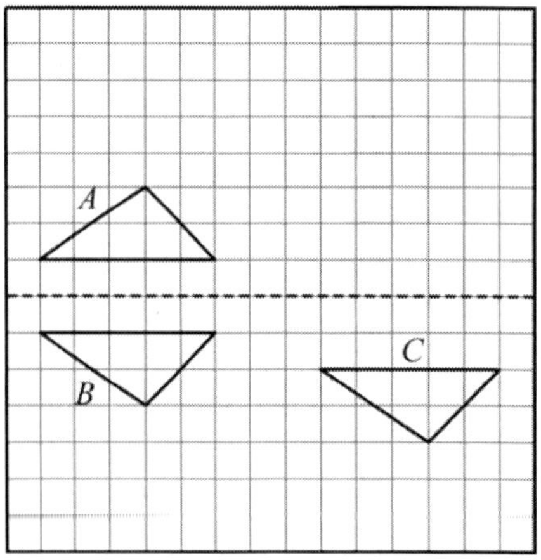

NOTES

Solution

Step 1
Identify the first transformation.

The first transformation is a flip. The original figure is one unit above the flip line, and it is flipped so that the image is now one unit below the flip line.

Each vertex of the image is exactly the same distance from the flip line as its corresponding vertex in the original shape.

Step 2
Identify the second transformation.

The second transformation is a slide. The figure slides horizontally to the right for eight units, or R8.

Step 3
Describe the third transformation.

To result in the image shown, the figure then slides downward for one unit, or D1.

Step 4
Describe the combined transformations.

To result in the image shown, the given shape flipped over the flip line, slid horizontally to the right for eight units, and then slid vertically downward for one unit.

Time to Try 2

Describe the transformations that occurred for the given shape in position *A* to result in the image in position *C*.

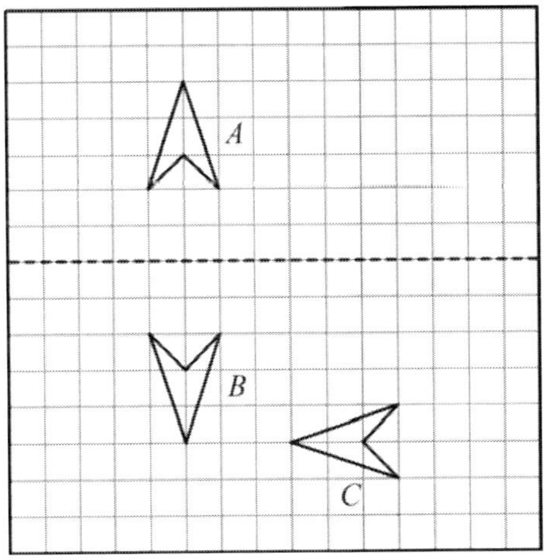

NOTES

DRAWING THE IMAGE RESULTING FROM MORE THAN ONE TRANSFORMATION

When you are given a combination of transformations, you can perform the transformations on the original shape and then draw the resulting image to show the final position.

Example

Draw the final position of the image after the following two successive transformations have been performed:

1. Slide the shape six units to the left (L6).
2. Slide the image five units down (D5).

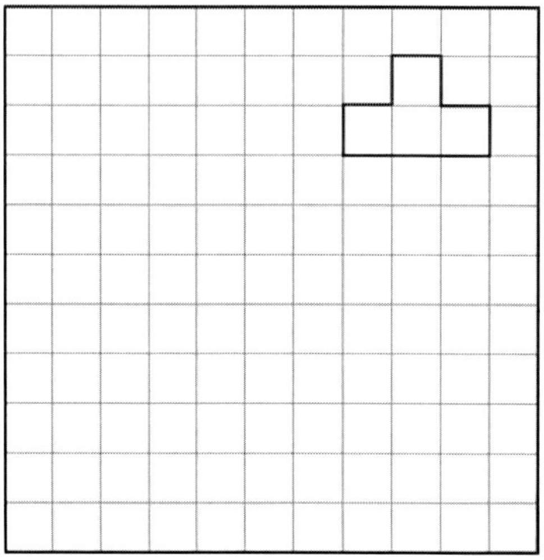

Solution

After the two slides, the final image should be in this position.

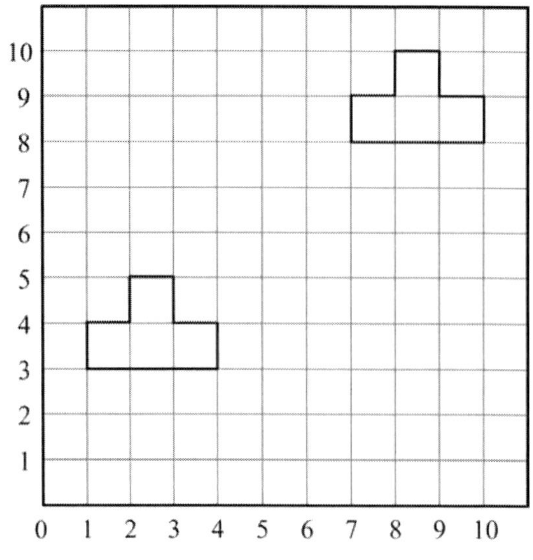

NOTES

Time to Try 3

Translate the given shape four units to the right (R4), and then flip the image across the dotted flip line.

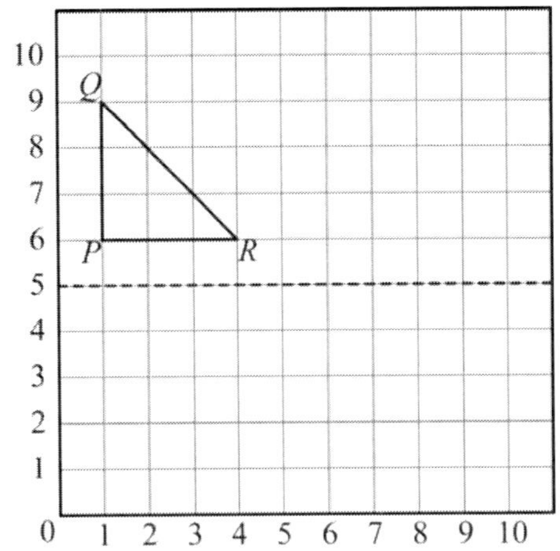

PRACTICE EXERCISES

Use the following information to answer the next question.

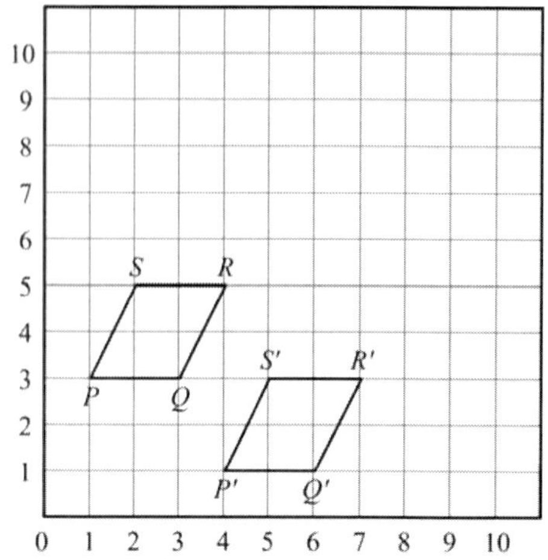

1. Describe the two successive transformations that occurred to shape *PQRS* for it to result in image *P'Q'R'S'*.

Use the following information to answer the next question.

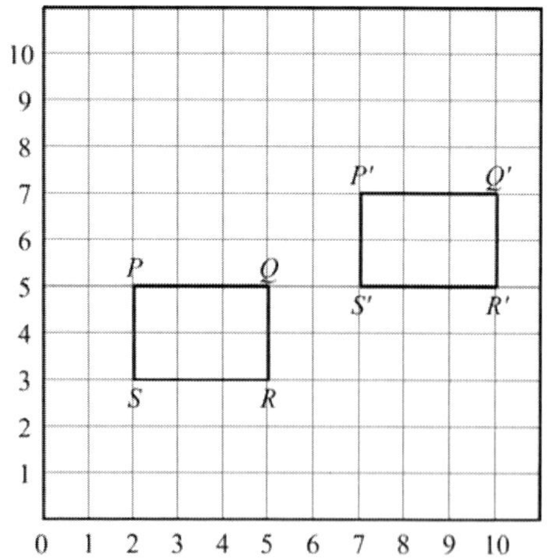

2. Describe the two successive transformations that occurred to shape *PQRS* for it to result in image *P'Q'R'S'*.

Use the following information to answer the next question.

- Slide the shape in position I five units downward (D5) to position II.
- Reflect the image in position II to create the image in position III.

3. Show the given transformations to the shape in position I by drawing and labelling the two positions.

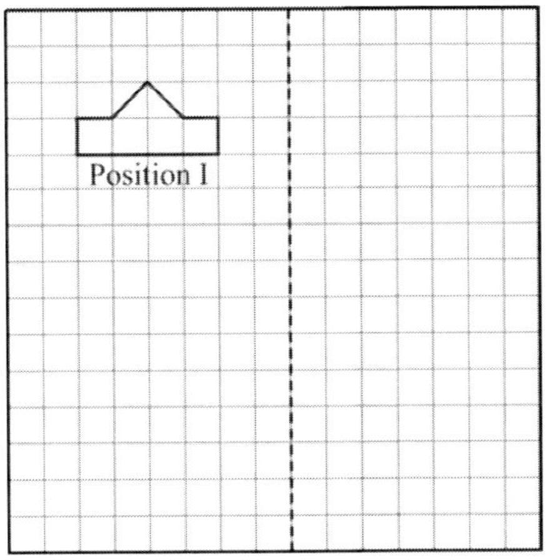

Use the following information to answer the next question.

- Reflect the shape in position *A* across the dotted flip line.
- Slide the reflected image four units to the right.
- Slide the translated image three units upward.

4. Draw the reflected image, and label its position as *B*. Then, draw the final translated image, and label its position as *C*.

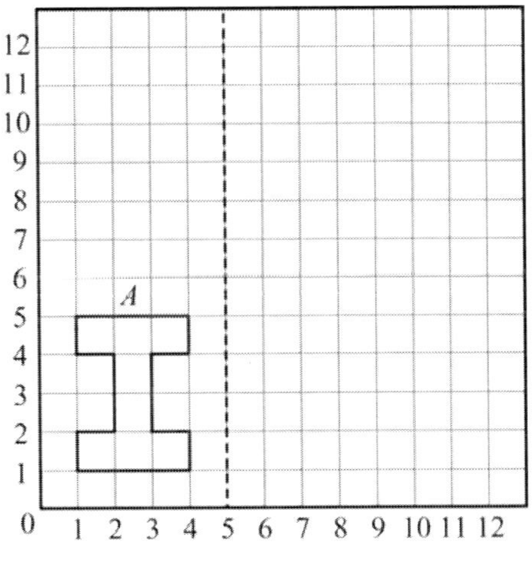

Use the following information to answer the next question.

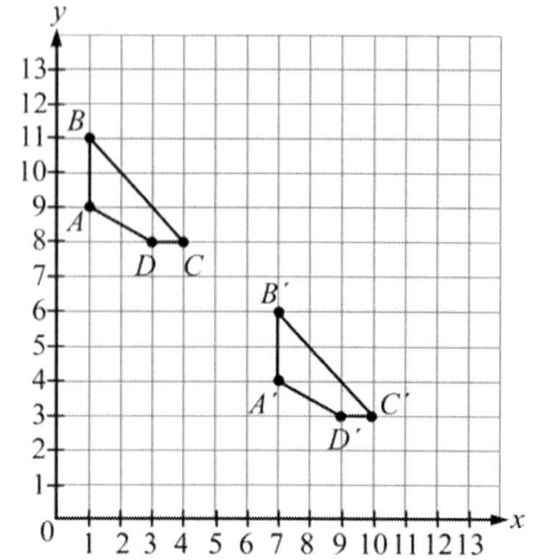

5. Describe the two successive transformations that occurred to shape *ABCD* for it to result in image $A'B'C'D'$.

Use the following information to answer the next question.

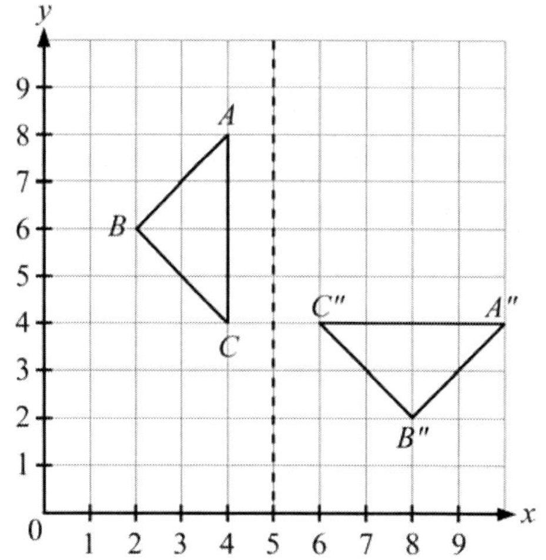

6. Describe the two transformations needed for figure *ABC* to move to the position shown by figure $A''B''C''$.

7. Reflect figure *ABCD* across the dotted flip line, and then slide the new image down three units. Draw and label both transformations.

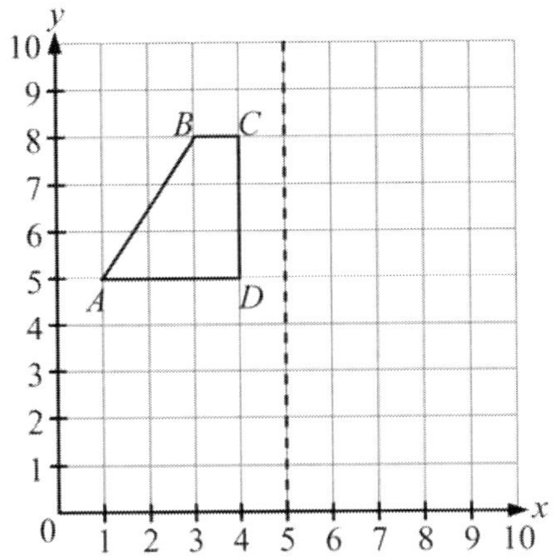

Lesson 2 DESIGNS CREATED FROM SUCCESSIVE TRANSFORMATIONS

NOTES

Designs can be created using successive transformations of a basic shape. The basic shape is called the original shape, and it becomes the first shape of the design. The original shape is then translated, reflected, or rotated to form the rest of the design. Any given design can be analyzed to determine the original shape and the transformations that created it.

CREATING DESIGNS

Start with a basic 2-D shape or pattern. Then, repeatedly slide, reflect, and rotate that shape or pattern to create your design.

Example

This is the original shape.

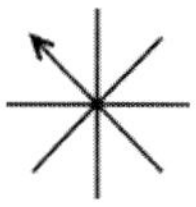

Use this shape to create a horizontal ($\rightarrow$) design that shows four 45° clockwise rotations.

Solution

Step 1
Start the design.

Use the original shape for the first shape of the design.

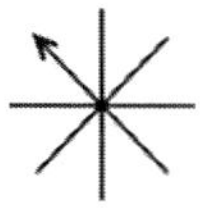

Step 2
Continue the design by rotating the shape four times.

Rotate the shape so that the arrow turns 45° to the right for each rotation.

Your design will look like this picture.

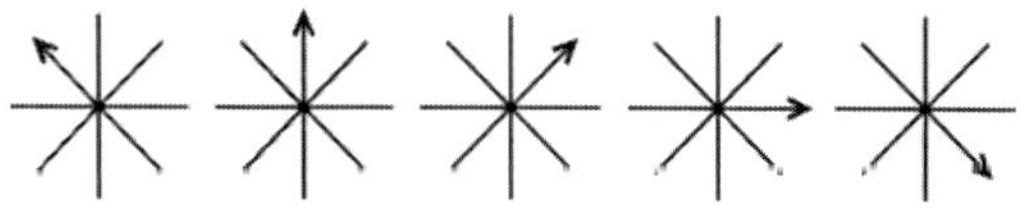

NOTES

Time to Try 1

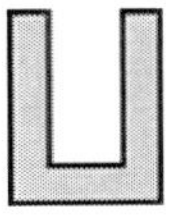

Use this original shape to create a design from the following translations.

- The first translation is an upward diagonal move (↗).
- The second translation is a downward diagonal move (↘).

ANALYZING DESIGNS

When a design is created, the basic shape used is referred to as the original shape. When analyzing designs, you need to identify the original shape that was used to create the design. You also need to describe the types of transformations that were applied to the original shape including successive translations (slides), reflections (flips), or rotations (turns).

Example

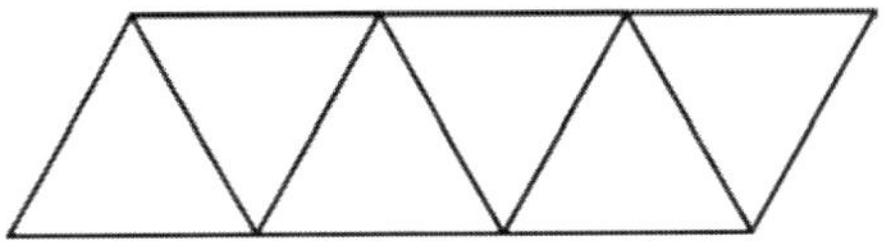

Identify the original shape, and describe the transformations used to create the design.

Solution

The original shape is an equilateral triangle.

The first triangle is rotated in a clockwise direction around the bottom right vertex. The next triangle is rotated counterclockwise around the top vertex. The pattern of rotation keeps repeating.

- Clockwise around the bottom right vertex
- Counterclockwise around the top vertex
- Clockwise around the bottom right vertex

NOTES

Time to Try 2

Identify the original shape, and describe the successive transformations used to create the given design.

PRACTICE EXERCISES

Use the following design to answer the next two questions.

1. Describe the original shape used for the given design.

2. Describe the transformations used to create the design.

Use the following information to answer the next question.

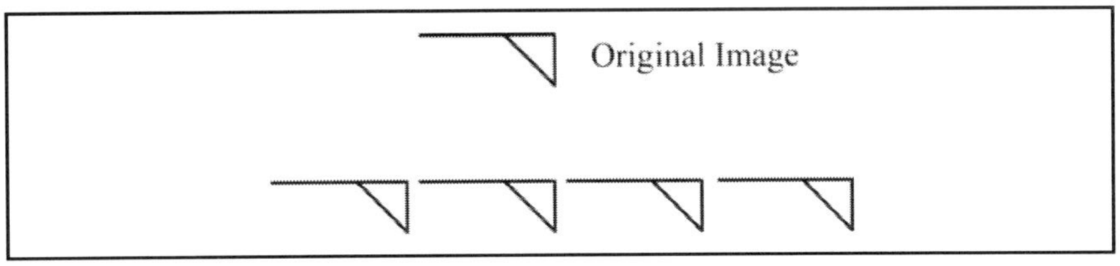

3. Describe the transformations used to create the given design.

Use the following information to answer the next question.

Sam and John are describing the transformations used to create the given pattern. Sam says that successive translations were used. John says that successive reflections were used.

4. Determine whether Sam or John described the transformations correctly, and justify your answer.

Use the following design to answer the next two questions.

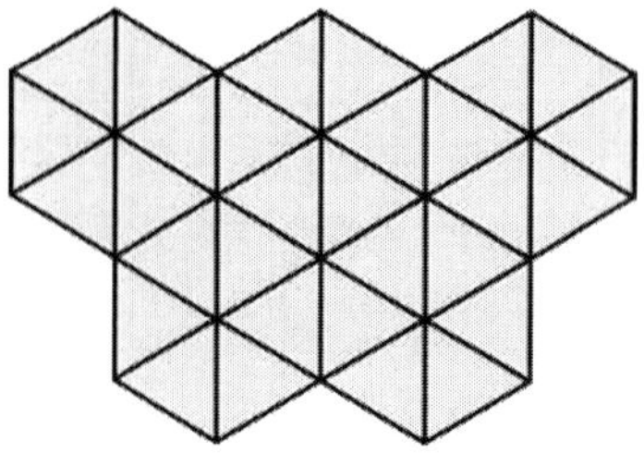

5. Describe the original shape used to create the given design.

6. Describe the transformations used to create the design.

Use the following design to answer the next question.

7. Describe the successive transformations used to create the given design.

Use the following design to answer the next question.

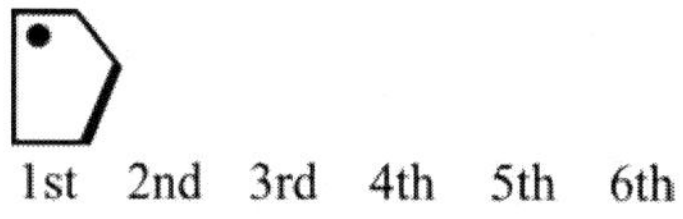

8. Draw the design that will be created when the given shape is flipped five times to the right (→).

REVIEW SUMMARY

- A transformation is a change in a figure that results in a different position or direction.
- Transformations include rotations (turns), translations (slides), and reflections (flips).
- After any transformation, the resulting image has the same size, area, angle sizes, and line lengths as the original shape.
- A rotation is a transformation in which a shape is turned about a point to create a congruent image of the original shape.
- Shapes can be rotated in either a clockwise or a counterclockwise direction.
- A translation is the movement of a figure in which each point of the figure is moved the same distance and in the same direction.
- A reflection is a transformation in which a figure is flipped along a line of reflection, called the mirror line or flip line, to create a mirror image of the original figure.
- A combination of transformations can take place. It is important to do each transformation separately to achieve the desired figure.

PRACTICE TEST

Use the following design to answer the next question.

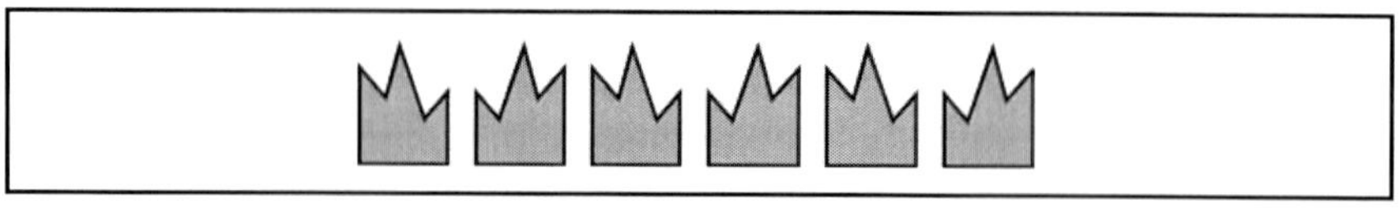

1. Describe the successive transformations used to create the given design.

Use the following design to answer the next two questions.

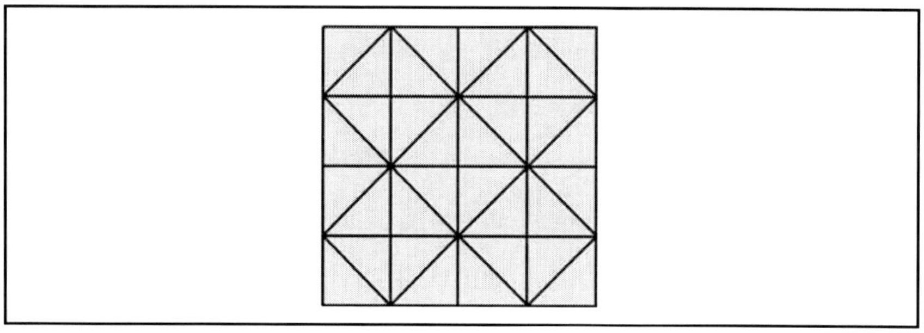

2. Identify the original shape used to create the given design.

3. Describe the successive transformations used to create the design.

Use the following information to answer the next question.

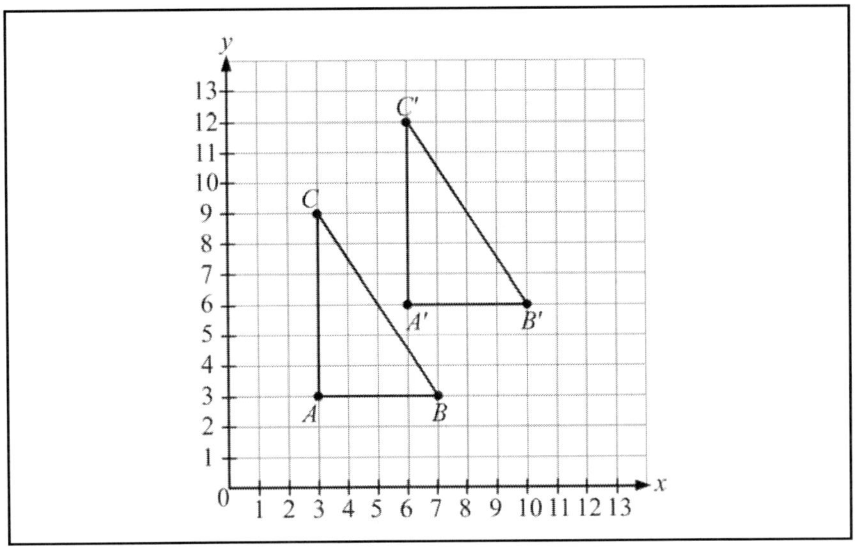

4. Describe the two successive transformations to shape *ABC* that resulted in image $A'B'C'$.

Use the following information to answer the next question.

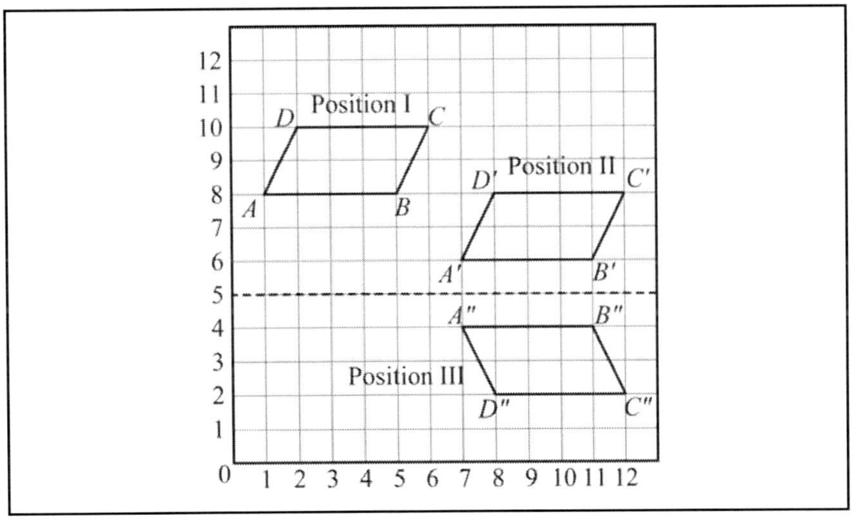

5. Identify the three transformations to shape *ABCD* in position I that resulted in image $A''B''C''D''$ in position III.

6. Perform the following three transformations on the given shape:

- Reflect the figure in position I across the dotted flip line. Draw the image, and label it as position II.
- Translate the reflected image four units to the right.
- Then, translate the image two units upward. Draw the image after the two translations, and label it as position III.

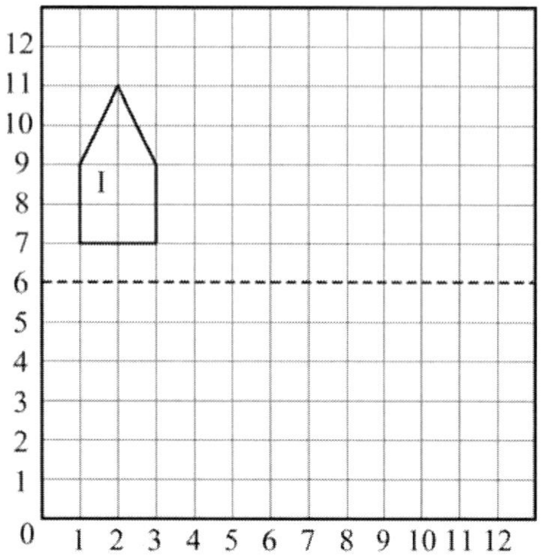

Use the following design to answer the next question.

7. Describe the successive transformations used to create the given design.

8. Translate shape *ABCD* to the right six units, and reflect the image across the line of reflection. Draw and label both transformations.

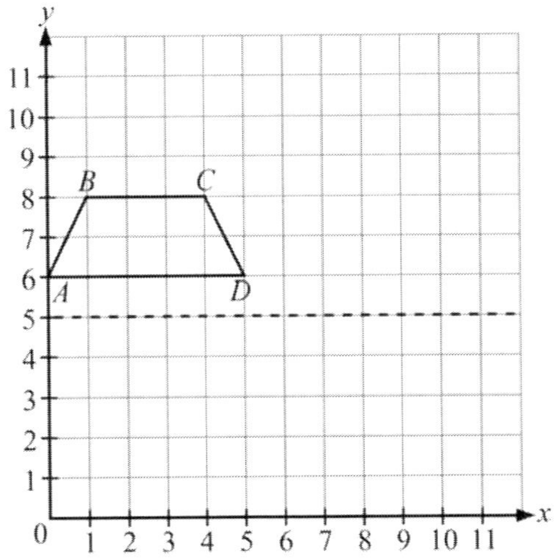

Use the following diagram to answer the next question.

Shape R can be moved to the position of image S using two combined transformations.

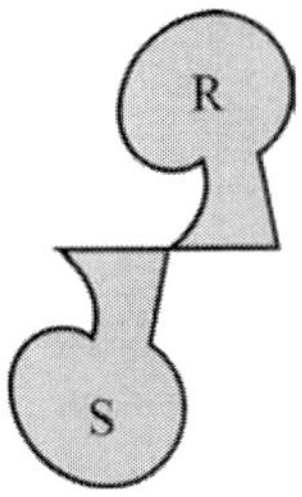

9. Describe two possible combined transformations that could move shape R to the position of image S.

Use the following diagram to answer the next question.

Ryan uses this original shape to create a horizontal design with three successive reflections.

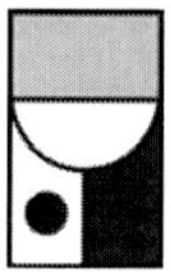

10. Draw Ryan's completed design.

FIRST QUADRANT OF A CARTESIAN PLANE

When you are finished this unit, you should be able to…

- Identify the x- and y-axes and the point of origin
- Plot points in the first quadrant given ordered pairs
- Write ordered pairs to correspond to plotted points
- Determine the distance between points along horizontal and vertical lines
- Identify points used to produce shapes or designs
- Draw shapes or designs given ordered pairs
- Perform a single transformation of a shape, and identify the coordinates of the vertices of the image
- Describe the positional change of the vertices of a given shape to the corresponding vertices of its transformed image

Lesson	Page	Completed on
1. Identifying Points and Distances Between Points	240	
2. Plotting Points in the First Quadrant	247	
3. Single Transformations in the First Quadrant	255	
Review Summary	262	
Practice Test	263	
Answers and Solutions	at the back of the book	

PREREQUISITE SKILLS AND KNOWLEDGE

Prior to starting this unit, you should be able to...

- Identify and describe a single translation, rotation, or reflection of a two-dimensional shape
- Perform and draw a single translation, rotation, or reflection of a two-dimensional shape

Lesson 1 IDENTIFYING POINTS AND DISTANCES BETWEEN POINTS

NOTES

A Cartesian plane is formed by the intersection of a horizontal and vertical number line at a 90° angle. The horizontal and vertical number lines are called the x-axis and y-axis, respectively. The lines intersect at a point called the origin (0, 0). A Cartesian plane is divided into four areas called quadrants, as shown in the given diagram.

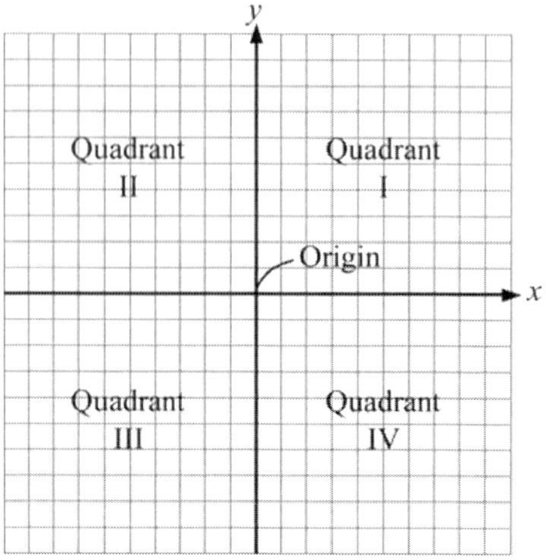

LABELLING THE FIRST QUADRANT OF A CARTESIAN PLANE

The following grid shows the first quadrant of a Cartesian plane. It is important to label the origin, the axes, and the intervals on the plane.

- The origin of the plane is located at the bottom left corner of the grid. The origin is labelled 0.
- The horizontal axis (→) has a directional arrow and is labelled x.
- The vertical axis (↑) has a directional arrow and is labelled y.
- This Cartesian plane uses an interval of 1 along the x- and y-axis (1, 2, 3…).

Ordered pair: a pair of numbers used to locate a point on a Cartesian plane

Coordinate: one of the numbers in an ordered pair

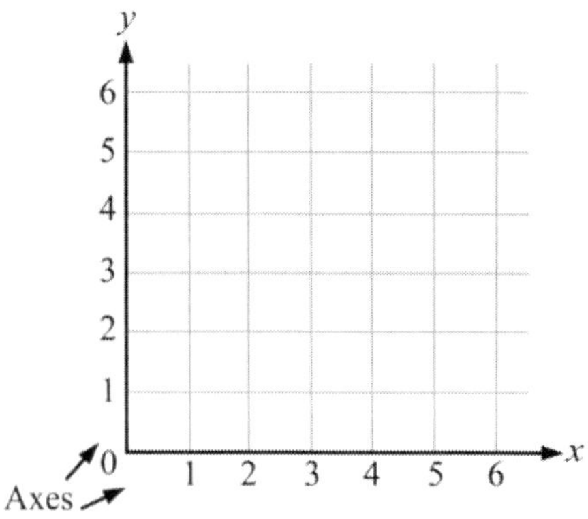

WRITING ORDERED PAIRS FOR PLOTTED POINTS

A point in the first quadrant of a Cartesian plane can be described by its location on the grid. Its location is written as an ordered pair or set of coordinates. An example of an ordered pair is (2, 6).

When you are writing the coordinates to describe the location of a given point, the order of the coordinates is very important. Always be sure to write the horizontal coordinate (x-axis) first and the vertical coordinate (y-axis) second—(horizontal, vertical) or (x, y).

Follow these steps to write an ordered pair describing the location of a point on the first quadrant:

- First, look along the x-axis to see how far the point is to the right of the 0 (origin).
- Write that number, followed by a comma.
- Next, look along the y-axis to see how far the point is up from the origin.
- Write that number to the right of the comma.
- Place brackets on each side of the numbers.

NOTES

x-coordinate: the first number in an ordered pair

y-coordinate: the second number in an ordered pair

You can remember that the x-axis coordinate comes before the y-axis coordinate just like x comes before y alphabetically.

Example

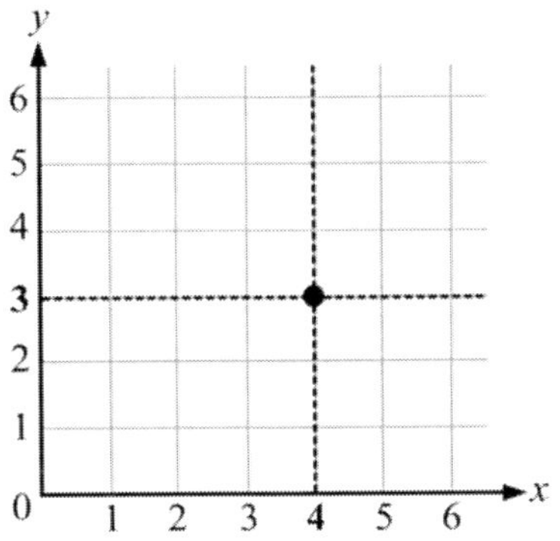

Write the coordinates of the given point.

Solution

1. Start at the origin (0).
2. Count along the x-axis to the line that has the given point.
   1, 2, 3, 4
3. The number 4 is the first number of the ordered pair.
4. Count up to the line that has the given point.
   1, 2, 3
5. The number 3 is the second number of the ordered pair.
6. Record the numbers within brackets separated by a comma.
   (4, 3)

The coordinates (ordered pair) of the given point are (4, 3).

NOTES

Time to Try 1

Write the ordered pairs for each of the three points.

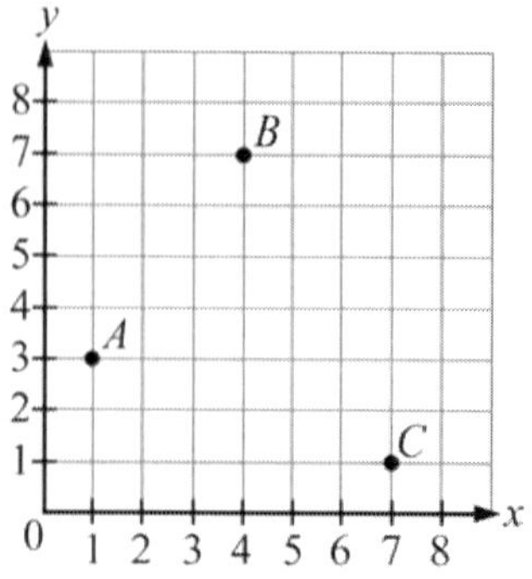

IDENTIFYING POINTS USED TO PRODUCE SHAPES OR DESIGNS

When shapes or designs are drawn in the first quadrant of a Cartesian plane, you can identify the points used to produce the shapes.

Example

Darian drew the following rectangle in the first quadrant. Identify the points used to produce the shape.

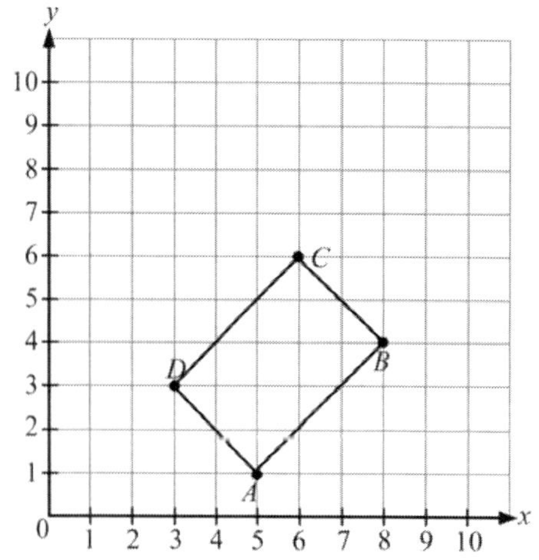

NOTES

Solution

- Point A is 5 units to the right of the origin and 1 unit above it. The ordered pair of point A is (5, 1).
- Point B is 8 units to the right of the origin and 4 units above it. The ordered pair of point B is (8, 4).
- Point C is 6 units to the right of the origin and 6 units above it. The ordered pair of point C is (6, 6).
- Point D is 3 units to the right of the origin and 3 units above it. The ordered pair of point D is (3, 3).

The ordered pairs used to produce the rectangle are $A(5, 1)$, $B(8, 4)$, $C(6, 6)$, and $D(3, 3)$.

Time to Try 2

Lara drew the following triangle in the first quadrant. Identify the points used to produce the shape.

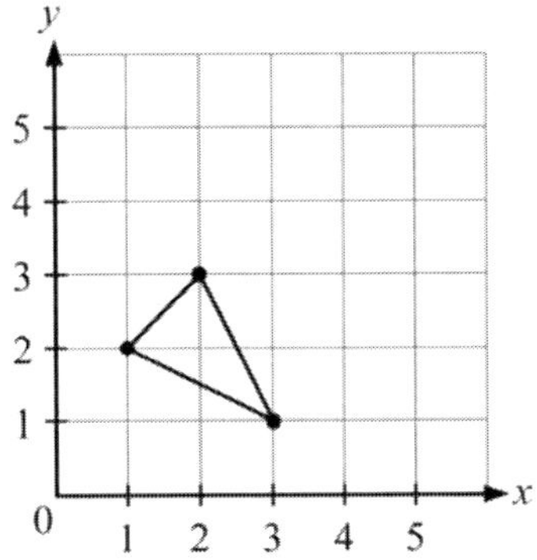

DETERMINING DISTANCES BETWEEN POINTS

To determine the distances between points along the horizontal and vertical lines in the first quadrant of a Cartesian plane, always start with the horizontal distance first. Count the number of spaces between the first point and the next point. Then, count the number of spaces between the points along the vertical lines.

Remember to look at the interval used. If the interval is 1, you count the spaces between the points by 1s. If the interval is 2, you count the spaces between the points by 2s, and so on.

NOTES

Example

What is the distance between points A and B?

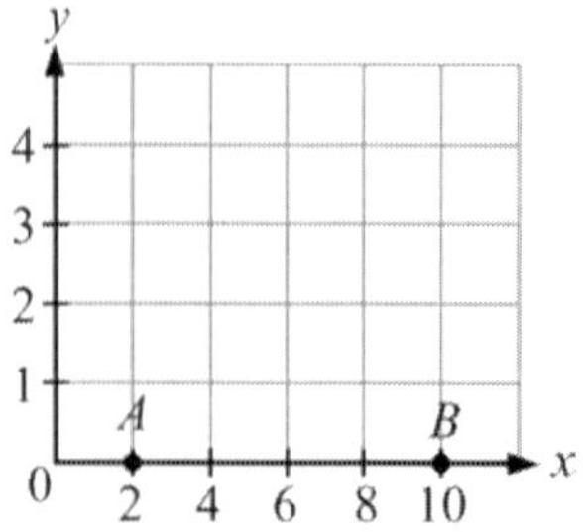

Solution

Step 1
Identify the interval.

The interval is 2.

Step 2
Determine the distance.

Count the number of spaces between the two points. There are four spaces. Start at 2 and count by 2s for four counts until you reach 10.

$\boxed{2}...2,4,6,8,...\boxed{10}$

The distance between point A and point B is 8 units.

Time to Try 3

Describe the distance between points A and C.

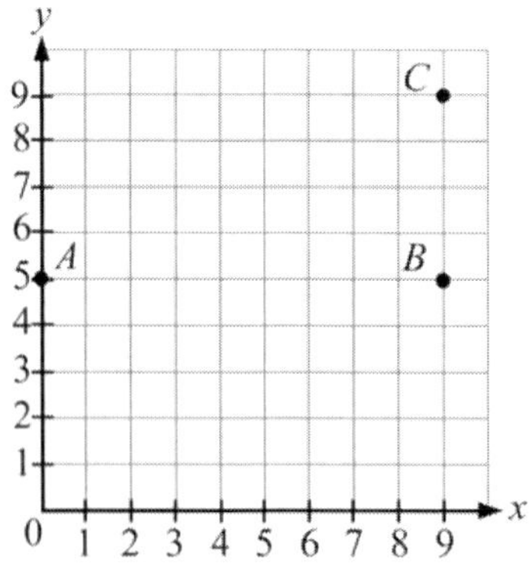

PRACTICE EXERCISES

Use the following information to answer the next question.

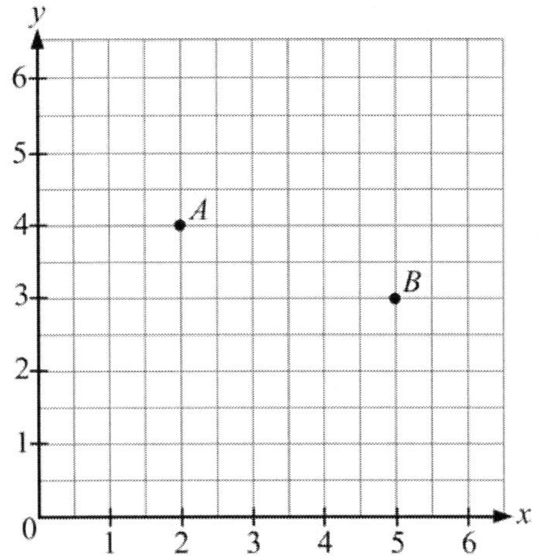

1. Identify the location of points A and B by writing their ordered pairs.

Use the following information to answer the next question.

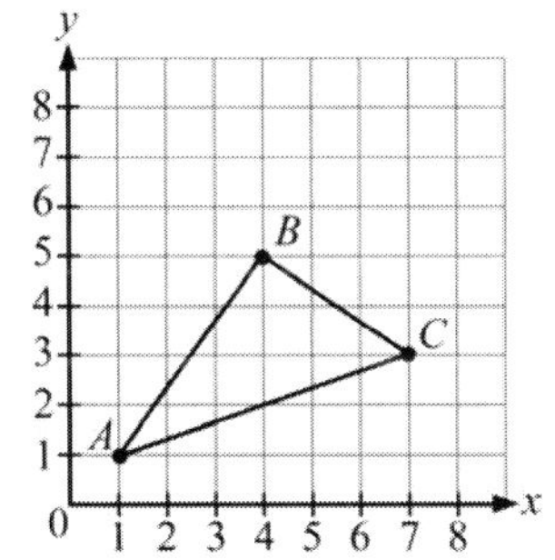

2. Identify the ordered pairs used to produce the triangle.

Use the following information to answer the next four questions.

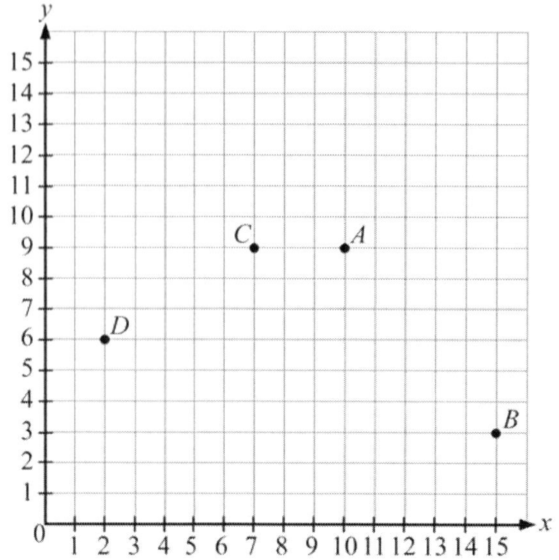

Describe the distance between each given set of points.

3. Points *A* and *C*

4. Points *D* and *A*

5. Points *C* and *B*

6. Points *D* and *C*

Lesson 2 PLOTTING POINTS IN THE FIRST QUADRANT

NOTES

PLOTTING POINTS WITH INTERVALS OF ONE

You can plot a point on a Cartesian plane if you know the point's coordinates. When given an ordered pair and asked to plot the point, follow these steps:

1. Look at the first number of the ordered pair. The first number tells you how far to the right of the origin (0) the point will be.
2. Look at the second number. The second number tells you how far up from the origin the point will be.
3. Draw the point where the two lines intersect.

Example

Given the ordered pairs (1, 2), (4, 6), and (7, 4), plot the points on the following Cartesian plane.

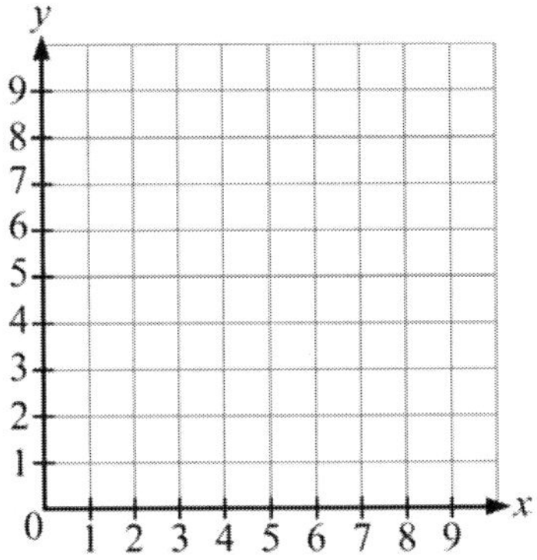

Solution

Step 1

Plot the ordered pair (1, 2).

- Start at the origin (0).
- Move to the right along the x-axis until you get to 1 (the first number in the ordered pair).
- Move up along the y-axis until you get to 2 (the second number in the ordered pair).
- Plot the point where these two lines intersect.

Step 2

Plot the ordered pair (4, 6).

- Start at the origin.
- Move to the right along the x-axis until you get to 4 (the first number in the ordered pair).
- Move up along the y-axis until you get to 6 (the second number in the ordered pair).
- Plot the point where these two lines intersect.

NOTES

Step 3

Plot the ordered pair (7, 4).

- Start at the origin.
- Move to the right along the x-axis until you get to 7 (the first number in the ordered pair).
- Move up along the y-axis until you get to 4 (the second number in the ordered pair).
- Plot the point where these two lines intersect.

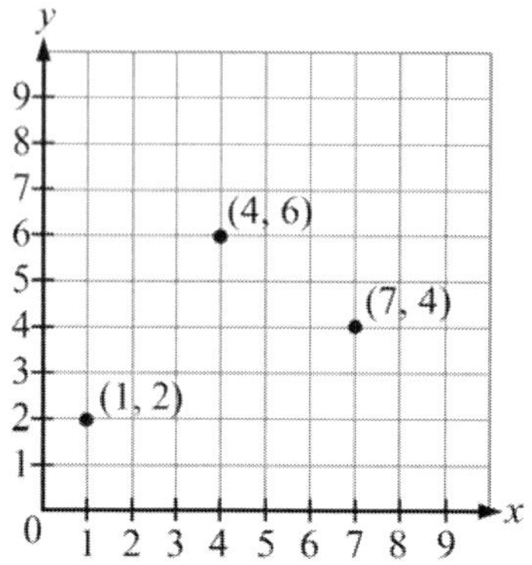

Time to Try 1

Plot the ordered pairs (1, 7), (3, 1), and (5, 3) on the following Cartesian plane.

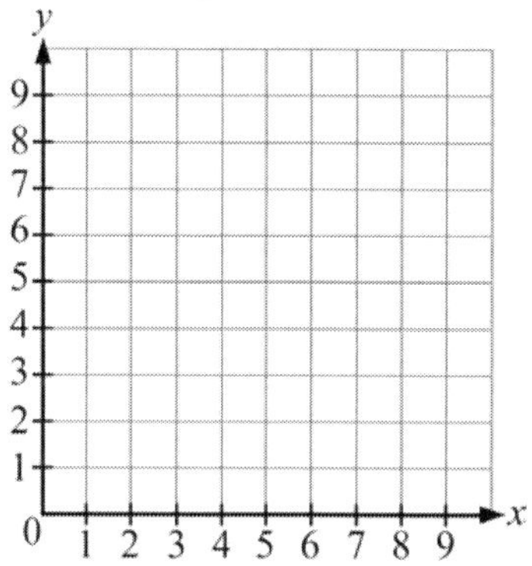

NOTES

PLOTTING POINTS WITH INTERVALS 0F 2, 5, OR 10

Not all Cartesian planes have axes with number intervals of one. Other commonly used intervals are 2, 5, and 10. It is always important to look at the number interval before plotting the ordered pair of a point.

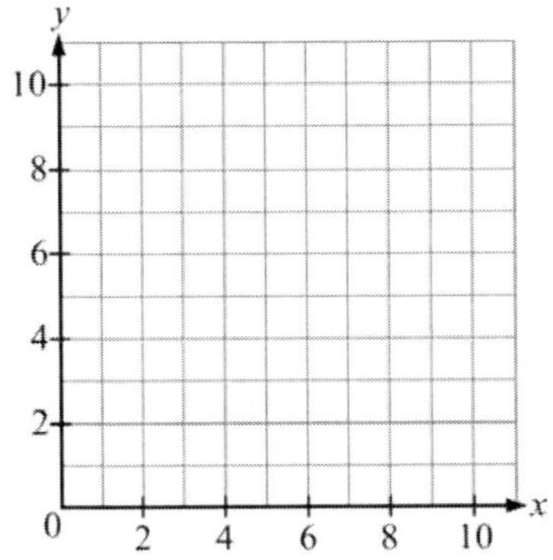

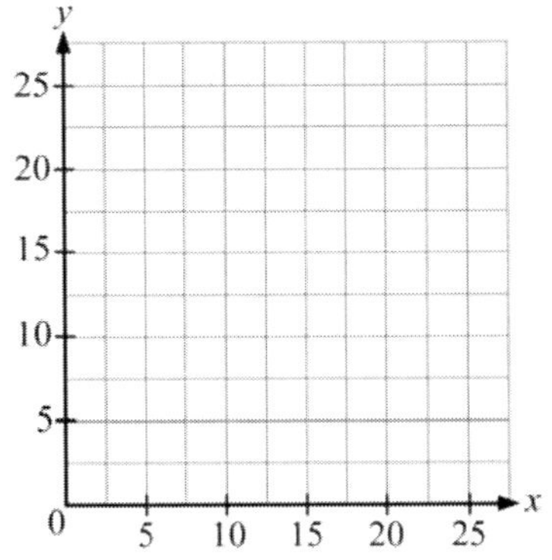

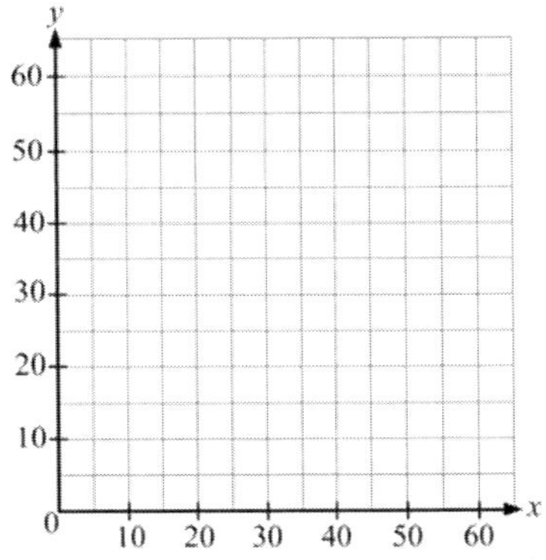

NOTES

Example

The following quadrant shows intervals of two. Plot the ordered pairs (1, 4), (4, 6), and (6, 5).

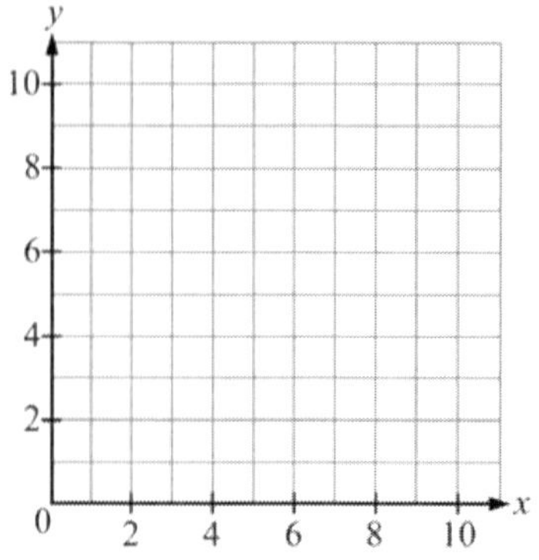

Solution

Step 1

Plot the ordered pair (1, 4).

- Start at the origin.
- Move to the right along the x-axis until you get to the line that represents 1 (the first number in the ordered pair). The 1 is between the origin (0) and the numbered line for 2.
- Move up along the y-axis until you get to 4 (the second number in the ordered pair).
- Plot the point where these two lines intersect.

Step 2

Plot the ordered pair (4, 6).

- Start at the origin.
- Move to the right along the x-axis until you get to 4 (the first number in the ordered pair).
- Move up along the y-axis until you get to 6 (the second number in the ordered pair).
- Plot the point where these two lines intersect.

NOTES

Step 3

Plot the ordered pair (6, 5).

- Start at the origin.
- Move to the right along the x-axis until you get to 6 (the first number in the ordered pair).
- Move up along the y-axis until you get to the line that represents 5 (the second number in the ordered pair). The 5 is between the numbered lines for 4 and 6.
- Plot the point where these two lines intersect.

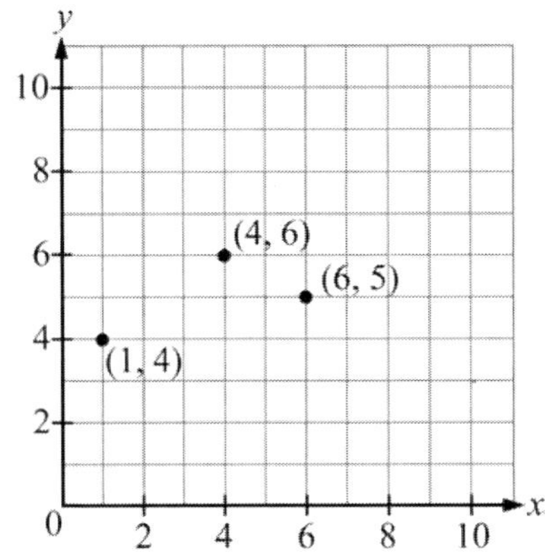

Time to Try 2

Plot the ordered pairs (2, 8), (3, 5), and (8, 3).

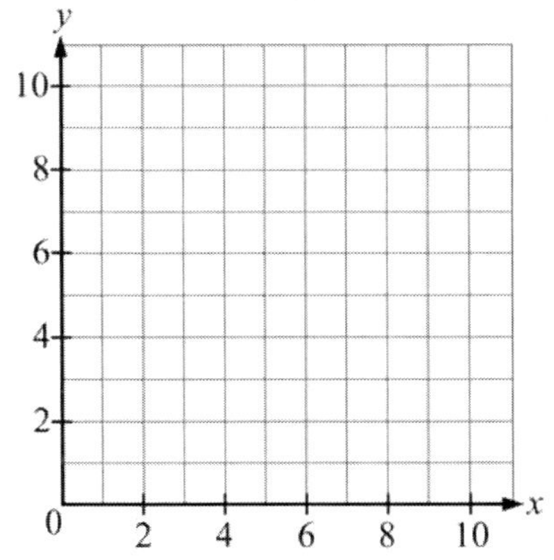

DRAWING SHAPES OR DESIGNS WHEN GIVEN ORDERED PAIRS

You can plot a set of ordered pairs and create a shape or design by connecting the points.

NOTES

Example

Plot the ordered pairs *P*(3, 9), *Q*(8, 9), *S*(3, 4), and *T*(8, 4).
Connect the points with line segments. Identify the shape produced.

Solution

Step 1

Plot the ordered pairs.

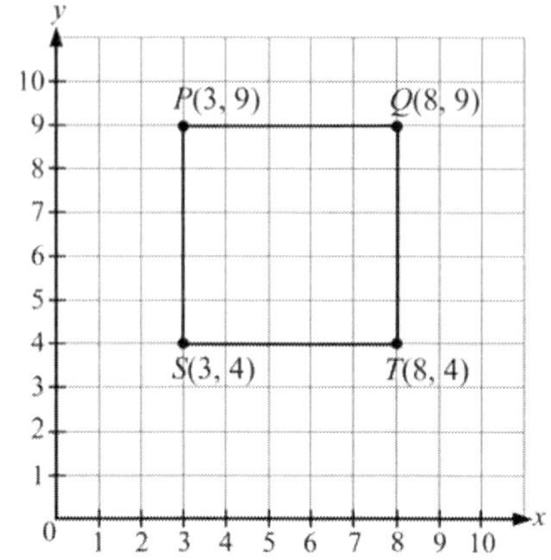

Step 2

The shape produced is a square.

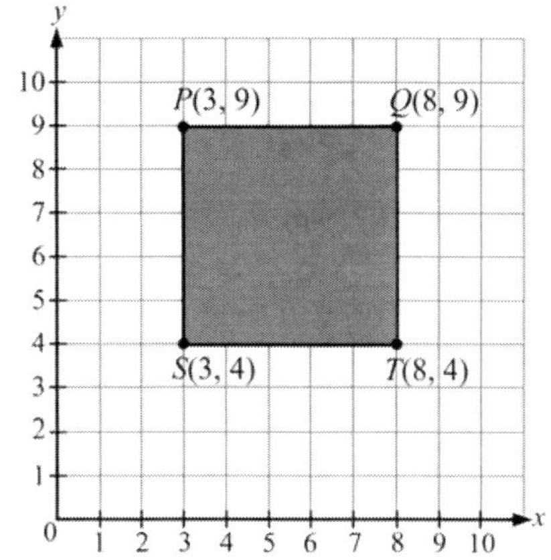

Time to Try 3

Plot the ordered pairs *A*(1, 7), *B*(2, 3), *C*(7, 3), and *D*(6, 7).
Connect the points with line segments. Identify the shape produced.

PRACTICE EXERCISES

1. Plot the ordered pairs (1, 3), (3, 5), (6, 6), and (8, 0). Label the plotted points from *A* to *D*, respectively.

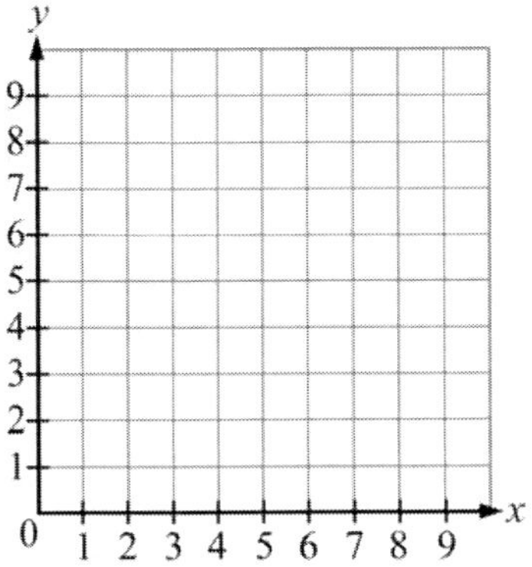

2. Plot the ordered pairs *A*(5, 4), *B*(2, 7), and *C*(8, 8).

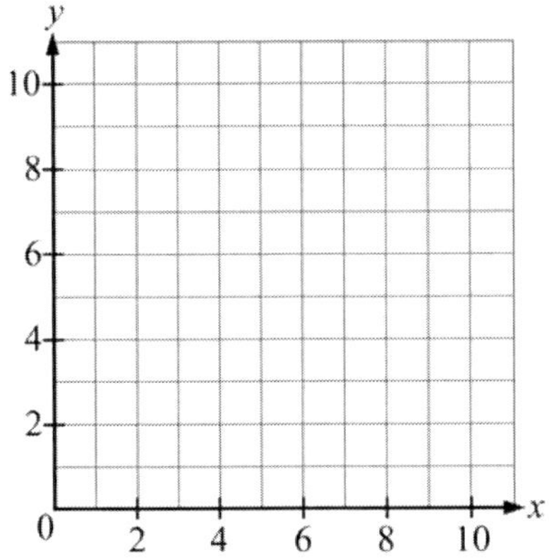

3. Plot the ordered pairs (5, 5), (10, 12), and (25, 25). Label the plotted points *A*, *B*, and *C*, respectively.

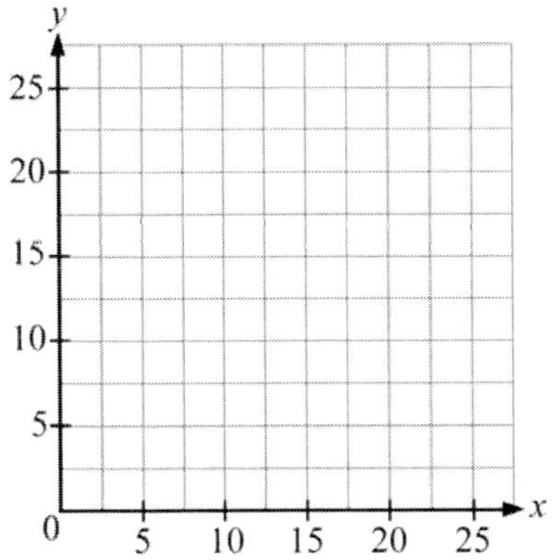

4. Plot the ordered pairs (2, 2), (3, 4), and (4, 2). Connect the points. Identify the shape produced.

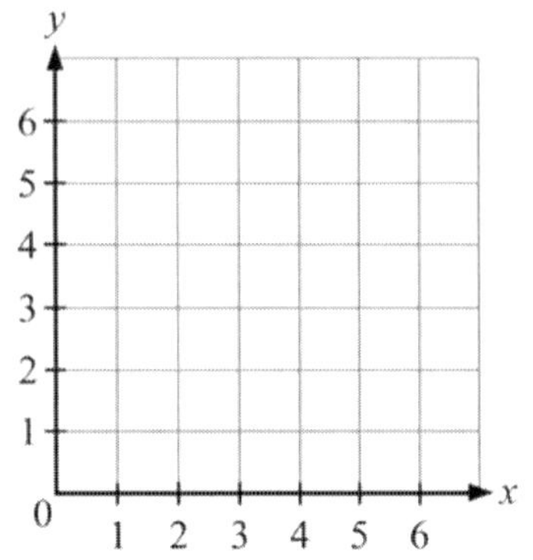

5. Plot the ordered pairs (2, 7), (5, 7), and (2, 4). Connect the points. Identify the shape produced.

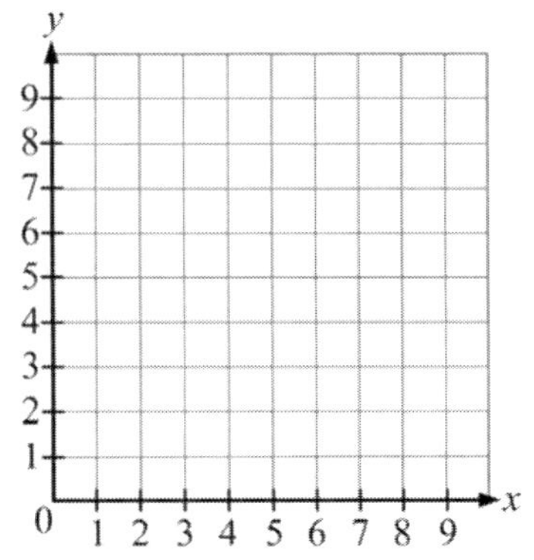

Lesson 3 SINGLE TRANSFORMATIONS IN THE FIRST QUADRANT

DESCRIBING POSITIONAL CHANGES OF VERTICES

NOTES

A Cartesian plane can be used to perform a single transformation of a 2-D shape. The positional change of the image can be described by comparing the coordinates (ordered pairs) of its vertices to the coordinates (ordered pairs) of the given shape.

Example

Sean drew a square on the coordinate grid in position *A*. He then translated the shape to the image seen in position *B*.

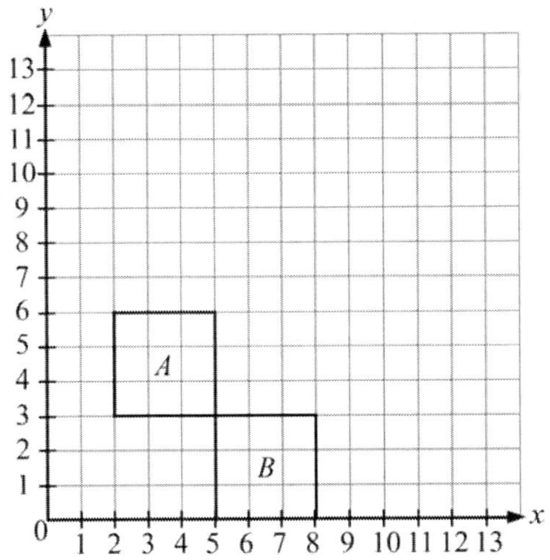

Identify the coordinates of the given shape in position *A*. Describe the positional change of each vertex from position *A* to position *B*.

Solution

Step 1
Identify the coordinates of shape *A*.

Starting at the top left corner of the square and moving in a counterclockwise direction, the coordinates of the vertices of the square are (2, 6), (2, 3), (5, 3), and (5, 6).

Step 2
Describe the positional change in vertices.
Position A $\rightarrow$ Position B

(2, 6) $\rightarrow$ (5, 3)
(2, 3) $\rightarrow$ (5, 0)
(5, 3) $\rightarrow$ (8, 0)
(5, 6) $\rightarrow$ (8, 3)

NOTES

Time to Try 1

Identify the coordinates of the given shape labelled Figure 1. Describe the positional change of each vertex from the shape labelled Figure 1 to its image labelled Figure 2.

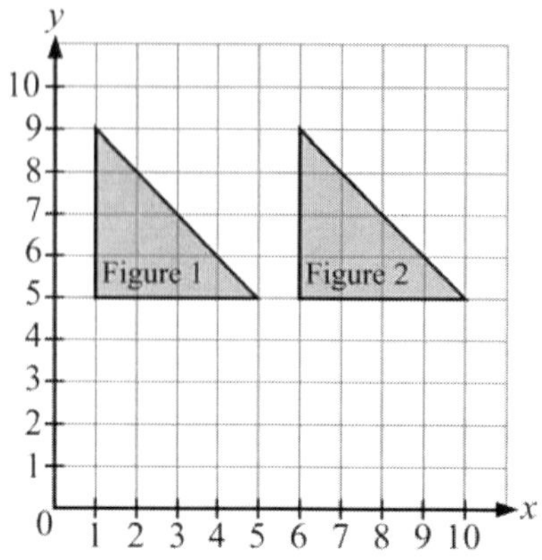

PERFORMING A SINGLE TRANSFORMATION

When performing a transformation in the first quadrant, you can count the number of spaces to determine the location of the transformed image. You can also use the ordered pairs (coordinates) of the shape to determine where to draw the transformed image.

When a shape is transformed, the corresponding vertices of the image are often labelled with primes. For example, vertex A in a given shape may be labelled A' in the image.

NOTES

Example

Flip the given 2-D shape labelled *A* across the dotted flip line, and draw its image in the new position, labelling it *B*.
Identify the coordinates of the vertices in the image.

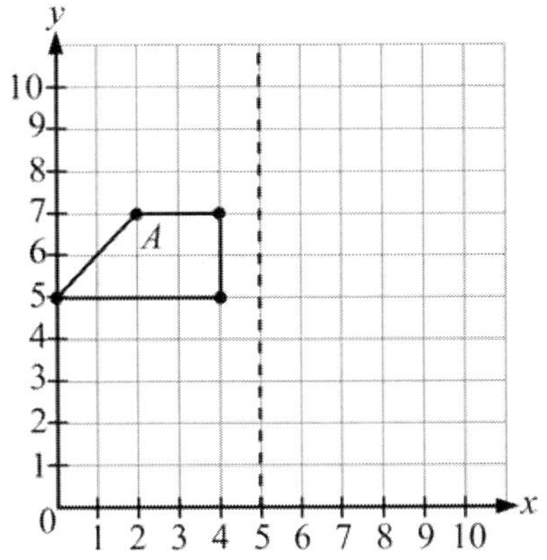

Solution

Step 1
Identify the distance of shape *A* from the dotted flip line.

Since the two vertices closest to the flip line (4, 5) and (4, 7) are one square to the left of the flip line, the flipped image (*B*) must be one square to the right of the flip line.

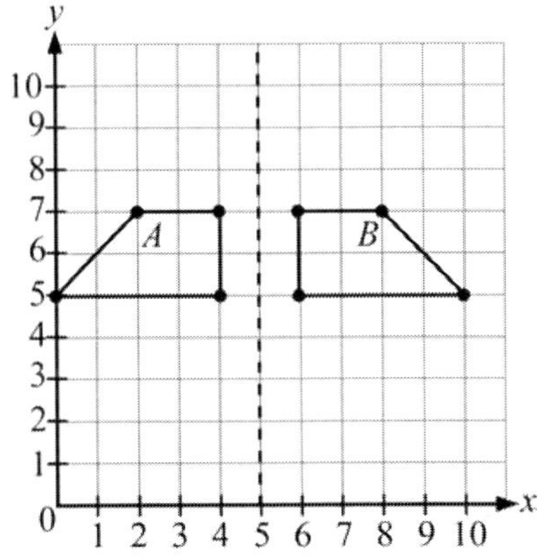

Step 2
Identify the vertices of the image labelled *B*.

Starting at the top left vertex and moving in a counterclockwise direction, the vertices of the reflected image are (6, 7), (6, 5), (10, 5), and (8, 7).

NOTES

Time to Try 2

Slide the given triangle three squares diagonally (↗), and draw the translated image. Identify the coordinates of the vertices in the image.

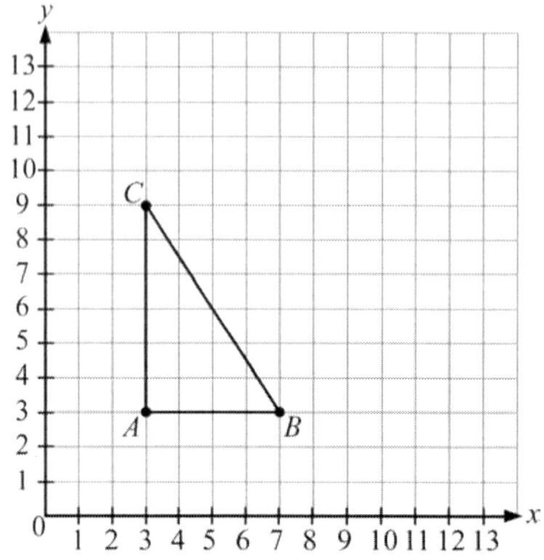

PRACTICE EXERCISES

Use the following information to answer the next question.

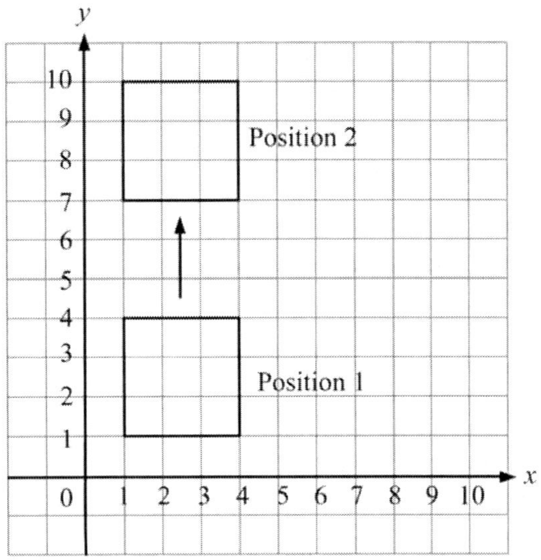

1. Identify the coordinates of the given shape in position 1. Describe the positional change of each vertex from position 1 to position 2.

Use the following information to answer the next question.

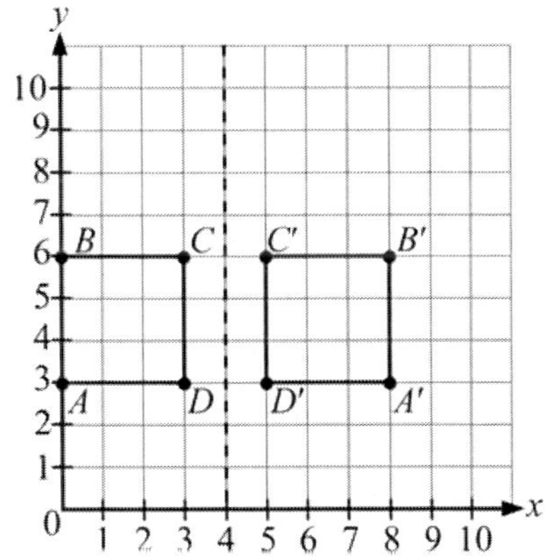

2. Identify the coordinates of the given shape *ABCD*. Describe the positional change of each vertex from shape *ABCD* to image $A'B'C'D'$.

Use the following information to answer the next question.

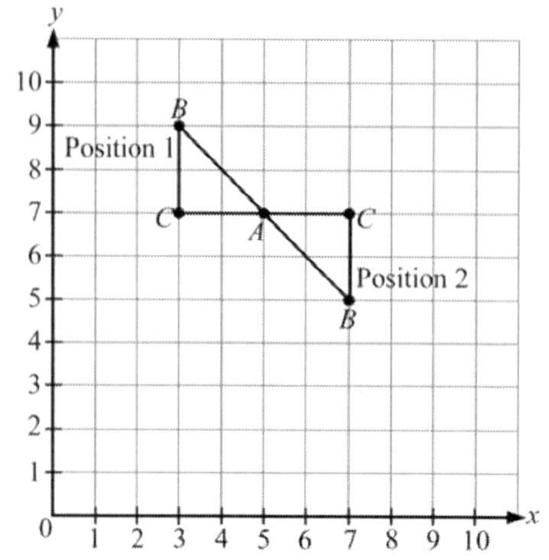

3. Identify the coordinates of triangle *ABC* in position 1. Describe the positional change of each vertex from the triangle in position 1 to its image in position 2.

4. Slide the given shape five squares diagonally (↗) and draw the image, labelling the vertices of the image as $A'B'C'D'$. Identify the coordinates of the vertices in the image.

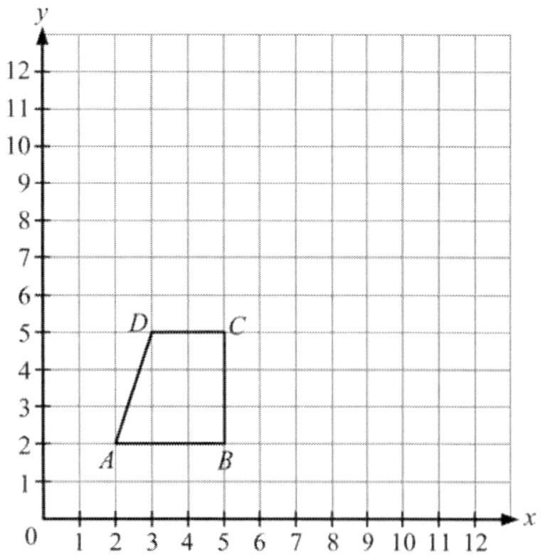

5. Reflect the given shape *PQRS* across the flip line and draw the image, labelling the vertices of the image as $P'Q'R'S'$. Identify the coordinates of the vertices in the image.

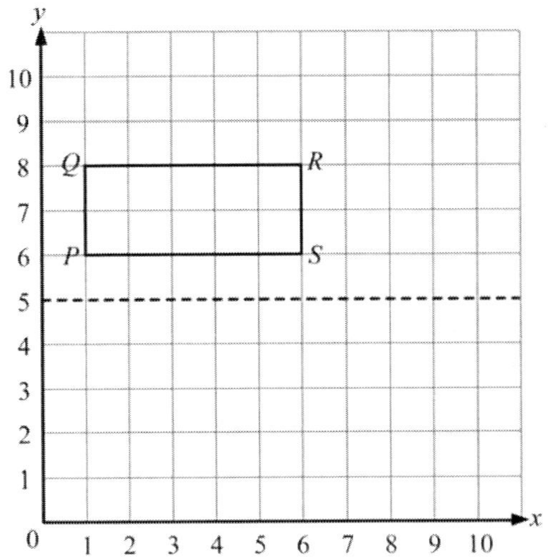

6. Rotate the given rectangle counterclockwise for a 90° turn. Let vertex *A* be the centre of rotation. Draw the rotated shape, labelling the vertices of the rotated rectangle. Identify the coordinates of the vertices in the rotated image.

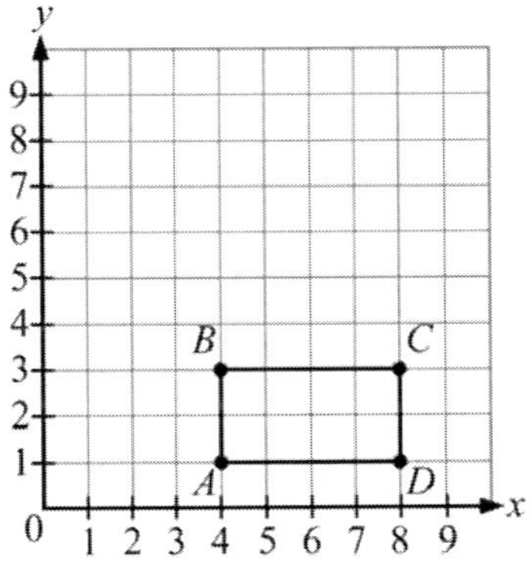

REVIEW SUMMARY

- A Cartesian plane is formed by the intersection of a horizontal and vertical number line at a 90° angle.
- The horizontal and vertical lines are called the x-axis and y-axis, respectively.
- The x-axis and the y-axis intersect at a point called the origin. The origin is the starting place for all coordinates or ordered pairs.
- The first number in an ordered pair tells you how far away from the origin to move on the x-axis.
- The second number tells you how far from the origin to move up the y-axis.
- Ordered pairs are written in brackets (x, y). The horizontal coordinate (x-axis) is listed first, and the vertical coordinate (y-axis) is listed second (horizontal, vertical).
- To determine the location of a point on a Cartesian plane and write its coordinates, first look along the x-axis to see how far to the right the point is located. Next, look along the y-axis to see how far up the point is located.
- You can plot points on a Cartesian plane if you are given the point's coordinates. Start from the 0 and count to the right to the first number of the ordered pair. Then count up to the second number in the ordered pair.
- Not all Cartesian planes have axes with number intervals of one. Other common intervals are two, five, and ten.
- You can draw shapes or designs in the first quadrant of a Cartesian plane, when given a set of ordered pairs.
- You can identify the coordinates of a shape drawn by you or someone else.
- You can perform a single transformation on a given 2-D shape and then identify the coordinates of the image drawn.
- You can perform a single transformation and then describe the positional change of vertices. To do this, compare the coordinates of the vertices of the shape to the corresponding vertices of the shape's image.

PRACTICE TEST

Use the following information to answer the next three questions.

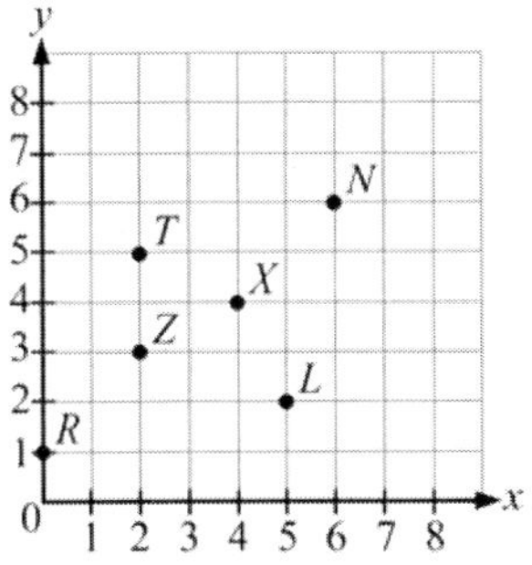

1. What point is located at (2, 5)?

2. Describe the distance between point *Z* and point *N*.

3. Write the ordered pair that shows the location of point *R*.

Use the following information to answer the next two questions.

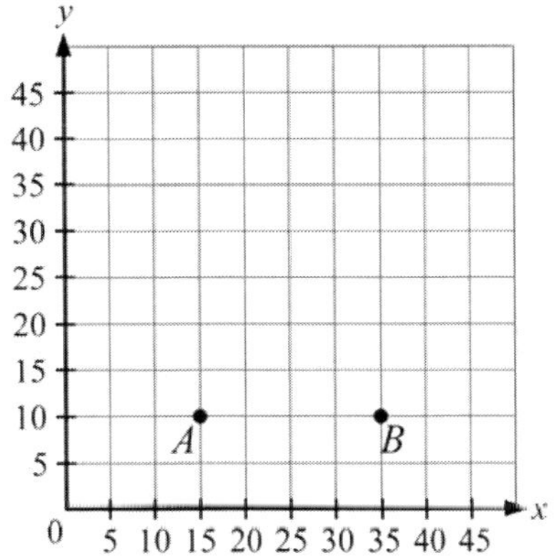

4. Write the ordered pairs that show the location of points *A* and *B*.

5. Describe the horizontal distance between point *A* and point *B*.

6. Plot the ordered pairs (2, 6), (7, 4), and (9, 3). Label the ordered pairs *A*, *B*, and *C*, respectively.

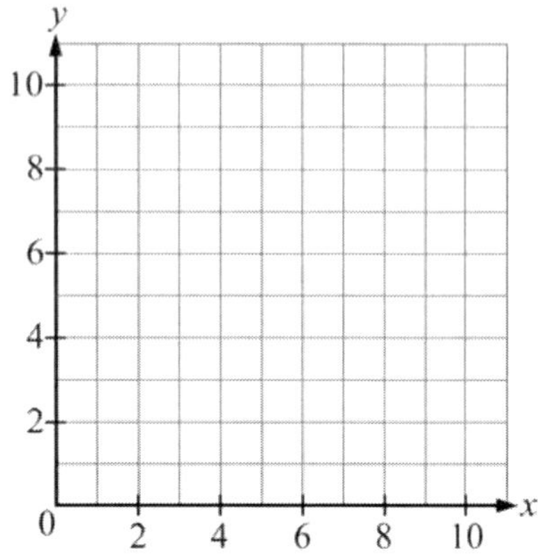

7. Plot the ordered pairs (10, 0), (30, 15), and (50, 40). Label the plotted points *A*, *B*, and *C*, respectively.

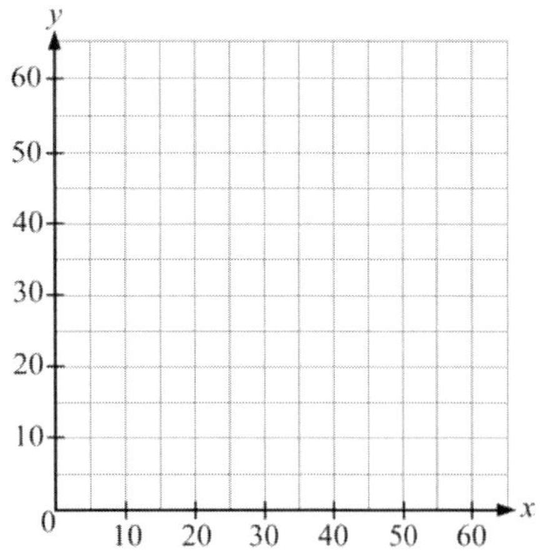

8. Plot and label the ordered pairs $A(4, 8)$, $B(7, 8)$, $C(10, 5)$, and $D(2, 5)$. Connect the points to produce a shape. Identify the shape formed by the given ordered pairs.

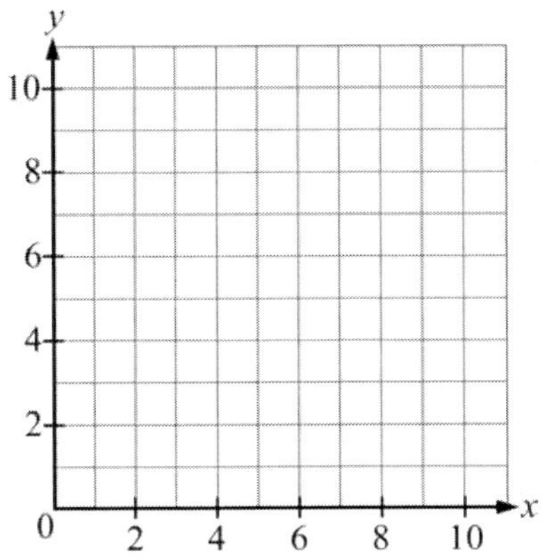

Use the following information to answer the next question.

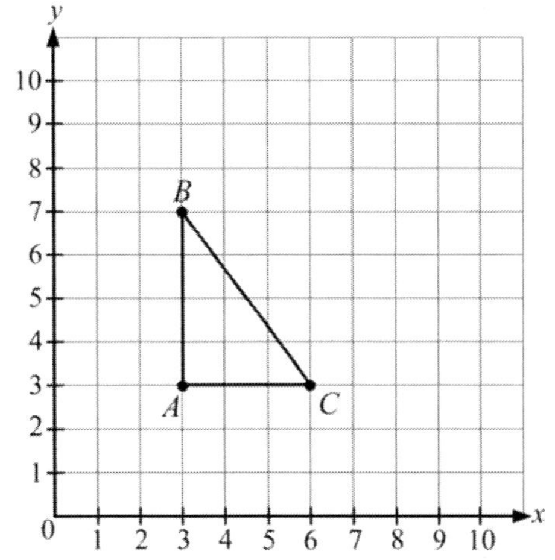

9. Identify the ordered pairs used to produce the triangle.

Use the following information to answer the next question.

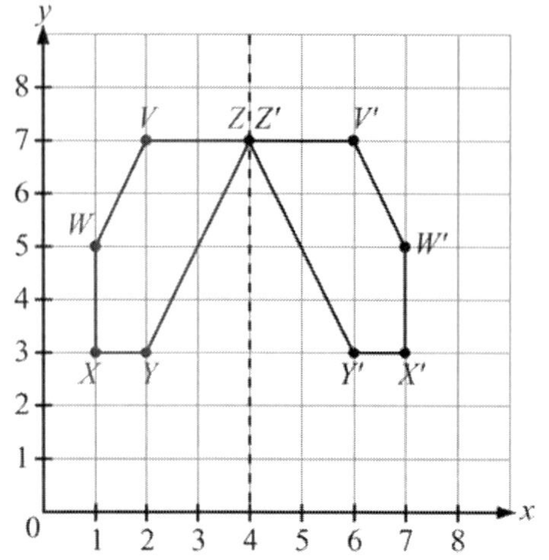

10. Identify the coordinates of shape *VWXYZ* on the left side of the flip line. Describe the positional change of each vertex from the shape on the left to its reflected image on the right.

11. Reflect shape *ABCD* across the dotted mirror line and draw the image, labelling the vertices of the image as $A'B'C'D'$. Identify the coordinates of the vertices in the image.

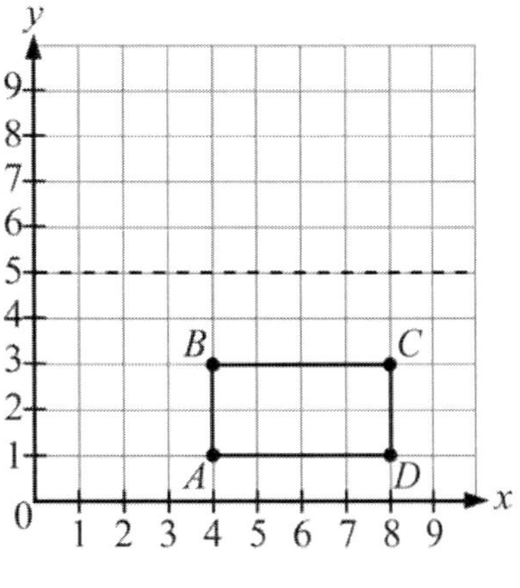

EXPRESSIONS AND EQUATIONS

When you are finished this unit, you should be able to…
- Apply the order of operations to solve multi-step problems
- Explain formulas for determining area and perimeter of rectangles as equations using letter variables
- Apply the commutative properties of addition and multiplication to equations using letter variables
- Express relationships between numbers within tables as expressions using letter variables
- Represent pattern rules (relationships between numbers) as expressions using letter variables
- Identify the unknown in a problem, express the problem as an equation, and then solve the problem created
- Create a problem for a given equation with one unknown
- Identify the unknown in a problem where the unknown could have more than one value, and then represent the problem as an equation
- Demonstrate the preservation of equality using balance scales and objects
- Use balance scales and objects to determine the value of a variable

Lesson	Page	Completed on
1. Order of Operations	268	
2. Generalizations Arising from Number Relationships	273	
3. Expressing Problems as Equations	280	
4. Preserving Equality	285	
Review Summary	291	
Practice Test	292	
Answers and Solutions	at the back of the book	

PREREQUISITE SKILLS AND KNOWLEDGE

Prior to beginning this unit, you should be able to…
- Explain the purpose of a letter variable in an equation
- Express a given problem as an equation with a letter variable representing the unknown
- Create a problem for a given equation involving one letter variable
- Solve problems involving single-variable one-step equations, such as $4 + a = 6$

Lesson 1 ORDER OF OPERATIONS

NOTES

There are specific rules to follow when solving expressions involving more than one operation.

1. Do all operations inside the brackets first.
2. Then, multiply and divide in order from left to right.
3. Then, add and subtract in order from left to right.

Example

Solve the expression $23-16\div 2$.

Solution

Since there are no brackets, start with step 2: divide $16\div 2=8$.

Go on to step 3 and subtract the quotient from 23: $23-8=15$.

$23-16\div 2=15$

Time to Try 1

Solve the expression $20-8+3$.

REASON FOR FOLLOWING THE ORDER OF OPERATIONS

It is important to follow the rules of operation when solving expressions with multiple operations so that everyone gets the same answer. If the order is changed, the answers will be different.

Example

When solving the expression $3+7\times 2$, Anna got an answer of 20 and Gordon got an answer of 17.

Who is correct? Justify your answer.

Solution

Step 1
Determine who is correct.

Gordon is correct.

NOTES

Step 2
Justify your answer.

Gordon followed the order of operations. He multiplied $7 \times 2 = 14$ first and then added 3 to the product: $3 + 14 = 17$.

Anna is incorrect because she did not follow the order of operations. She added $3 + 7 = 10$ first and then multiplied the product by 2.

Time to Try 2

When solving the expression $50 \div (2 + 3)$, Cora got an answer of 28 and Ben got an answer of 10. Who is correct? Justify your answer.

BRACKETS

Brackets are used to indicate which operations in the expression must be done first. If there are two or more operations inside the brackets, follow the order of operations.

- Multiply and divide in order from left to right.
- Then, add and subtract in order from left to right.

Example

Explain the order of operations needed to solve $150 \div (6 + 3 \times 8) - 5$.

Solution

First operation	Multiplication inside brackets: $3 \times 8 = 24$	$150 \div (6 + 24) - 5$
Second operation	Addition inside brackets: $6 + 24 = 30$	$150 \div 30 - 5$
Third operation	Division: $150 \div 30 = 5$	$5 - 5$
Fourth operation	Subtraction: $5 - 5 = 0$	$150 \div (6 + 3 \times 8) - 5 = 0$

NOTES

Time to Try 3

Explain the order of operations needed to solve $15 \div (16 \div 2 - 3)$.

USING BASIC CALCULATORS COMPARED WITH SCIENTIFIC CALCULATORS

Basic four-function calculators are different from scientific calculators. If you are using a basic calculator to solve an expression, follow the order of operations as though you were using a paper and pencil strategy or mental strategy.

1. brackets
2. multiply and divide from left to right
3. add and subtract from left to right

If you are using a scientific calculator, however, the calculator is programmed to use the order of operations. Type in the expression just as it appears, and the calculator will automatically perform the operations in the correct order to give the correct answer.

Example

Kate used a four-function calculator to solve $12 - 6 \div 2$. How should she use the calculator to get the correct answer?

Solution

Since Kate is using a basic four-function calculator, she needs to follow the order of operations.

- Since there are no brackets, go to step 2 and type in $6 \div 2 =$. The answer 3 will appear on the screen.
  $6 \div 2 = 3$
- Go to step 3 and type in $12 - 3$. The answer 9 will appear on the screen.
  $12 - 3 = 9$

When Kate follows the order of operations, $12 - 6 \div 2 = 9$.

NOTES

Time to Try 4

To solve $10+2\times 5$, Darin took his calculator and typed in $10+2\times 5=$. The answer that appeared on the screen was $10+2\times 5=20$.

What kind of calculator did Darin use? Justify your answer.

PRACTICE EXERCISES

1. When solving the expression $26-12\times 2$, Jesse got an answer of 2 and Becky got an answer of 28. Who is correct? Justify your answer.

2. Explain how you would use a four-function calculator to solve $14+6\div 2$.

3. Explain the order of operations needed to solve $(2\times 5)\times 3\div 15$.

4. When solving the expression $4+(2\times 3-2)\div 2$, Dayton got an answer of 4 and Shaylah got an answer of 6. Who is correct? Justify your answer.

5. Explain the order of operations needed to solve $(5\times 2)\div(4+1)$.

Choose mental math, a calculator, or paper and pencil to solve the following expressions:

6. $220 \div (4 + 3 \times 6)$

7. $9 + 3 \times 4 - 6$

8. $234-2\times 9$

Lesson 2 GENERALIZATIONS ARISING FROM NUMBER RELATIONSHIPS

NOTES

WRITING AND EXPLAINING FORMULAS FOR AREA AND PERIMETER OF RECTANGLES

When writing or explaining formulas for area and perimeter of rectangles, the relationships between the lengths and widths of a rectangle are being expressed.

The relationships between the length and width of a rectangle can be expressed using letter variables in formulas. The following standard variables are the first letters of the words:

- The variable l represents the length.
- The variable w represents the width.
- The variable A represents the area.
- The variable P represents the perimeter.

Example

The following rectangle has a length of 5 cm and a width of 3 cm.

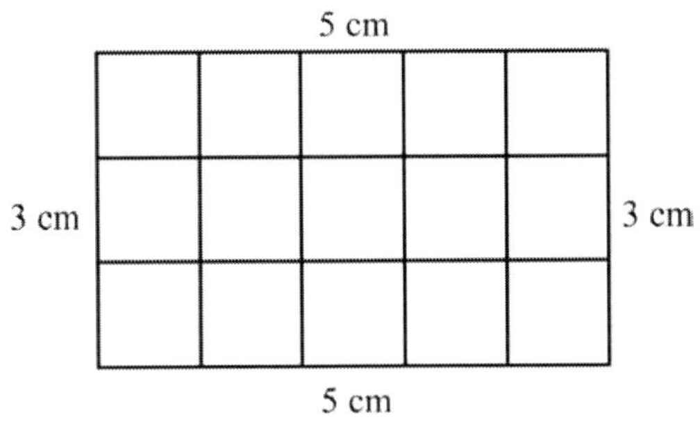

Write and explain a formula that can be used to determine the area of the rectangle.

Solution

Step 1
Write a formula.

Using the variables l for length, w for width, and A for area, the formula that explains how to determine the area of the rectangle is $A = l \times w$.

NOTES

Step 2
Explain the formula.

To use the area formula, substitute the length (5 cm) and the width (3 cm) of the rectangle into the formula and then multiply the two dimensions to determine the number of square centimetres needed to cover the surface of the rectangle.

$A = l \times w$

$A = 5 \times 3$

$A = 15 \text{ cm}^2$

Time to Try 1

Write and explain a formula that can be used to determine the perimeter of the same rectangle.

USING EQUATIONS TO ILLUSTRATE THE COMMUTATIVE PROPERTY OF ADDITION AND MULTIPLICATION

The commutative property of addition means that numbers can be added in any order and the sum will be the same. Adding $3+5$ gives the sum of 8. Adding $5+3$ still gives the sum of 8.

The commutative property of multiplication means that numbers can be multiplied in any order and the product will be the same. Multiplying 2×7 gives the product of 14. Multiplying 7×2 still gives the product of 14.

Equations can be written using variables to illustrate the relationships between the numbers for the commutative properties of addition and multiplication. Justify these equations by providing examples of how the equations work.

NOTES

Example

Using the variables x and y, write an equation to illustrate the commutative property of addition. Justify your answer with an example.

Solution

Step 1
Write an equation.

This equation shows that the two numbers can be added in any order: $x + y = y + x$. The variable x represents one number and the variable y represents a different number.

Step 2
Justify your answer with an example.

Let x represent the number 10 and y represent the number 5. Since $x + y = y + x$, then:

$$10 + 5 = 5 + 10$$
$$15 = 15$$

Time to Try 2

Using the variables a and b, write an equation to illustrate the commutative property of multiplication. Justify your answer with an example.

NOTES

USING EXPRESSIONS TO DESCRIBE RELATIONSHIPS IN A TABLE

You can describe the relationship between numbers presented in a table by writing an expression using a variable.

Example

In	Out
12	5
26	19
7	0
19	12

Express the relationship between the In and Out numbers as an expression using the variable n to represent any In number.

Solution

Step 1
Determine the relationship.

Each number in the Out column is 7 less than its corresponding number in the In column.

Step 2
Write an expression using the variable n.

The relationship between the In and Out numbers can be expressed as $n-7$, where the variable n represents any In number.

Time to Try 3

Term	1	2	3	4
Term Number	8	16	24	32

Express the relationship between the given term numbers as an expression, using the variable x to represent any term number.

NOTES

USING EXPRESSIONS TO REPRESENT PATTERN RULES

The relationship between numbers presented in a sequence can be described by writing an expression using a variable. The written expression is the pattern rule used to make the pattern of numbers. The variable used will represent any number in the pattern.

Example

Adam uses a particular pattern rule to make this sequence of numbers: 1 875, 375, 75, 15, 3.

Represent the pattern rule as an expression using the variable k to represent any number in the pattern.

Solution

Step 1
Determine how each term changes.

Each term is divided by 5 to get the next term.

Step 2
Represent the rule as an expression.

The pattern rule can be represented by the expression $k \div 5$.

Time to Try 4

Jillian uses a particular pattern to make this sequence of numbers: 3, 7, 15, 31, 63.

Represent the pattern rule as an expression using the variable x to represent any number in the pattern.

PRACTICE EXERCISES

1. Using a letter variable, write and explain a formula for finding the area of any square.

2. If $24 \times 6 = 6 \times h$, what is the value of the variable *h*? Explain your answer.

3. If $m + 19 = 19 + 83$, what is the value of the variable *m*? Explain your answer.

Use the following information to answer the next question.

Term	1	2	3	4	5
Term Number	2	10	20	40	80

4. Express the relationship between the given term numbers as an expression using the variable *a* to represent any given number.

Use the following information to answer the next question.

In	Out
3	7
5	11
8	17
1	3

5. Express the relationship between the In and Out numbers as an expression using the variable k.

6. Using the letter variable a, represent the pattern rule for the sequence 91, 101, 111, 121 as an expression.

7. Using the letter variable h, express the pattern rule for the sequence 2, 5, 14, 41 as an expression.

Lesson 3 EXPRESSING PROBLEMS AS EQUATIONS

NOTES

To express a problem as an equation and then to solve that problem, follow these steps:

1. Identify the unknown.
2. Choose a letter variable to represent the unknown.
3. Express the problem as an equation.
4. Solve the problem concretely (using models like base ten blocks or connecting cubes), pictorially (drawing diagrams of pictures), or symbolically (using numbers and symbols).

Example

Ben and Tom have the same number of shirts. Ben has two short-sleeved shirts and five long-sleeved shirts. Tom has four short-sleeved shirts. How many long-sleeved shirts does Tom have?

Using the letter variable m, write an equation to express the problem. Solve the problem symbolically and pictorially.

Solution

Step 1
Identify the unknown.

The unknown is the number of long-sleeved shirts Tom has. This number will be represented by the variable m.

Step 2
Express the problem as an equation.

The expression that shows the number of shirts Ben has is $2+5$.

The expression that shows the number of shirts Tom has is $4+m$.

Since the boys have the same number of shirts, the two expressions are equal to each other.

The problem expressed as an equation is $2+5=4+m$.

Step 3
Solve the problem symbolically.

$$2+5=4+m$$
$$7=4+m$$
$$7-4=m$$
$$3=m$$

Since $m=3$, Tom had three long-sleeved shirts.

NOTES

Step 4
Solve the problem pictorially.

Use pictures of shirts to help you understand and solve the problem.

Ben's Shirts	Tom's Shirts
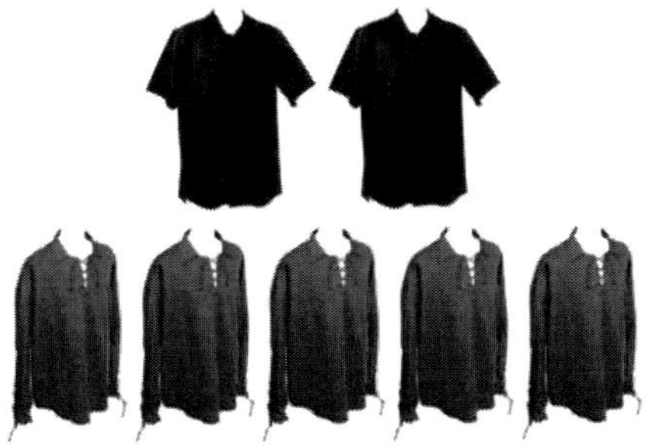 $2+5=7$	 $4+m=7$

Think of a number that can be added to 4 to have a sum of 7.
The number 3 can be added to 4 to equal 7.

Tom had three long-sleeved shirts:

Time to Try 1

A coach is buying one pair of socks for each of the seven boys and eight girls on her basketball team. The socks come in packages of three pairs. How many packages of socks does the coach need to buy?

Using the letter variable *m*, write an equation to express the problem. Solve the problem symbolically and concretely.

NOTES

CREATING A PROBLEM FOR A GIVEN EQUATION

When given an equation, try creating your own problem to fit the equation.

The first step in writing the problem is to determine what the variable (the unknown) represents. Once what it represents is understood, it will be easier to create a problem.

Example

Create a problem that can be solved using the equation $28 \div g = 4$.

Solution

Step 1
Determine the unknown.

Division can be explained as *total* ÷ *number of groups* = *number in each group*. The variable *g* represents the number of groups.

Step 2
Create a problem.

Write a problem where you start with a total of 28 items.
The items are placed in an unknown number of equal groups (*g*), with four items placed in each group.

Dylan was organizing his collection of 28 model cars by placing them in equal groups. Each group had 4 model cars. How many groups did Dylan make?

Step 3
Solve the problem.

$$28 \div g = 4$$
$$28 \div 7 = 4$$
$$g = 7$$

Dylan made seven groups of cars.

Time to Try 2

Write a problem that can be solved using the equation $200 = 5C$.

NOTES

VARIABLES AS CHANGING QUANTITIES

When there is more than one variable in an equation, the values for the variables may change, depending on the problem.

For example, consider the equation $a + n = 4$, when $a = 1$ and $n = 3$. After replacing a with 1 and n with 3, the equation becomes $1 + 3 = 4$.

However, if $a = 2$, then $n = 2$, the equation becomes $2 + 2 = 4$.

Example

Carla has a total of five sports cards. Some of the cards are hockey cards and some of the cards are soccer cards. How many hockey and soccer cards could Carla have?

Write an equation to express the problem. Let h represent the number of hockey cards and s represent the number of soccer cards. Explain how the variables could have more than one value.

Solution

Step 1

Determine the equation.

The equation that represents the problem is $h + s = 5$.

Step 2

Explain how the values can change.

The value of s will change depending on the value of h.

- If Carla has one hockey card, then she has four soccer cards.
- If Carla has two hockey cards, then she has three soccer cards.
- If Carla has three hockey cards, then she has two soccer cards.
- If Carla has four hockey cards, then she has one soccer card.

Time to Try 3

Dylan has four balloons. Some are red and some are blue. How many red and blue balloons could Dylan have?

Write an equation to express the problem. Let r represent the number of red balloons and b represent the number of blue balloons. Explain how the variables could have more than one value.

PRACTICE EXERCISES

1. Bernie weighs 70 kg, and Ronald weighs k kg. Together, they weigh 130 kg. How much does Ronald weigh? Write an equation to express the problem. Solve the problem symbolically and concretely.

Use the following information to answer the next question.

Water is filling a pool at a rate of 150 L per minute. There will be 2 500 L of water in the pool when it is filled. How many minutes will it take to fill the pool?

2. Jan writes the expression $150 \times m$ and the equation $150 \times m = 2\ 500$ to represent the problem. The unknown in both the expression and equation is expressed as the variable m. Explain what the variable m represents in both the expression and the equation.

In the following equations, determine whether the variable (or variables) has one value or could have more than one value. Explain your answer.

3. $18 = l \times w$

4. $3 \times n = 10 + 2$

Use the following information to answer the next question.

There are four seats in each row of a bus except the front row, which has three seats. Determine the number of rows in the bus if there are 51 seats in total.

5. Express the problem as an equation where the variable is represented by the letter r.

6. Write a problem that can be solved using the equation $12 \times n = 312$. Solve the problem.

Lesson 4 PRESERVING EQUALITY

NOTES

The equal symbol (=) indicates that the left side and right side of an equation are balanced. For example, in the equation 3 + 3 = 4 + 2, the numbers on each side of the equal sign (=) have the same value. Both expressions have a value of 6.

Preservation of equality refers to keeping things equal or balanced. To preserve equality, it is necessary to perform the same operation to both sides of the equation.

BALANCING OBJECTS TO DEMONSTRATE EQUALITY

Equality in an equation can be illustrated using a balance scale and objects. In order to balance a scale and preserve equality, the objects on the left side of the scale must have a combined value (mass) equal to the combined value (mass) of the objects on the right side of the scale.

APPLYING PROPERTIES OF EQUALITY

When balancing objects on a scale to solve an equation, apply the properties of equality.

1. Addition: When the same number is added to both sides of an equation (or balance scale), both sides remain equal.
2. Subtraction: When the same number is subtracted from both sides of an equation (or balance scale), both sides remain equal.
3. Multiplication: When both sides of an equation (or balance scale) are multiplied by the same number (except zero), the sides remain equal.
4. Division: When each side of an equation (or balance scale) is divided by the same number (except zero), the sides remain equal.

Example

This balance scale shows one block and three balls on the left side of the scale balancing five balls on the right side. The balance scale illustrates the equation $x + 3 = 5$, where x represents the mass of the block.

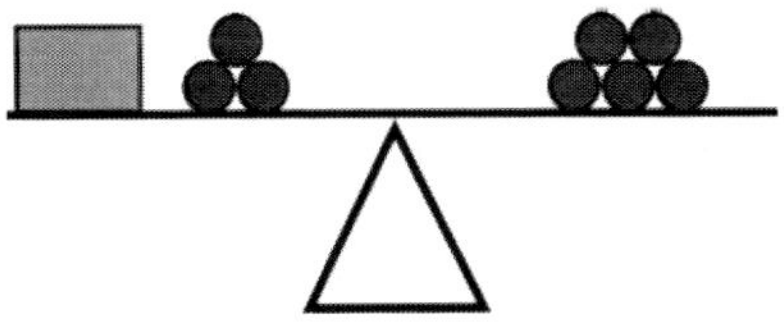

Determine the value of x by applying the property of subtraction.

NOTES

When a balance scale is level, each side has the same mass.

Solution

The goal is to arrive at a situation where there is only the block on the left side of the scale and an equivalent number of balls on the right side of the scale.

Remove three balls from each side of the scale, and you are left with the block balancing two balls.

$$x+3=5$$
$$x+3-3=5-3$$
$$x=2$$

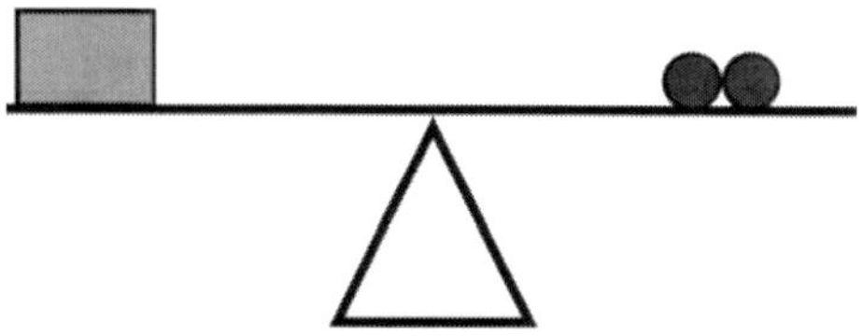

$x = 2$

The solution is 2; the block has a mass that is equal to the mass of two balls.

Time to Try 1

This balance scale shows that two blocks have the same mass as eight balls. The scale illustrates the equation $2x=8$ where x represents the mass of one block.

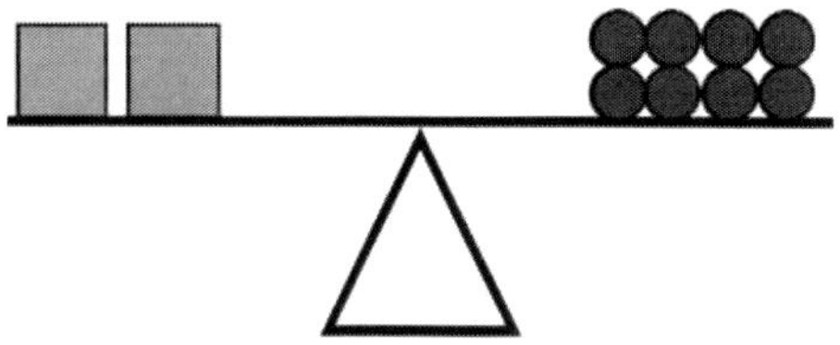

$2x = 8$

Determine the value of x by applying the property of division.

PRACTICE EXERCISES

Use the following information to answer the next question.

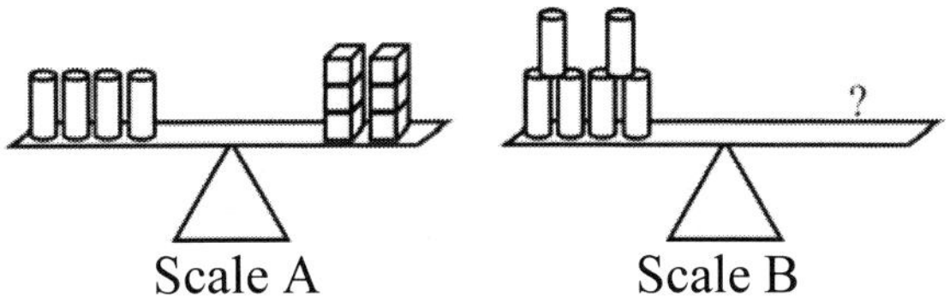

1. In order for scale B to be balanced, how many blocks would have to be placed on the right side of the scale? Assume that each cylinder has the same mass and that each block has the same mass. Explain your answer.

Use the following information to answer the next question.

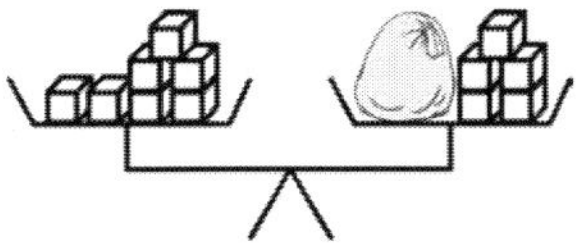

This balance scale shows seven cubes on the left side of the scale balancing a bag and five cubes on the right side. The balanced scale illustrates the equation $7 = x + 5$, where x represents the mass of the bag.

2. Apply the property of subtraction to determine the value of x (the mass of the bag).

Use the following information to answer the next question.

Cube = 5 g
Cylinder = 4 g

This balance scale shows a rectangular block on the left side of the scale balancing a cube and three cylinders on the right side of the scale.

3. Use the given masses of the cube and cylinder to determine the mass of the rectangular block on the left side of the scale.

Use the following information to answer the next question.

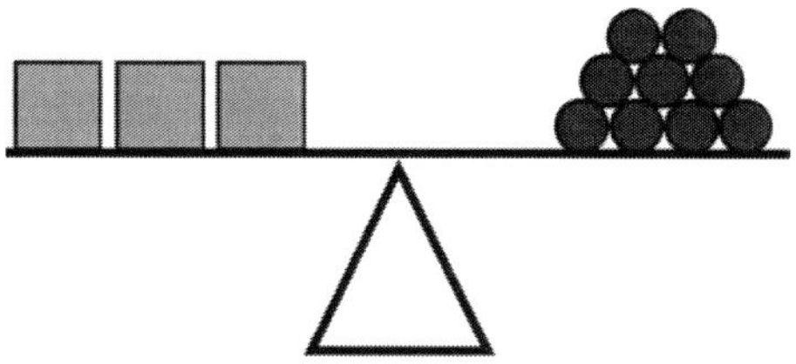

$3x = 9$

This balance scale shows three blocks on the left side of the scale balancing nine balls on the right side. The balanced scale illustrates the equation $3x = 9$, where x represents the mass of one block.

4. Determine the value of x.

Use the following information to answer the next question.

Rectangular block = 30g

This balance scale shows six cubes on the left side of the scale balancing one rectangular block on the right side of the scale. The balanced scale illustrates the equation $6x = 30$, where x represents the mass of one block.

5. Determine the value of x.

Use the following information to answer the next question.

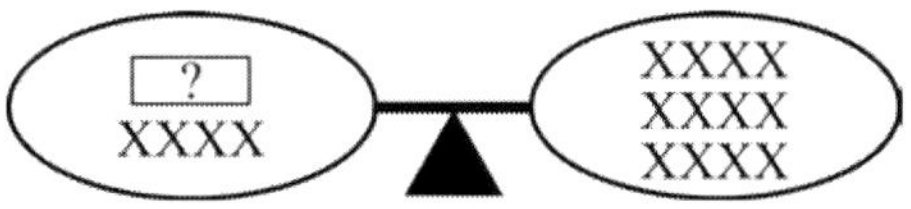

This balance scale shows that one rectangle and four Xs on the left side of the scale balance 12 Xs on the right side. The scale illustrates the equation $m + 4 = 12$, where m represents the mass of the rectangle.

6. If each X on the balance scale is equal to 2 g, then what is the mass of the rectangle?

Use the following information to answer the next question.

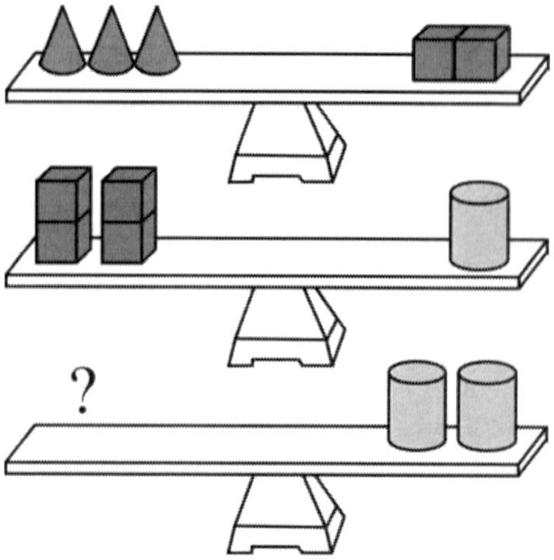

This balance scale is used to show equality of 3-D objects.

7. If three cones balance two cubes and four cubes balance one cylinder, how many cones are needed to balance two cylinders?

REVIEW SUMMARY

- The standard order of operations is as follows:
  1. Perform calculations inside brackets first.
  2. Calculate multiplication and division facts, working from left to right.
  3. Calculate addition and subtraction facts, working from left to right.
- The relationship between the length and width of a rectangle can be explained by writing formulas for perimeter or area using letter variables.
- Pattern rules can be expressed as expressions using letter variables.
- The relationships between terms and term numbers in tables can be described using expressions with letter variables.
- The commutative property of addition and multiplication can be illustrated by using variables.
  Addition: $a + b = b + a$ Multiplication: $a \times b = b \times a$
- In some equations, a variable will represent an unknown quantity with only one value.
- In other equations, like formulas, a variable will represent a changing quantity where one value is dependent on the other value.
- A balance scale and objects can be used to illustrate preservation of equality.
- A balance scale and objects can be used to determine the value of a variable in an equation.
- In order to balance a scale, the objects on the left side of the scale must have a combined value equal to the combined value of the objects on the right side of the scale.
- Mass is commonly used to illustrate equality of objects on a balance scale.
- To preserve equality, whatever operations are performed on the left side must also be performed on the right side.
  1. When the same number is added to both sides of a scale or equation, both sides remain equal.
  2. When the same number is subtracted from both sides of a scale or equation, both sides remain equal.
  3. When both sides of a scale or equation are multiplied by the same number (except zero), the sides remain equal.
  4. When both sides of a scale or equation are divided by the same number (except zero), the sides remain equal.

PRACTICE TEST

Use the following information to answer the next question.

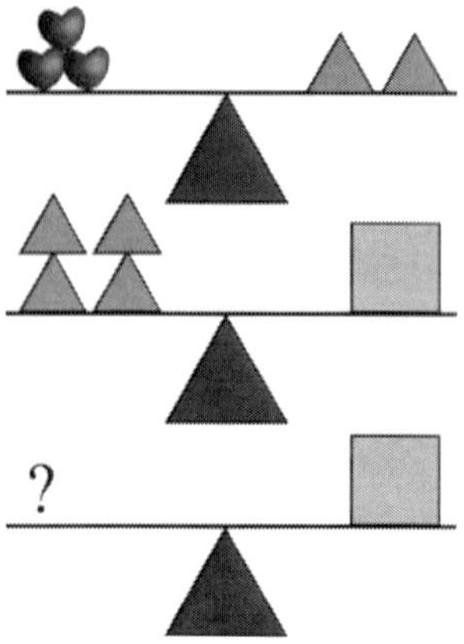

This set of balance scales shows that three hearts balance two triangles, and four triangles balance one square.

1. How many hearts will balance one square? Explain your work.

Solve the next three problems using the order of operations. Show your work.

2. $4 \times 3 + 15 \div 5$

3. $4 \times 10 \div 5 + 2$

4. $(3 + 2 \times 7) - 12$

Use the following information to answer the next question.

Daria has five hair bands. Some are pink and some are green. How many pink and green hair bands could Daria have?

5. Write an equation to express the problem. Let p represent the number of pink hair bands and g represent the number of green hair bands. Explain how the variables could have more than one value.

6. Create a problem that can be solved using the equation $4g = 52$.

7. Carl weighs 60 kg, and his older brother weighs x kg. Together, they weigh 150 kg. How much does the older brother weigh? Write an equation to express the problem and then solve the problem symbolically and concretely.

8. Using the variables m and z, write an equation to illustrate the commutative property of multiplication. Justify your answer with an example.

Use the following information to answer the next question.

Term	1	2	3	4	5
Term Number	1	3	9	27	54

9. Express the relationship between the given term numbers as an expression using the variable a to represent any given number.

10. Using a letter variable, represent the pattern rule for the series 112, 56, 28, 14 as an expression.

Use the following information to answer the next question.

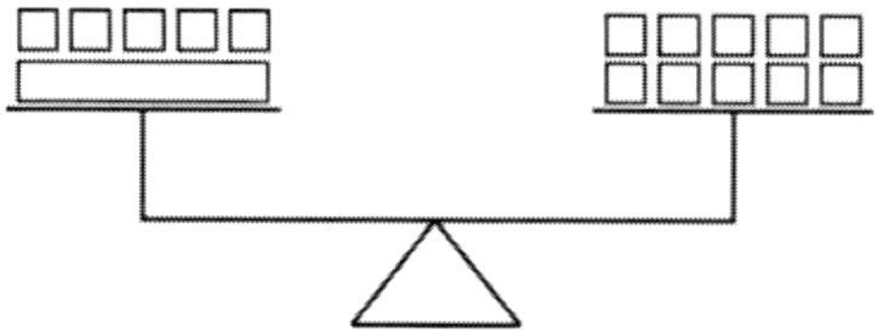

This balance scale shows that the mass of one rectangle and five squares on the left side of the scale balances the mass of 10 squares on the right side. The scale illustrates the equation $x+5=10$, where x represents the mass of the rectangle.

11. Use the property of subtraction to determine the mass of the rectangle.

DATA ANALYSIS

When you are finished this unit, you will be able to…

- Identify common attributes of line graphs
- Determine whether a given set of data is continuous or discrete
- Create a line graph
- Interpret a line graph and draw conclusions
- Select an appropriate method for collecting data and justify your choice
- Collect data by administering a questionnaire, performing an experiment, or using a database or electronic media
- Draw conclusions based on collected data
- Select an appropriate type of graph for displaying a set of collected data and justify the choice
- Graph collected data and interpret the results to solve problems

Lesson	Page	Completed on
1. Line Graphs	296	
2. Creating and Interpreting Line Graphs	303	
3. Selecting, Justifying, and Using Appropriate Methods of Collecting Data	310	
4. Selecting and Justifying an Appropriate Graph to Display Data	318	
5. Graphing Collected Data and Interpreting the Resulting Graph	323	
Review Summary	327	
Practice Test	328	
Answers and Solutions	at the back of the book	

PREREQUISITE SKILLS AND KNOWLEDGE

Prior to beginning this unit, you should be able to…

- Distinguish between first-hand and second-hand data
- Construct and interpret double bar graphs
- Draw conclusions from a given double bar graph
- Solve a given problem by constructing and interpreting a double bar graph

Lesson 1 LINE GRAPHS

NOTES

A line graph uses points and lines to show how values change over a period of time.

DISPLAYING DISCRETE AND CONTINUOUS DATA

Discrete data is data that has a limited number of values.
For example, the months of the year, days of the week, or the number of students in a class. Discrete data is always presented as whole numbers. There are no values between the whole numbers. For example, there is no day between Monday and Tuesday, and there is no whole number between 25 and 26 students.

If the data on the x-axis and the y-axis is discrete data, the points plotted are usually not connected with a line.

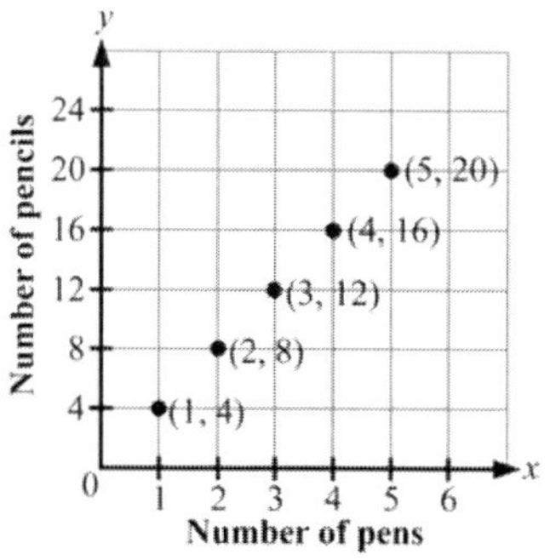

A line graph plots continuous data as points and then joins the points with a line. Continuous data is data that changes as time goes on. There can be values between the plotted points. For example, weather patterns, temperatures, or a tree's height keep changing over time, so they are examples of continuous data.

A broken-line graph is a graph that shows the shape of the relationship between the values of the x- and y-axis with line segments connecting the points. Broken line graphs are almost always discrete data.

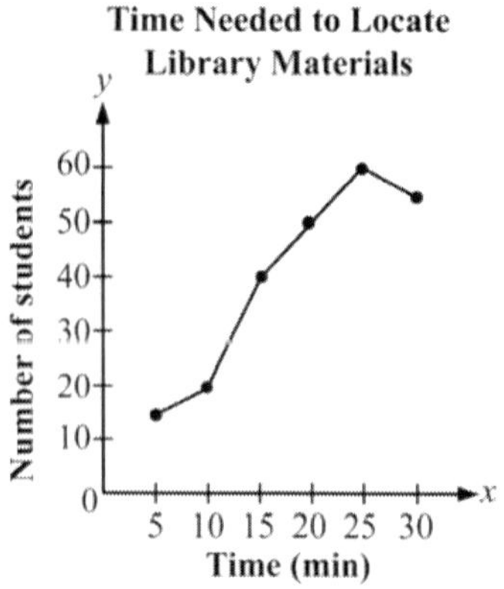

NOTES

A continuous line graph is a graph that shows the relationship of two items or events that happen at the same rate over time. The points are connected with a straight continuous line.

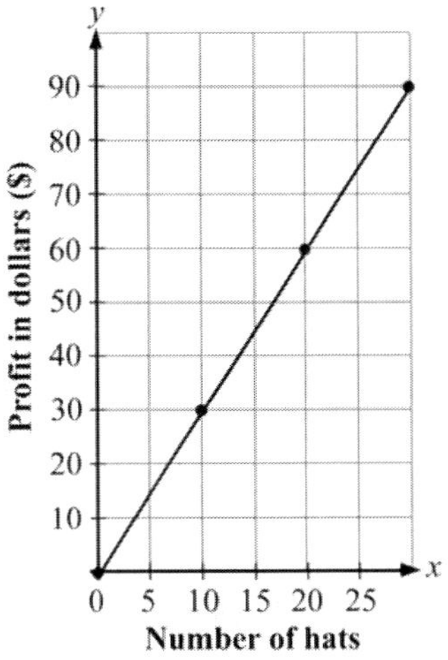

Example

Classify each set of data as discrete or continuous.

a) The number of suitcases lost by an airline.

Solution

Discrete: The number of suitcases lost must be a whole number.

b) The height of a child over a number of years.

Solution

Continuous: The height of a child can take on many possible values between the designated years.

c) The number of pumpkins grown.

Solution

Discrete: The number of pumpkins must be a whole number.

Time to Try 1

A five-question quiz is given in Math class. Each question receives one mark if it is correctly answered. Is the number of correct answers on a student's quiz an example of discrete or continuous data? Explain.

NOTES

COMMON ATTRIBUTES OF LINE GRAPHS

When creating a line graph to display data, there must be an appropriate title, labelled axes, and scales with intervals that suit the range and distribution of the data. These attributes allow people to read and interpret the graph accurately. Consider the following graph:

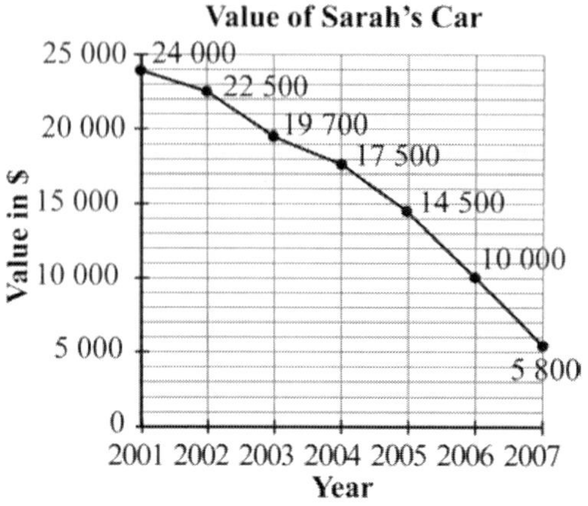

The Title

The title offers a short explanation of what the graph is about. By reading the title, the reader should immediately be able to identify the main point of the data presented.

The title of the given graph is "Value of Sarah's Car." This title tells the reader that the graph contains information about the value of Sarah's car.

Scale

A **scale** is a series of marks on a graph at set **intervals** that allow the reader to judge the value of a variable. On a graph, a scale is shown with a sequence of markers. For example,

Axes: the plural form of axis.

Each number has a marker on the line. This line can run along the x-axis or the y-axis. The numbers on the line are in increasing order. The interval of the scale depends on the data being recorded. The interval could increase by ones, fives, tens, hundreds, or any other number that is appropriate for the graph.

NOTES

x-Axis

The *x*-axis runs horizontally. Typically, the *x*-axis consists of numbers representing different time periods.

The label of the *x*-axis in the given graph is "Year" and the scale measures different years, from 2001 to 2007, in intervals of one year.

y-Axis

The *y*-axis runs vertically. Typically, the *y*-axis consists of numbers representing the item being measured. The *y*-axis usually starts at 0 and can be divided into as many intervals as required.

The label of the *y*-axis in the given graph is "Value in \$." The *y*-axis represents the value of Sarah's car in dollars and the scale increases in intervals of \$5 000. This interval is appropriate as it reflects the range of the data. When determining the interval, the lowest and highest numerical amounts should be considered. In this example, the range of data is from \$24 000 to \$5 800.

Points

The points or dots on the line graph show the measured data.

Line

The line connecting the points provides an estimate of the values between the points.

Example

What attribute is missing from this line graph? Explain your answer.

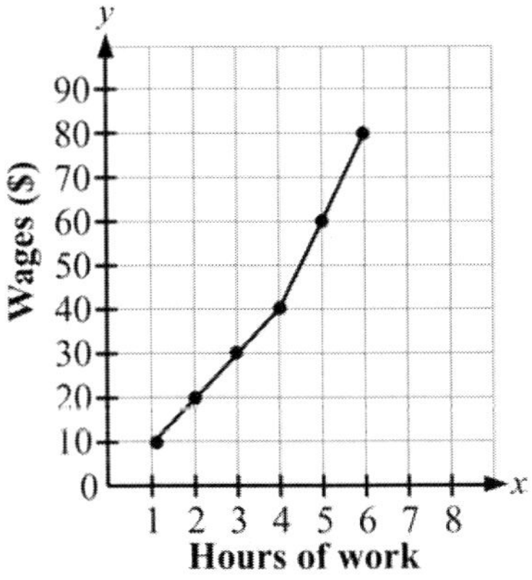

Solution

The title is missing. Without a title, we do not know who or what the graph is about.

NOTES

Time to Try 2

What attributes are missing from this graph? Explain your answer.

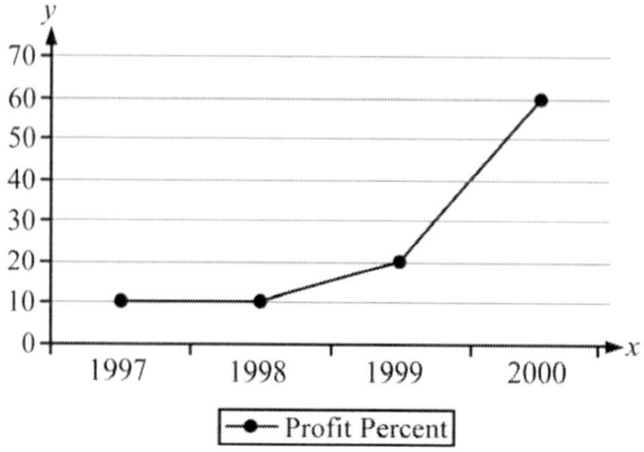

PRACTICE EXERCISES

Classify each set of data as discrete or continuous. Justify your answer.

1. The height of an apple tree over a one-year period.

2. The number of multiplication facts answered correctly on a quiz.

Use the following information to answer the next question.

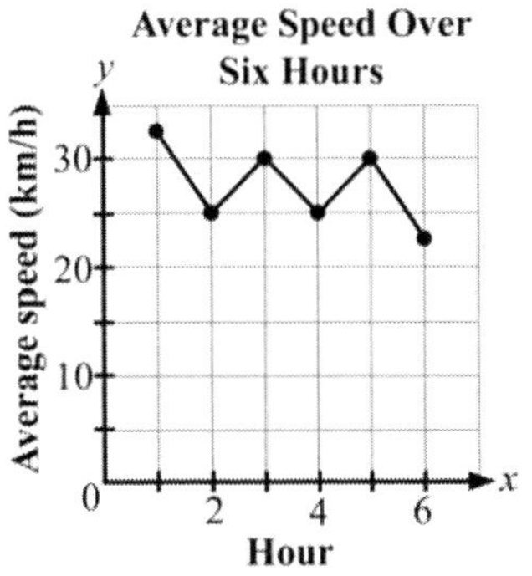

3. Determine whether the data displayed is continuous or discrete. Explain your answer.

Use the following information to answer the next question.

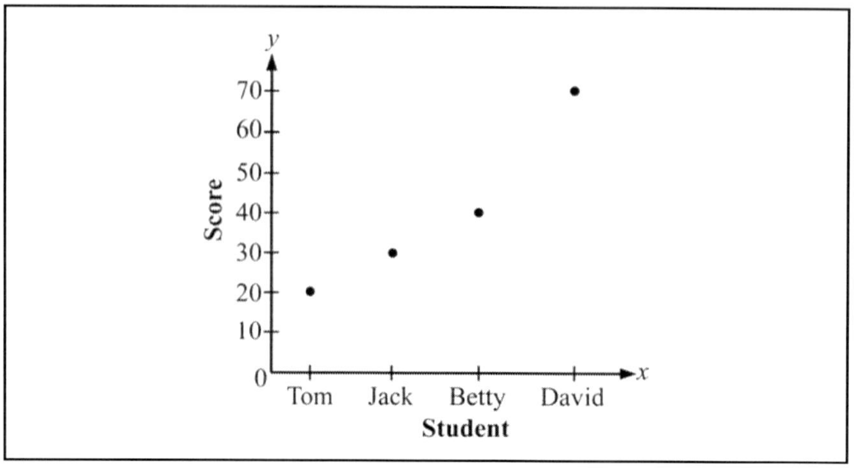

4. What attribute is missing from the graph shown? Explain your answer.

Use the following information to answer the next question.

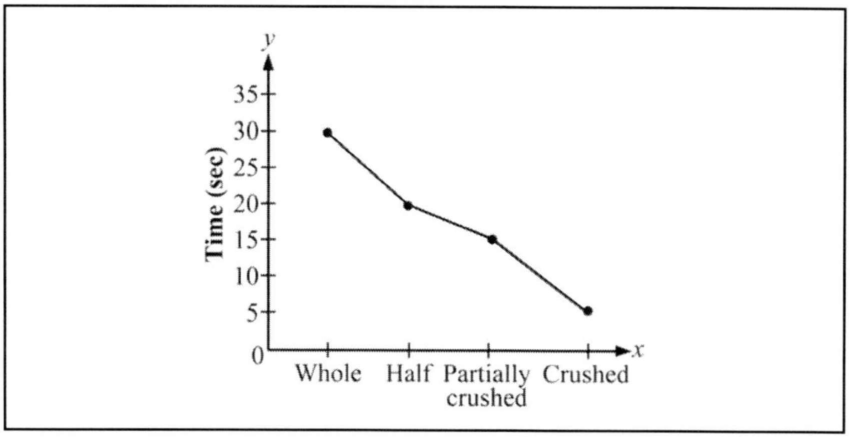

5. What attributes are missing from the graph shown? Explain your answer.

Lesson 2 CREATING AND INTERPRETING LINE GRAPHS

CREATING A LINE GRAPH

To create a line graph from a given set of data, follow these steps:

1. Draw and label the horizontal and vertical axes using a scale that includes all the information. Place any units of measurement in brackets.
2. Plot the data on the graph using a dot for each piece of data (ordered pair).
3. Connect the dots using straight lines if the data is continuous data.
4. Give the graph a title.

Example

Create a line graph from this set of data:

Number of Hours Worked	Wages Received ($)
2	10
4	20
6	30
8	40

Solution

Step 1

Draw and label the x- and y-axes. Be sure to use a ruler to ensure precise measurements.

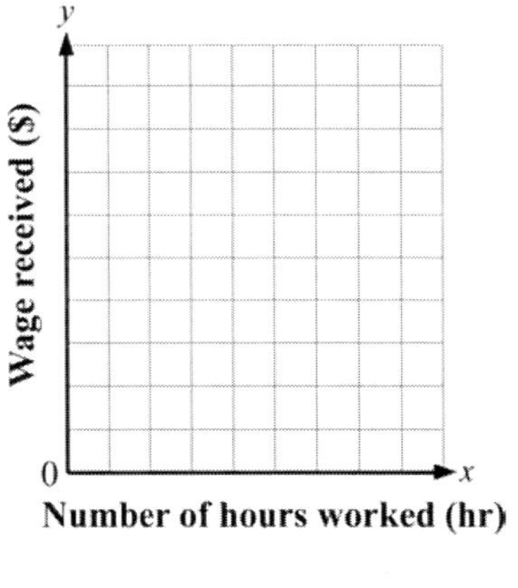

Step 2

Plot the ordered pairs on the graph using dots.

(2, 10), (4, 20), (6, 30), (8, 40)

NOTES

An ordered pair is a pair of numbers of the form (x, y) that give the location of a point on a coordinate plane.

The first number (x) in the ordered pair describes the horizontal distance.

The second number (y) in the ordered pair describes the vertical distance.

NOTES

Step 3
Connect all the dots on the graph with straight lines.

Step 4
Give the graph a title.

A Pay Scale for Hours Worked

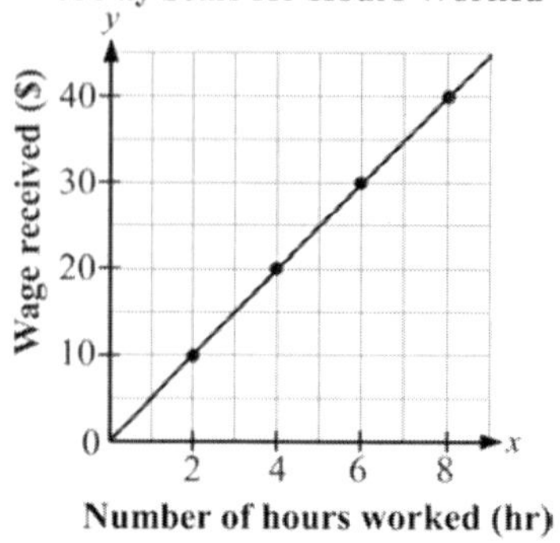

Time to Try 1

This table shows the average market price of television sets from 2000 to 2005.

Year	Price ($)
2000	1 200
2001	1 100
2002	1 400
2003	1 000
2004	1 100
2005	1 500

Create a line graph to show the changes in the cost of televisions over the years.

NOTES

Remember:
1. Use a ruler to draw and label the horizontal and vertical axes using a scale that includes all the data.
2. Plot the information on the graph using a dot for each ordered pair.
3. Connect each dot using a straight line if the data is continuous.
4. Write a title above the graph.

INTERPRETING LINE GRAPHS TO DRAW CONCLUSIONS

When interpreting graphs, remember to look at all the information given. A conclusion, which is the answer to a question that can be answered by looking at the information on the graph, can be drawn based on the data.

Example

The following line graph shows the number of football viewers on Monday nights from 2001 to 2006.

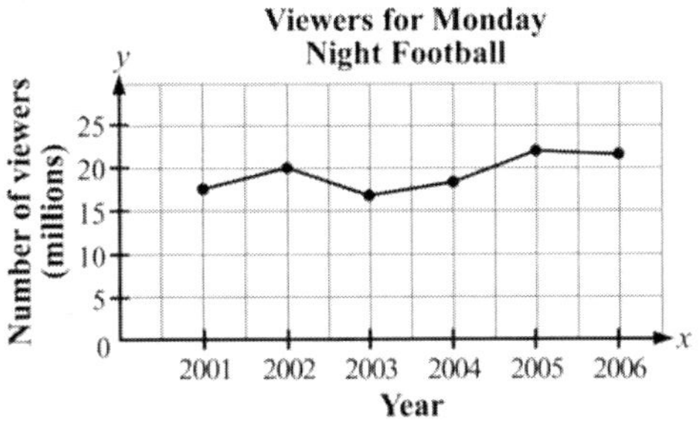

What conclusion can be drawn about the number of people who watched Monday night football from 2001 to 2006?

Solution

Because the number of viewers for each given year is just below, above, or on the 20 million line, it can be concluded that about 20 million people watched the Monday night football games from 2001 to 2006.

NOTES

Time to Try 2

Chuck made a line graph to show the wages he makes compared with the hours he works at his job.

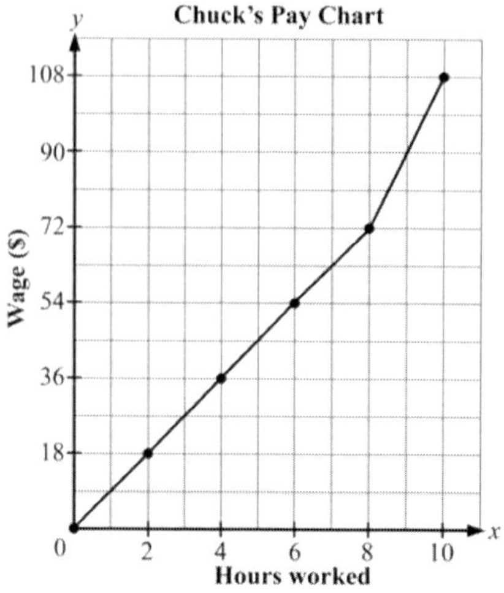

Write two conclusions that can be drawn from this data.

PRACTICE EXERCISES

Use the following information to answer the next question.

Heart rate is a measure of the number of heartbeats per minute. The following table shows the average number of heartbeats per minute for an adult male.

Time (min)	Heartbeats
0	0
1	70
2	140
3	210
4	280

1. Create a line graph for the data presented in the table.

Use the following information to answer the next question.

This line graph shows the populations of hawks and squirrels in a particular northern forest over a period of ten years.

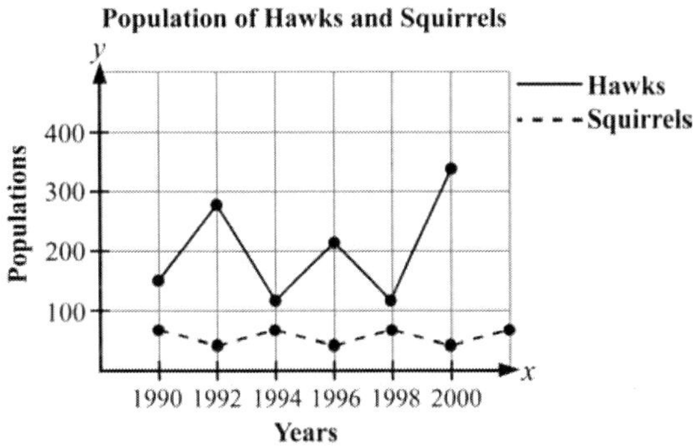

2. Write two conclusions that can be made about the populations of hawks and squirrels.

Use the following information to answer the next question.

When Charlie noticed that the tap outside his back door was leaking, he placed a bucket beneath the tap. For four days in a row, he measured the amount of water that dripped into the bucket. Here are his results.

Day	1	2	3	4
Amount of Water (L)	2	4	6	8

3. Use the data in Charlie's table of values to create a line graph.

Use the following information to answer the next question.

This line graph shows the number of people (patrons) who visited a particular water park every day for one week.

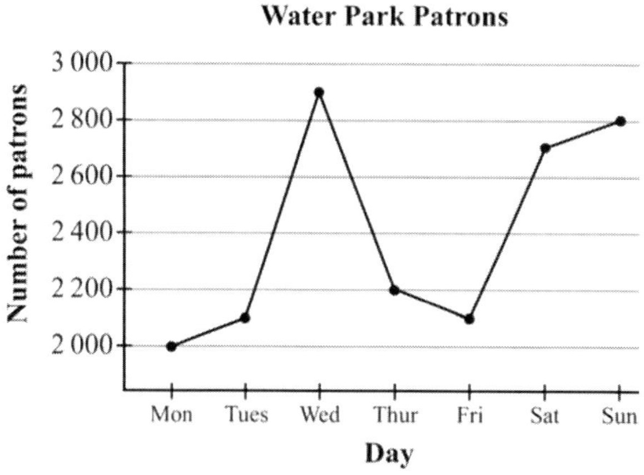

4. What conclusions can be drawn from this graph?

Use the following information to answer the next question.

The following graph shows the cost of seven tickets purchased by Mr. Barker for a particular music festival.

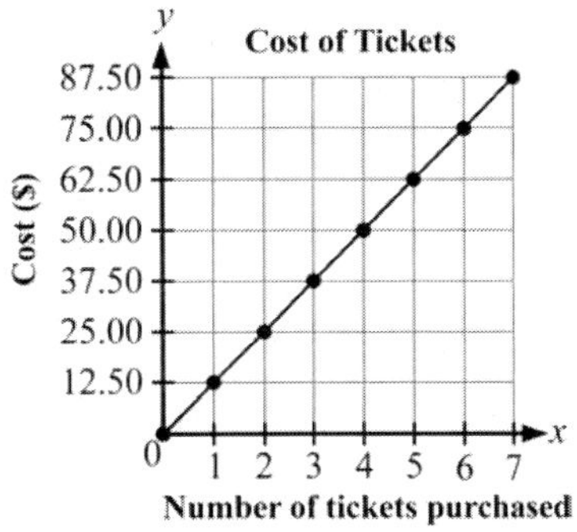

5. What conclusion can be drawn from the data?

Lesson 3 SELECTING, JUSTIFYING, AND USING APPROPRIATE METHODS OF COLLECTING DATA

NOTES

There are two types of data that can be collected: primary data (sometimes called first-hand data) and secondary data (sometimes called second-hand data). When collecting primary data, it is useful to make observations, perform experiments, conduct surveys, or design questionnaires. When collecting secondary data, it is useful to read books and newspapers, access databases (information stored on a computer), or use electronic networks (Internet).

QUESTIONNAIRES

A questionnaire or survey collects data in the form of answers to questions. A survey is the best way to collect data in order to find out the opinions of a group of people. Questionnaires are surveys that are given to the selected sample and collected right away. There are different types of surveys, such as telephone or mail surveys, personal interviews, or questionnaires.

DESIGNING AND ADMINISTERING A QUESTIONNAIRE

To design a questionnaire, first decide what you want to learn; this will determine who will be surveyed and what kind of questions they will be asked.

1. The key to designing a good survey question is phrasing the survey question in such a way that you get the desired data from the participants.
   - Use clear and concise vocabulary.
   - Have the same meaning for each participant.
   - Have one clear response from each participant.

2. Questions can be asked in the way of multiple-choice, numerical response, or short answer. Responses to multiple choice questions are easy to tabulate and interpret, and therefore are the most frequently used.

3. Organize your choices to make it easier to record and interpret your results. To record results, use a tally chart or a table. If you have access to a computer, you could use a spreadsheet to record the results.

4. Determine the number of people needed to answer the given question for the results to be reliable. The number of people that you ask to participate can influence the outcome of your questionnaire. Usually, the more people asked, the more reliable the results.

NOTES

Example

Maria wants to use a questionnaire to collect data from the Grade 6 students at her school about what their favourite breakfast food is. The question she plans to ask is, "What is your favourite breakfast food?"

Explain how Maria can design and administer her questionnaire to collect the data she wants.

Solution

Step 1
Offer choices.

Maria could offer choices such as bacon and eggs, pancakes, cereal, toast, or other.

Step 2
Determine how to collect data.

Depending on how many Grade 6 students are in her school, Maria could ask for a show of hands in each class or she could give the students a written questionnaire that they could fill out and return to her.

Step 4
Organize a method for recording results.

If Maria asked for a show of hands, she could use a tally chart to record the responses.

Favourite Breakfast Food	Number of Students
Bacon and eggs	卌 卌 卌 I
Pancakes	卌 卌 卌
Cereal	卌 卌 IIII
Toast	卌 卌 III
Other	IIII

NOTES

Time to Try 1

Jared wants to use a questionnaire to collect data from the Grade 5 and 6 students at his school about what their favourite weekend activity is. The question he plans to ask is, "What is your favourite weekend activity?"

Explain how Jared can design and administer his questionnaire to collect the data he wants.

PERFORMING EXPERIMENTS

If you collect data by doing an experiment, you carry out an action once or several times to gather results for a question that was asked. The more you repeat the given action, the more reliable the results will be.

Data is collected through observations of the action performed. Observations can be measured by using equipment like a thermometer, a scale, a die, or a spinner. Often, a table, graph, or chart is used to record the data observed. Organizing the results in tables, charts, or graphs can make it easier to interpret the data and then draw conclusions from it.

Example

Rhonda wants to answer the question, "Will a coin land on heads more than on tails if I flip it?"

Explain how Rhonda can perform an experiment to answer her question.

Solution

Step 1

Determine the number of times the action will be repeated.

Since there are only two outcomes, landing on heads or tails, tossing the coin 25 times should give reliable results.

NOTES

Step 2
Determine how to record the observations.

Rhonda can use a tally chart to record the way the coin lands for each of the 25 tosses.

Heads	Tails
𝍸 𝍸 ||	𝍸 𝍸 |||

Step 3
Draw a conclusion from the results.

Based on the results, Rhonda can conclude that the coin is no more likely to land on heads than on tails. Landing on heads or landing on tails is equally likely.

Time to Try 2

Jayden asks himself the question, "How many times can I kick the soccer ball into the air without letting it land on the ground?" Explain how Jayden can perform an experiment to answer his question.

USING A DATABASE OR ELECTRONIC MEDIA TO COLLECT DATA

Using a computer and the Internet is an excellent and fast way to find information.

Computers can be used in many ways to help collect and record data. You can use word processing programs to print out questionnaires and e-mail to send them out. You can use programs to create and print out tables and charts for recording results. You can use spreadsheets to organize all the results to make it easier to draw conclusions.

A database is a collection of information that is stored in a computer system. Databases are designed to store, manage, and retrieve information.

It is appropriate to use a database as a source of information unless the information being sought is too current. For example, a question regarding current affairs and/or pop culture might not be answerable from a database.

NOTES

Instead, you might need to access a specific website that discusses the information in question. The Internet provides large amounts of secondary data about all kinds of topics. Two examples of informative sites for students to use are the Guinness World Records site and the Statistics Canada site.

The key to effectively using databases and electronic media as a source of data is using appropriate search strategies. Sometimes, information can be missed if you are unsure how to find it.

Here is some information that can make using the computer easier.

- **Pull-down menus** are lists that are on many websites. If you click on a pull-down menu, a list of choices will appear. Move your mouse down the list until you highlight the item or topic you want. Click and release to go there.
- **Icons** are mini pictures or symbols that appear along the top and bottom of your screen. For example, on most computers there is a small picture of a printer at the bottom of the screen. If you click on this icon of the printer, your document or website information will be printed.
- **Passwords** are required on many websites, usually because the information is not free or contains personal information, such as online banking.
- **Keyword searches** are used on search engines such as Google, Yahoo!, or Ask.com. In order to search for your keywords, type the words in the designated space and click search. For example, "North American mountain lion" will narrow the list of websites on your topic to a certain species of lion. Typing too few keywords, like lion, for instance, would bring up so many results that you might find it confusing. Parents can be a great help when you are trying to decide on keywords for information searches.
- **Entry menu features** are all the menus, icons, and hyperlinks that are found on the home page. When you visit a website for the first time, the home page shows lists of topics that are covered within that website.

NOTES

Example

Suppose you are gathering information on population growth in Alberta. Your task is to determine how the population of Alberta has changed since 1900. Explain how a database could help you collect data.

Solution

A database would be the best way to gather information; it would be accurate and available. Using the Online Reference Centre, you could search the following databases:

- AlbertaSource.ca
- The Alberta Online Encyclopedia
- Peel's Prairie Provinces
- The Canadian Encyclopedia

Within these databases, you could search under the following terms:

- population growth
- demographics
- Alberta

Time to Try 3

What search term(s) would you use if you were trying to answer the following question: "How has the average temperature in Canada changed over the last 100 years"?

SELECTING AN APPROPRIATE METHOD TO ANSWER A QUESTION AND JUSTIFYING YOUR CHOICE

Choosing an appropriate method for collecting data depends on the type of question that you are trying to answer and the type of data you are trying to collect. You need to choose a method of collection that will give you the best and most reliable data.

NOTES

Example

Amanda wants an answer to the question, "How many times can I continuously toss a ball up in the air without dropping it?"

How should Amanda collect the data to answer her question? Justify your choice of method.

Solution

Step 1
Select a method of collecting data.

Amanda should conduct an experiment to answer her question.

Step 2
Justify your choice of method.

An experiment is an activity where you carry out an action once or several times to gather results, which then become the data. The action Amanda needs to carry out is to repeatedly toss the ball up into the air and then catch it. The results will be the number of times she catches the ball until she drops it. She can repeat the experiment several times and compare the results in order to draw a conclusion.

Time to Try 4

Trace wants an answer to the question, "What kind of cereal do my classmates like to eat for breakfast?"

How should Trace collect the data to answer his question? Justify your choice of method.

PRACTICE EXERCISES

Decide whether or not it would be appropriate to consult a database to answer the following four questions. Explain your answer.

1. How has the average price of a Canadian home changed in the last 50 years?

2. What are the top 10 songs currently being played on the radio?

3. What city in Canada recorded the warmest daytime temperature this week?

4. How has consumer spending changed over the last 50 years in Canada?

For each of the following questions, state whether a questionnaire, experiment, or database would be most appropriate for collecting the data needed. Justify your answer.

5. "How many jumping jacks can I do in a row?"

6. Morris asked 10 teachers the question, "Which type of book do you prefer to read: adventure, biography, mystery, science fiction, or thriller?"

Lesson 4 SELECTING AND JUSTIFYING AN APPROPRIATE GRAPH TO DISPLAY DATA

NOTES

Once your data has been collected, you need to decide how it will be best displayed. Although different types of graphs can be used to represent the same data, one type of graph is often more appropriate than another.

Following is a review of some graphs:

BAR GRAPH

A bar graph is best used to display facts about countable data using vertical or horizontal bars. These bars show the connections between two or more things and make it easy to compare data groups.

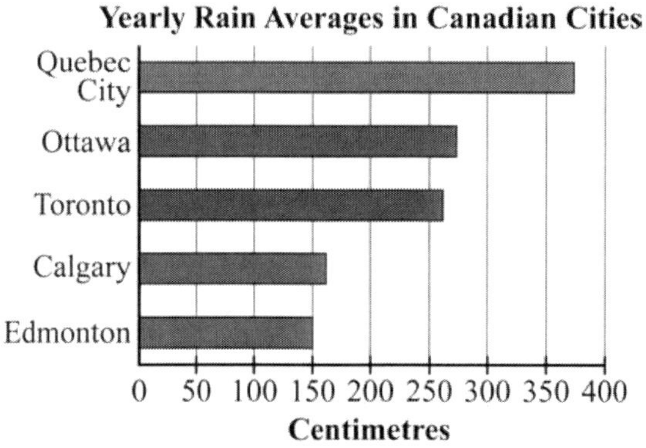

DOUBLE BAR GRAPH

A **double bar graph** is best used to show comparisons between and within sets of similar data. A double bar graph uses two connected bars to compare separate pieces of information within the same category.

A double bar graph has two bars for each category.

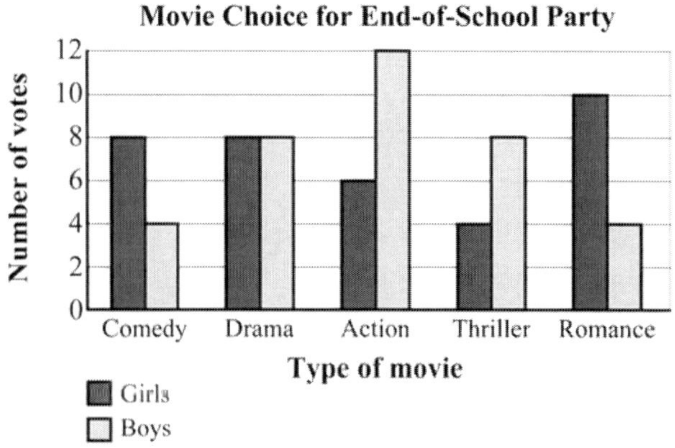

NOTES

LINE GRAPH

A line graph is best used to show changes over a period of time. A line graph is formed by line segments that join or a straight line that joins points representing the data.

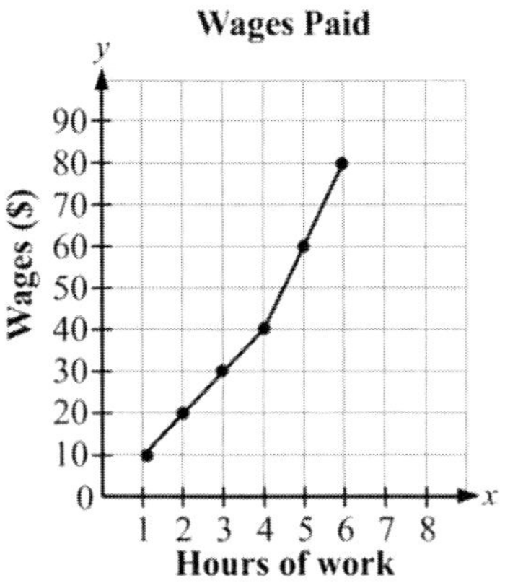

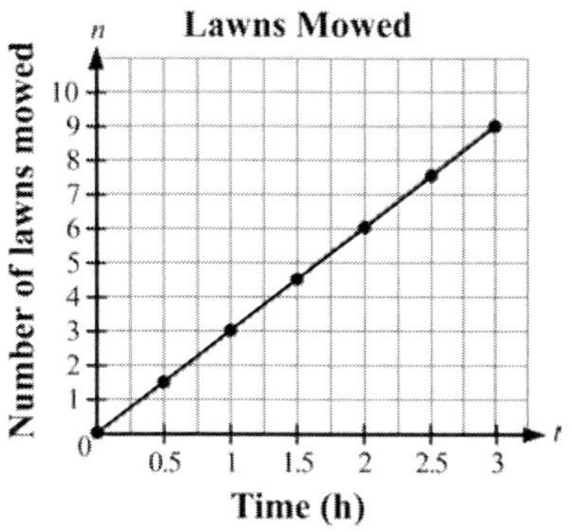

Example

Bob collected the following data from his schoolmates:

Favourite Food	Votes
Pizza	25
Hot Dogs	19
Steak	15
Hamburgers	21
Tacos	12

Bob could not decide which of the following graphical representations best represented his data.

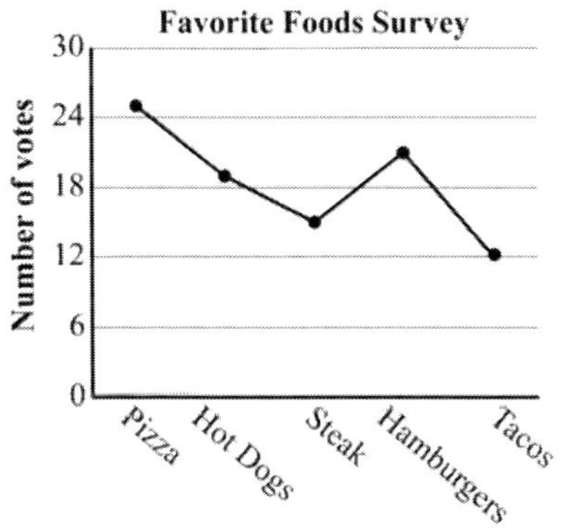

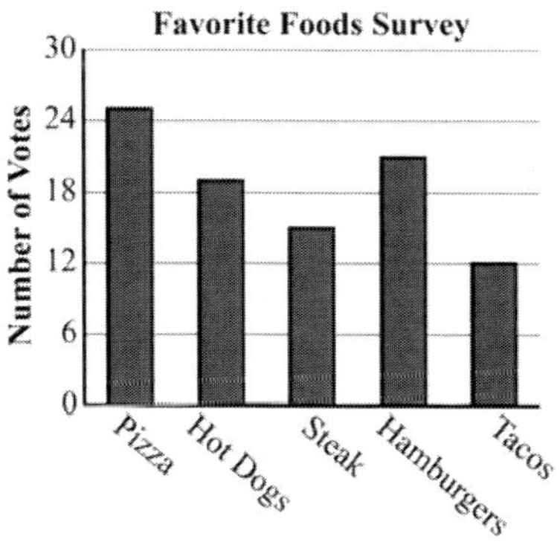

Which graph best represents the data? Justify your answer.

NOTES

Solution

The vertical bar graph is the best choice because it represents the data collected in a clear and precise manner. It is easy to compare the number of votes for each choice of food.

The line graph is not a good choice for this type of data because there are no changes over time.

Time to Try 1

The table below shows daily temperatures for New Westminster, recorded for 6 days, in degrees Celsius.

Temperatures in New Westminster (°C)	
Day	**Temperature**
1	21
2	24
3	18
4	25
5	17
6	17

Which graph would best represent the data? Justify your answer.

PRACTICE EXERCISES

Use the following information to answer the next question.

Students vote for their favourite fun day activity by writing the name of one of the four given activities on a slip of paper.

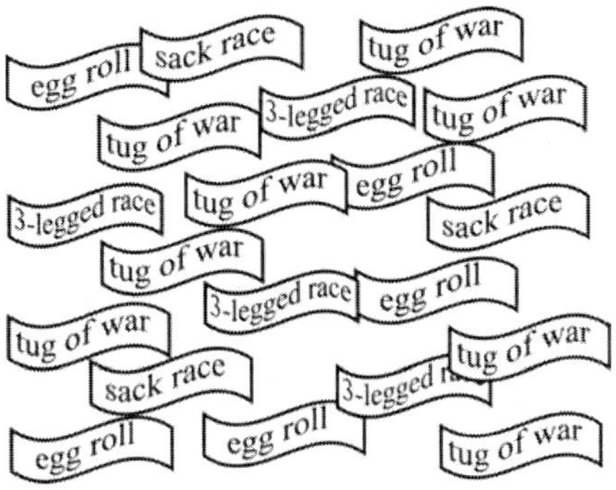

1. What type of graph would be most appropriate to display the given data? Justify your choice of graph.

Use the following information to answer the next question.

Bean Plant Growth	
Date	**Plant Height (cm)**
July 4	3
July 11	7
July 18	9
July 25	15
August 1	23

2. What type of graph would be most appropriate to display the given data? Justify your choice of graph.

Use the following information to answer the next question.

Five students were each given a bag of marbles. Each student first predicted the number of marbles in the bag and then counted the actual number of marbles.

Student	Prediction	Actual
Sam	35	52
Kate	40	47
Bob	50	49
Reece	25	52
Kel	38	50

3. What type of graph would be most appropriate to display the given data? Justify your choice of graph.

Use the following information to answer the next question.

Heart Rate of a Grade 6 Student	
Time of Day	**Beats per Minute**
9:00 A.M.	69
9:30 A.M.	73
10:00 A.M.	88
10:30 A.M.	120
11:00 A.M.	77
11:30 A.M.	75
12:00 P.M.	70

4. What type of graph would be most appropriate to display the given data? Justify your choice of graph.

Lesson 5 GRAPHING COLLECTED DATA AND INTERPRETING THE RESULTING GRAPH

NOTES

To display certain kinds of data, one form of graph is often more appropriate than another. Before choosing a particular form, think about what you want to communicate. When you are deciding how to display your data, you must think about what you want to show, who will be using your graph, and what might affect the appearance of your graph. Data is easier to understand if it is organized and presented appropriately. Once you have collected and graphed your data, you can interpret the resulting graph to solve problems or draw conclusions.

Interpret: To draw meaning from.

Example

The school cafeteria is considering cutting back on the number of choices given for lunch. Larry collected data about the favourite lunch snacks of ninety-two students at his school to help the school make their decision.

Favourite Food	Number of Votes
Pizza	25
Hot Dogs	19
Steak	15
Hamburgers	21
Tacos	12

Use a bar graph to graph the collected data. Draw a conclusion about the choices of food made.

Solution

Step 1
Graph the data.
The graph could look like this.

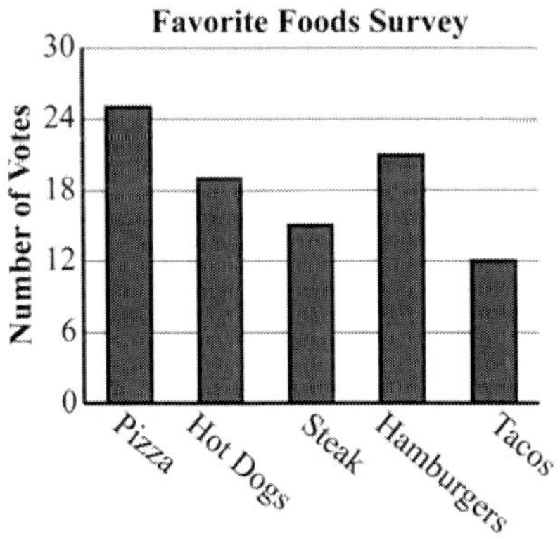

NOTES

Step 2
Draw a conclusion about the choices of food made.
Following are two examples of conclusions you can draw.

- Since tacos and steak got the lowest number of votes, you can conclude that if the cafeteria wanted to stop selling some of the foods, it should stop selling tacos and steak.
- Since almost half the students chose either pizza or hamburgers as their favourite foods, you can conclude that if the cafeteria only sold two items, it should sell pizza and hamburgers.

Time to Try 1

Sam conducted a questionnaire to determine the type of movie the class should view at the year-end party.

- The choices of types of movies were comedy, drama, action, thriller, and romance.
- Girls: 8 chose comedy and 8 chose drama; 6 chose action, 4 chose thriller, 10 chose romance
- Boys: 4 chose comedy and 4 chose romance; 8 chose drama and 8 chose thriller, 12 chose action

Use a double bar graph to graph the collected data. Which type of movie would be the best choice for the year-end party?

PRACTICE EXERCISES

Use the following information to answer the next question.

At the beginning of April, Ethan handed questionnaires to the 25 students in his class. This is the question he asked: "Did you choose to drink white or chocolate milk with your lunch?" He asked the students to keep track of their milk choices for the months of April, May, and June.

This is the data he collected:

- April: White milk was chosen by 10 students and chocolate milk was chosen by 15 students.
- May: White milk was chosen by 20 students and chocolate milk was chosen by 5 students.
- June: White milk was chosen by 12 students and chocolate milk was chosen by 13 students.

1. Use a double bar graph to graph the collected data. Overall, what milk was chosen the most?

Use the following information to answer the next question.

Kenton planted a marigold seed. Once it sprouted, he measured the height of the plant at the end of each week.

At the end of week 1, the plant was 25 mm tall; at the end of week two, the plant was 50 mm tall. At the end of week 3, the plant was 60 mm tall, and at the end of week 4, the plant was 90 mm tall.

2. Use a line graph to graph the collected data. Interpret the data displayed to draw a conclusion about how the plant grew over the four-week period.

Use the following information to answer the next question.

The Number of Days Grayson was Absent for Each Grade												
Grade	1	2	3	4	5	6	7	8	9	10	11	12
Days Absent	4	5	3	4	3	1	0	2	1	0	0	1

3. Display the data in a line graph. The biggest drop in absences was between which two consecutive grades?

REVIEW SUMMARY

- A line graph is useful for displaying data that changes continuously over time.
- Continuous data is data that can take on any value within a range. On a line graph, the data points are connected with a continuous line (continuous line graph) or with line segments (broken-line graph).
- Connecting the data points with a line illustrates the pattern that may have occurred between the actual measurements that were recorded.
- Discrete data is data that can only take on certain values. On a line graph, the data is usually presented as a series of points.
- When you create a line graph, be sure that you have an appropriate title, labels for your axes, and scales with intervals that suit the range and distribution of the data.
- When collecting and recording data, you can design questionnaires, perform experiments, access databases, or use electronic networks like the Internet.
- Choosing an appropriate method for collecting data depends on the type of question that you are trying to answer.
- A questionnaire is a research instrument consisting of one or more questions that are asked to a group of people.
- An experiment is an activity completed in order to predict the chances of an event occurring in the future.
- A database is a collection of information that is stored in a computer system. Electronic media involves the use of the Internet.
- Different forms of graphs (like bar graphs, double bar graphs, broken-line graphs, and continuous line graphs) may be used to represent the same data; however, one form is often more appropriate than another. Use the graph that is best suited to the type of data you have collected.
- You can graph the data you collected and then interpret the resulting graph to solve problems or draw conclusions.

PRACTICE TEST

1. Describe the common attributes of graphs that must be present in order for the reader to understand what the graph is about.

Use the following information to answer the next question.

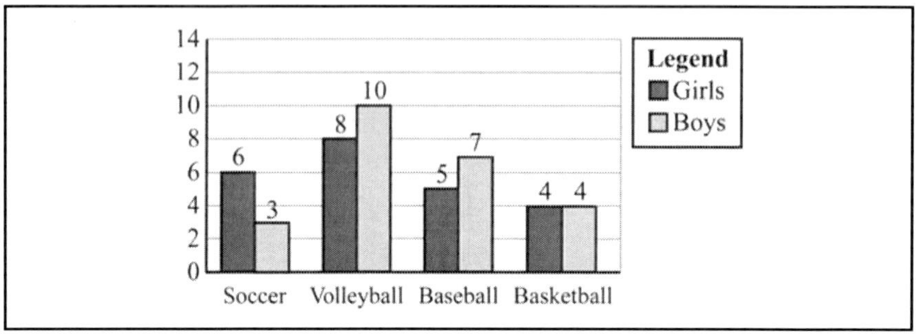

2. What attributes are missing from this graph? Explain your answer.

Describe the data in the next two questions as discrete or continuous. Justify your answer.

3. The time it took a runner to run 15 kilometres.

4. The number of trophies a hockey player won over a ten-year career.

Use the following information to answer the next question.

Brent was saving his money to buy some CDs. This graph shows the amount of money he had saved by the end of each week for five weeks.

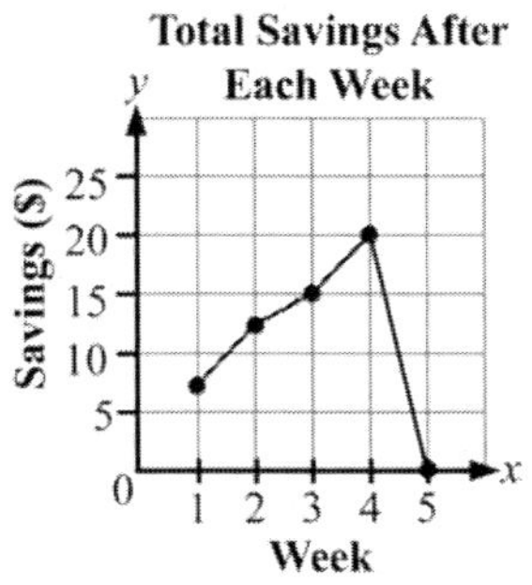

5. What conclusion can be drawn from the data?

Use the following information to answer the next question.

Ron is interested in what happens to butterflies during the winter months in Canada. He decides to collect data for the question, "What happens to butterflies during the cold winter months?"

6. Select the most appropriate method of collecting the data Ron wants. Justify your answer.

Use the following information to answer the next question.

Charlie checked the outdoor temperature four times during one day. Following are his results:

Time	Temperature (°C)
6:00 A.M.	13
11:00 A.M.	20
4:00 P.M.	27
9:00 P.M.	23

7. What type of graph would be most appropriate to display the collected data? Justify your answer.

Use the following information to answer the next question.

Sam observed a tortoise moving over a period of eight minutes. He noticed that the tortoise moved at exactly the same rate for each minute it moved.

At the end of three minutes, the tortoise had moved 30 cm. At the end of eight minutes, the tortoise had moved 80 cm.

Minute	Distance Travelled (cm)
0	0
3	30
8	80

8. Create a line graph for the data given.

Use the following information to answer the next question.

Tom wants to go to a movie on Saturday. He wonders which movie was the best box office movie over the past week.

9. Explain if it is appropriate for Tom to use a database as a source of data.

PROBABILITY

When you are finished this unit, you should be able to…

- Identify all the possible outcomes of a probability experiment
- Find the favourable outcomes of a probability experiment
- Find the theoretical probability of outcomes in a probability experiment
- Find the experimental probability of outcomes in a probability experiment
- Explain the difference between experimental and theoretical probabilities
- Compare experimental results with the theoretical probability of outcomes in a probability experiment

Lesson	Page	Completed on
1. Theoretical Probability	334	
2. Experimental Probability	341	
Review Summary	347	
Practice Test	348	
Answers and Solutions	at the back of the book	

PREREQUISITE SKILLS AND KNOWLEDGE

Prior to beginning this unit, you should be able to…

- Describe the chance of a single outcome happening using words such as *impossible*, *possible*, and *certain*
- Carry out a probability experiment a number of times, record the results, and explain the results
- Compare the chance of two possible outcomes happening using words such as *less likely*, *equally likely*, and *more likely*
- Design and carry out probability experiments with one outcome less likely to happen than another, with one outcome equally as likely to happen as another, and with one outcome more likely to happen than another

Lesson 1 THEORETICAL PROBABILITY

NOTES

Theoretical probability is the chance an outcome might happen in perfect conditions.

FAVOURABLE AND POSSIBLE OUTCOMES

The favourable outcomes are the number of results you want in a probability experiment. The possible outcomes are the number of all the results that can happen in the experiment.

You can use a ratio fraction to show the probability of an outcome. The numerator of the fraction is the number of favourable outcomes. The denominator is the number of all the possible outcomes.

$$\frac{\text{Numerator}}{\text{Denominator}} = \frac{\text{total number of favourable outcomes}}{\text{total number of possible outcomes}}$$

Theoretical probability can be expressed in words, such as *one out of three,* or as a probability fraction, such as $\frac{1}{3}$.

ALL POSSIBLE OUTCOMES

You can find the number of possible outcomes in a probability experiment by counting them:

- the number of sides of a coin
- the number of parts on a spinner
- the number of faces of a die used

Each side, part, or face is a possible outcome.

The denominator of the probability fraction will always be equal to the number of all the possible outcomes

Example

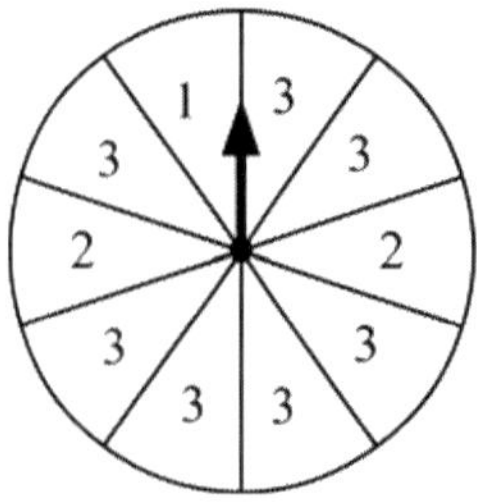

Brendan spins the spinner once. How many possible outcomes are there?

NOTES

Solution

Count the number of parts in the spinner. There are 10 parts, so there are 10 possible outcomes.

Even though there are only three different numbers used on the parts (1, 2, or 3), the spinner can stop on any of the ten parts, even if the numbers on the parts are the same.

Time to Try 1

Kate flips a coin in a probability experiment.

How many possible outcomes are there? Explain your answer.

ONE FAVOURABLE OUTCOME

When there is only one favourable outcome in a probability experiment, the numerator of the probability fraction will always be 1.

Example

What is the theoretical probability of rolling a six on a regular six-sided die?

Solution

Step 1

Find the favourable outcome.

Each six-sided die has 1 six on it. Therefore, there is 1 favourable outcome.

NOTES

Step 2
Find the possible outcomes.

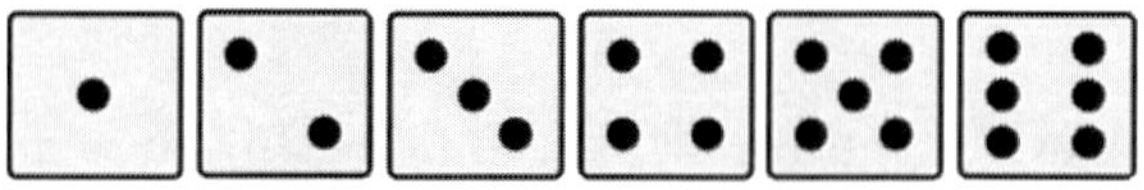

The die has six sides, and the sides are numbered from one to six. The number of all the possible outcomes is 6.

Step 3
Write the theoretical probability.

$$\text{Theoretical probability} = \frac{\text{number of favourable outcomes}}{\text{number of possible outcomes}}$$

The theoretical probability of rolling a six is 1 out of 6, or $\frac{1}{6}$.

Time to Try 2

What is the theoretical probability of rolling a two on a regular 10-sided die?

MORE THAN ONE FAVOURABLE OUTCOME

There can be more than one favourable outcome in a probability experiment. The numerator of the probability fraction will be equal to the number of favourable outcomes. The probability fraction can be simplified by reducing it to its lowest form.

Example

A spinner has 12 parts. Each part of the spinner contains one of the following numbers:
2, 2, 3, 4, 4, 4, 5, 6, 7, 7, 7, and 8

When the spinner is spun, what is the theoretical probability of the spinner stopping on four?

Solution

Step 1
Find the favourable outcomes.

There are 3 parts with number four. Therefore, there are 3 favourable outcomes.

NOTES

Step 2
Find the possible outcomes.

There are 12 numbered parts on the spinner, so the number of all the possible outcomes is 12.

Step 3
Write the theoretical probability.

The theoretical probability of the spinner landing on a four is 3 out of 12, or $\frac{3}{12}$.

Step 4
Simplify the probability fraction.

Simplify the probability fraction $\frac{3}{12}$ by dividing both the numerator and denominator by 3.

$$\frac{3}{12} = \frac{3 \div 3}{12 \div 3} = \frac{1}{4}$$

The theoretical probability of $\frac{3}{12}$ can be expressed as one out of four, or $\frac{1}{4}$.

Time to Try 3

A die has 12 faces. The faces are labelled with the following numbers:
2, 2, 2, 3, 3, 4, 4, 4, 5, 5, 6, and 6

When the die is rolled, what is the theoretical probability of rolling a five?

PREDICTING THEORETICAL PROBABILITY

You can predict the chances an outcome has for a probability experiment. This prediction is the theoretical probability. You can make this prediction if you know how many favourable and possible outcomes exist in the experiment.

NOTES

You can predict how many times one outcome may happen if you repeated the experiment. Make an equivalent fraction to the theoretical probability fraction. The denominator of the equivalent fraction should be the number of times you repeated the experiment.

Example

This spinner has four equal parts. Each part has a number.

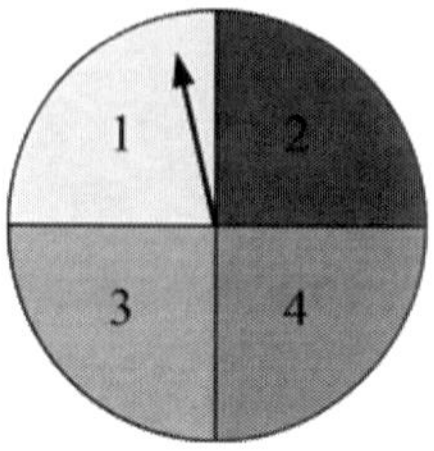

Jake spins the spinner 20 times. About how many times will it stop on the part with the number one?

Solution

Step 1

Find the theoretical probability.

The theoretical probability of the spinner stopping on one after 1 spin is 1 out of 4, or $\frac{1}{4}$.

Step 2

Write an equivalent fraction.

Since the spinner will be spun 20 times, write a fraction equivalent to $\frac{1}{4}$ with a denominator of 20.

$$\frac{1}{4} = \frac{1 \times 5}{4 \times 5} = \frac{5}{20}$$

A good prediction would be that out of 20 spins, the spinner will stop on one 5 times.

Time to Try 4

The faces of a standard six-sided die are numbered from one to six. Audrey rolls the die 30 times. Predict the number of times Audrey may roll a five.

PRACTICE EXERCISES

Use the following information to answer the next question.

Janice rolls an eight-sided die. The faces of the die have the following numbers: 6, 8, 8, 9, 9, 9, 9, and 9

1. How many possible outcomes are there? Explain your answer.

Use this information to answer the next four questions.

Don made a spinner with 12 equal parts. Each part was a particular colour (red, orange, black, or brown) and held a particular number.

Red	Orange	Black	Brown
3	6	7	2
4	7	8	5
9	9	4	
		9	

If Don spins the spinner once, what is the theoretical probability of the spinner stopping on each of the following parts?

2. A part with the number 3

3. A part with the number 9

4. A brown coloured part

5. A black coloured part

Use this spinner to answer the next two questions.

6. Henri spins this spinner 40 times. Predict the number of times the spinner will stop on a number containing a 7.

7. Out of 30 spins, predict the number of times the spinner will stop on an even number.

Lesson 2 EXPERIMENTAL PROBABILITY

NOTES

You can find experimental probability by doing an experiment. Experimental probability is based on the results of the experiment. You can use a ratio fraction to show the experimental probability of an outcome of an experiment. The numerator of the fraction is the number of times a particular outcome happened. The denominator is the number of all the times the event was tested.

$$\text{Experimental probability} = \frac{\text{number of times outcome occurs}}{\text{number of times event tested}}$$

Experimental probability can be expressed in words, such as *one out of three*. This means that the outcome happened once, and the event was tested three times. The experimental probability is more often expressed as a ratio or probability fraction, such as $\frac{1}{3}$.

Example

Mark wants to know how many times a coin would land heads up if he flips it 20 times. When he performs the experiment, the coin lands heads up 9 times. What was the experimental probability of the coin landing heads up?

Solution

Step 1
Identify the number of times the outcome happened.

The coin landed heads up 9 times.

Step 2
Identify the number of times the event happened.

Mark flipped the coin 20 times.

Step 3
Find the experimental probability.

The experimental probability of the coin landing heads up was 9 out of 20, or $\frac{9}{20}$.

NOTES

Time to Try 1

Jenn rolls a standard six-sided die 60 times. The die lands with three face-up 17 times. What is the experimental probability of rolling a three?

Theoretical probability is what should happen.

Experimental probability is what does happen.

DISTINGUISHING BETWEEN EXPERIMENTAL AND THEORETICAL PROBABILITIES

Theoretical probability is the outcome you would expect under perfect conditions. It is the chance of an event happening. You can calculate it without doing an experiment.

Experimental probability is the outcome you find by doing an experiment. It is the number of times an event happens in real life. You can calculate it using the results of the experiment.

The more you repeat the trials in an experiment, the more the experimental probability will be like the theoretical probability. But, the probabilities may never be equal.

COMPARING EXPERIMENTAL RESULTS WITH THEORETICAL PROBABILITY

It is a good idea to find the theoretical probability of the outcome you want before carrying out a probability experiment. After the experiment, you can then compare the actual results (experimental probability) to the prediction (theoretical probability).

NOTES

Example

Sarah rolled a standard six-sided die 60 times. The following chart shows how many times the die landed on each of the six numbers.

Number on Cube	Number of Times
1	12
2	5
3	13
4	10
5	17
6	3

Which number on the die had the same experimental probability as its theoretical probability?

Solution

Step 1
Find the theoretical probability.

There are six different numbers on the die. The theoretical probability of landing on any one of the numbers is $\frac{1}{6}$.

Step 2
Find the experimental probability.

The experimental probability is the same as the theoretical probability when it is $\frac{1}{6}$.

The denominator of the experimental probability is 60 (all of the rolls). The numerator must be 10 (the number of times the die landed on a particular number).

$$\frac{1}{6} = \frac{1 \times 10}{6 \times 10} = \frac{10}{60}$$

Step 3
Find which number had the same theoretical and experimental probabilities.

The cube landed 10 times on the number 4. This is shown by the probability $\frac{10}{60} = \frac{1}{6}$. The number four has the same theoretical probability as experimental probability.

NOTES

Time to Try 2

Kirby tossed a two-sided coin 14 times to see how many times it would land on heads. It landed on heads 9 times and on tails 5 times.

Compare the experimental probability of landing on heads to the theoretical probability of landing on heads.

USING TECHNOLOGY TO CONDUCT PROBABILITY EXPERIMENTS

You can use computers to conduct probability experiments. Many Internet websites offer probability experiments in which the actions of spinning pointers, flipping coins, and rolling dice is simulated. A *simulated action* is a model or a copy of a real action.

You can use spreadsheets to record and organize the results of an experiment. You can also use calculators to quickly perform calculations.

Follow the same procedure when you use simulated actions as when you use actual manipulatives in a probability experiment:

- Calculate the theoretical probability and make a prediction.
- Conduct the experiment.
- Find the experimental probability.
- Compare the experimental results with the theoretical probability.

PRACTICE EXERCISES

Use the following information to answer the next question.

Michelle has a spinner with six equal parts. Each part is a different colour.

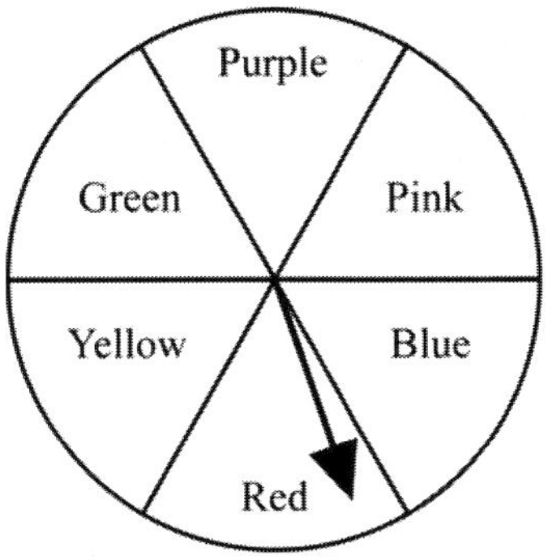

1. Michelle spins the spinner 100 times. The spinner stops on the red section 19 times. What is the experimental probability of the spinner landing on the red section?

2. Zeke tosses a penny in the air 25 times. It lands heads up 14 times. What is the experimental probability of the penny landing with heads face up?

3. Shawn tosses a two-sided coin 30 times. She wants to see how many times it will land on tails. The coin lands on heads 14 times and on tails 16 times. Compare the experimental results with the theoretical probability of the coin landing on tails.

Use the following information to answer the next question.

Ryder spins this spinner 20 times. He wants to see how often it will stop on a number divisible by two. The spinner lands on a number divisible by two 8 times.

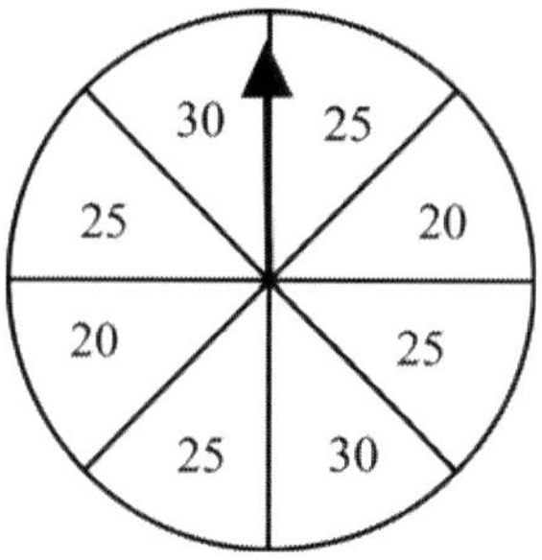

4. Compare the experimental results with the theoretical probability of the spinner stopping on a number divisible by two.

REVIEW SUMMARY

- A favourable outcome is the number of results you want in a probability experiment.
- Possible outcomes are the results that can happen in a probability experiment.
- Theoretical probability is the chance an outcome might happen in perfect conditions.
- Express theoretical probability as a ratio fraction.

  $$\text{Theoretical probability} = \frac{\text{total number of favourable outcomes}}{\text{total number of possible outcomes}}$$

- The experimental probability depends on the results of an experiment. It is found by carrying out the experiment.
- Express the experimental probability as a ratio fraction.

  $$\text{Experimental probability} = \frac{\text{number of times outcome happens}}{\text{number of times event tested}}$$

- You can use theoretical probability to predict an outcome of an experiment. Then, you can compare the results to the prediction.
- The more you repeat the trials, the more the experimental and theoretical probabilities will be alike.

PRACTICE TEST

1. Rhonda predicts that the probability of rolling a nine on a regular 12-sided die is $\frac{1}{12}$. Is this an example of theoretical probability or experimental probability? Explain your answer.

2. Mark tosses a penny in the air 10 times. The penny lands heads up three times, and Mark records these results. Is this an example of theoretical probability or experimental probability? Explain your answer.

If Don rolls a standard six-sided die once, what is the theoretical probability that he will roll each of the following outcomes?

3. A number greater than three

4. A number less than three

Use the following information to answer the next three questions.

Allie rolls a four-sided die 12 times and recorded the number of times each number is rolled.

Number on Die	Number of Rolls
1	3
4	3
6	4
9	2

5. What is the experimental probability of rolling a six?

6. Compare the experimental results with the theoretical probability of rolling a nine.

7. Explain how the theoretical and experimental probabilities for rolling a one are the same.

Use the following information to answer the next three questions.

Josh makes a spinner out of a pencil and a paper clip.

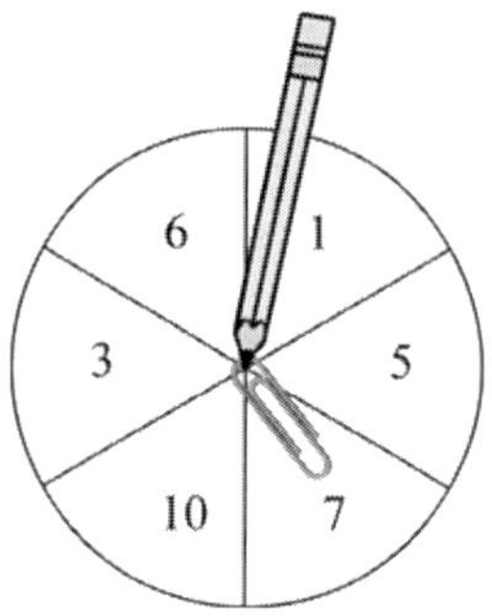

8. Josh spins the spinner 60 times. Predict the number of times the spinner will stop on the number seven.

9. When Josh spins the spinner 18 times, it lands on the number three 5 times. What was the experimental probability of landing on three?

10. When Josh spins the spinner 24 times, the spinner landed on the number six 8 times. Compare the experimental results with the theoretical probability of landing on the number six.

DEVELOPING AND APPLYING FORMULAS FOR PERIMETER, AREA, AND VOLUME

When you are finished this unit, you will be able to…

- Explain, using models, how the perimeter of any polygon can be determined
- Generalize a formula for determining the perimeter of polygons
- Explain, using models, how the area of any rectangle can be determined
- Generalize a formula for determining the area of rectangles
- Explain, using models, how the volume of any right rectangular prism can be determined
- Generalize a formula for determining the volume of right rectangular prisms
- Solve problems involving the perimeter of polygons, the area of rectangles, and the volume of right rectangular prisms

Lesson	Page	Completed on
1. Determining the Perimeter of Any Polygon	352	
2. Determining the Area of Any Rectangle	361	
3. Determining the Volume of Any Right Rectangular Prism	366	
Review Summary	372	
Practice Test	373	
Answers and Solutions	at the back of the book	

PREREQUISITE SKILLS AND KNOWLEDGE

Prior to starting this unit, you should be able to…

- Estimate perimeter, using referents for cm and m
- Measure and record perimeter in cm and m
- Construct different shapes for a given perimeter
- Select and justify referents for cm^2 and m^2
- Estimate and determine area
- Construct different rectangles for a given area
- Select and justify referents for cm^3 and m^3
- Estimate and determine volume
- Construct right rectangular prisms for a given volume

Lesson 1 DETERMINING THE PERIMETER OF ANY POLYGON

NOTES

A polygon is a closed figure made out of three or more straight sides. The perimeter of a polygon is the total distance all around the shape of the polygon.

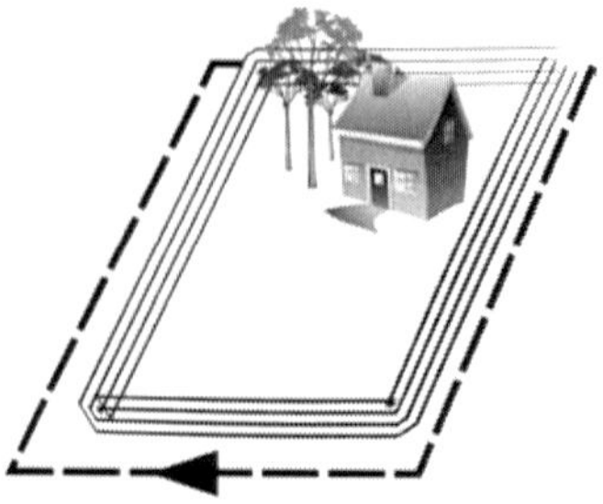

Perimeter

USING MODELS TO EXPLAIN THE PERIMETER OF IRREGULAR POLYGONS

On the geoboard shown, the space between two dots represents a 1 cm unit.

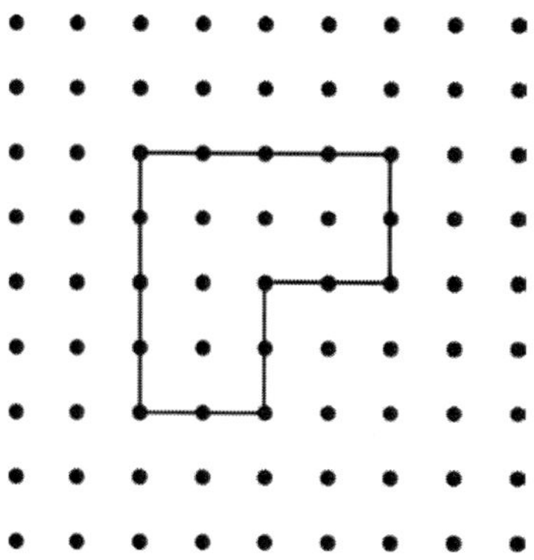

To find the perimeter of the outlined shape on the geoboard, count the number of centimetre units in each of the six sides and add them together. The numbers may be added in any order.

$4+4+2+2+2+2=16$

The perimeter of the polygon is 16 cm.

NOTES

Example

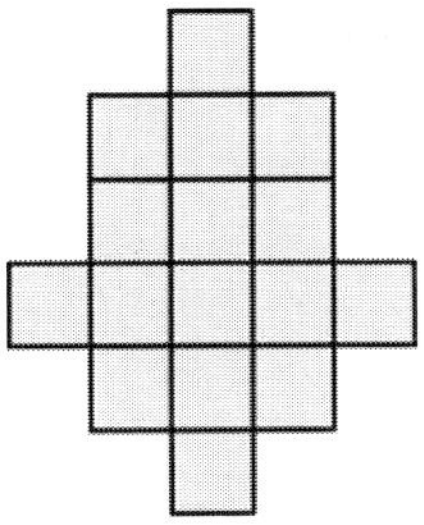

Explain how you can determine the perimeter of the given shape.

Solution

Count the number of units all around the outside edge of the shape. It may be helpful to make a tick on each unit as you count it so that you do not miss a unit or accidentally count a unit twice.

There are 22 units around the outside edge. The perimeter of the given shape is 22 units.

Time to Try 1

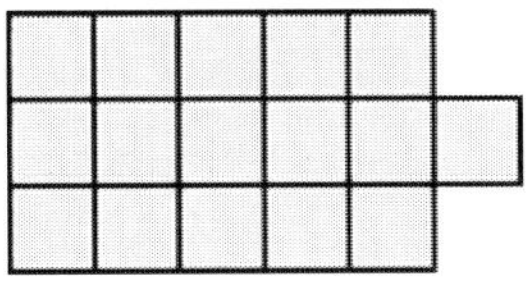

Explain how you can determine the perimeter of the given shape.

GENERALIZING A RULE (FORMULA) FOR THE PERIMETER OF IRREGULAR POLYGONS

Since perimeter is the distance all around a polygon, the letter s can represent each side length of the polygon.

Since you need to add all the side lengths to determine perimeter, you can develop a formula for perimeter by adding the side lengths for each shape.

If the shape has six sides like the diagram shown on the previous page, the formula could be $P = s + s + s + s + s + s$.

The number of s's used in the formula depends on the number of sides in the polygon being measured.

NOTES

SOLVING PROBLEMS INVOLVING PERIMETER OF IRREGULAR POLYGONS

Recall that when adding the lengths of each side, the numbers can be added in any order. Rearranging the numbers can make it easier to add.

Example

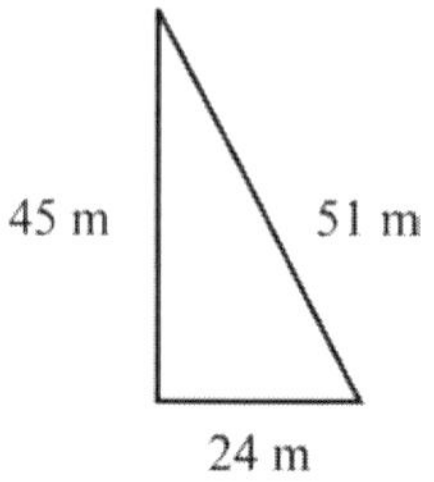

Use a formula to find the perimeter of the triangle.

Solution

Step 1
Develop a formula.

Since there are three sides in the triangle, you need three s's in the formula. The formula could be written as $P = s + s + s$.

Step 2
Apply the formula.

Substitute the numbers for the three side lengths into the formula. The numbers can be added in any order.

$$\begin{aligned} P &= 45 + 51 + 24 \\ &= 120 \end{aligned}$$

The perimeter of the triangle is 120 m.

Time to Try 2

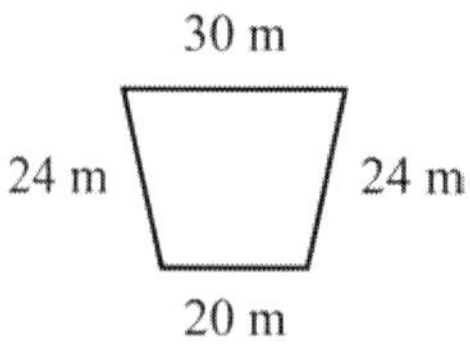

Use a formula to find the perimeter of the given trapezoid.

NOTES

USING MODELS TO EXPLAIN THE PERIMETER OF RECTANGLES AND SQUARES

A rectangle is a four-sided figure that has two sets of sides that are opposite each other and equal in length.

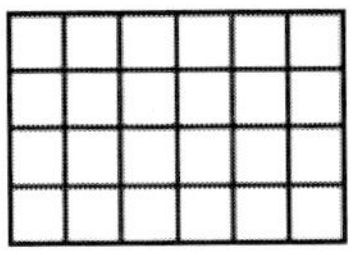

To find the perimeter of the given rectangle, you can count the number of 1 cm lengths around the outer edges of the rectangle (1, 2, 3, …18, 19, 20).

A quicker method is to count the number of centimetre units that make the length (6 cm units) and the number of units that make the width (4 cm units). Add the two lengths and two widths together. The numbers may be added in any order.
$4+6+4+6=20$

The perimeter of the rectangle is 20 cm.

A square is a four-sided figure that has four sides that are all the same length.

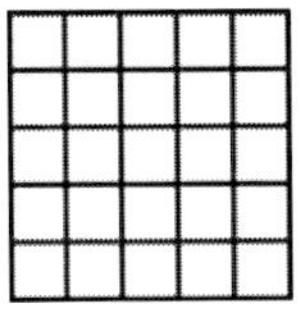

To find the perimeter of the given square, you can count the number of 1 cm lengths all around the edge (1, 2, 3,… 18, 19, 20), add the length of the square to itself four times ($5+5+5+5=20$) or multiply the length by 4 ($5\times4=20$).

The perimeter of the square is 20 cm.

NOTES

Example

Four rectangles are shown on the following grid.

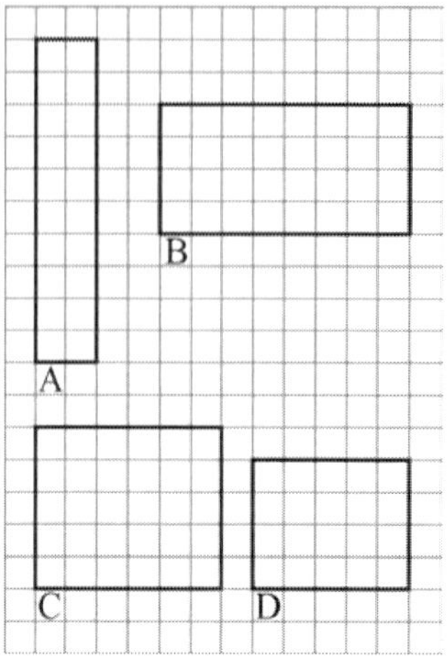

Which rectangle has a perimeter of 22 units?

Solution

One method of finding the perimeter is to count the number of units for the length and the width, and then add the two lengths and two widths.

Rectangle A: $10+10+2+2=24$
Rectangle B: $8+8+4+4=24$
Rectangle C: $5+5+6+6=22$
Rectangle D: $5+5+4+4=18$

Rectangle C has a perimeter of 22 units.

Time to Try 3

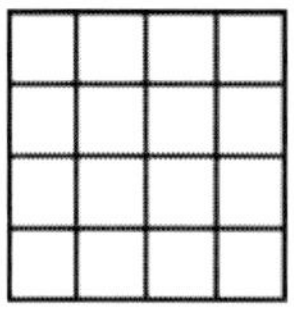

Explain how you can determine the perimeter of the square.

NOTES

GENERALIZING A RULE (FORMULA) FOR PERIMETER OF RECTANGLES AND SQUARES

Since the edges of a rectangle are referred to as lengths and widths, you can let the letter l represent the length of the rectangle and the letter w refer to the width of the rectangle.

Since you need to add the two lengths and widths together to find the perimeter, you can develop a formula like $P = l + w + l + w$.

Because the two lengths are the same and the two widths are the same, you can multiply each length and width by 2 and then add the products, like $P = (2 \times l) + (2 \times w)$. You can also add the length and width together and then multiply the sum by 2, like $P = 2(l + w)$.

Since all the side lengths are the same for a square, you can develop a formula like $P = l + l + l + l$ or $P = s + s + s + s$. You can also develop a formula like $P = 4 \times l$ or $P = 4s$, in which you multiply the side length by 4.

SOLVING PROBLEMS INVOLVING THE PERIMETER OF RECTANGLES AND SQUARES

Recall that when adding the lengths of each side, the numbers can be added in any order. Rearranging the numbers can make it easier to add.

Example

This figure represents the size and shape of Maria's garden.

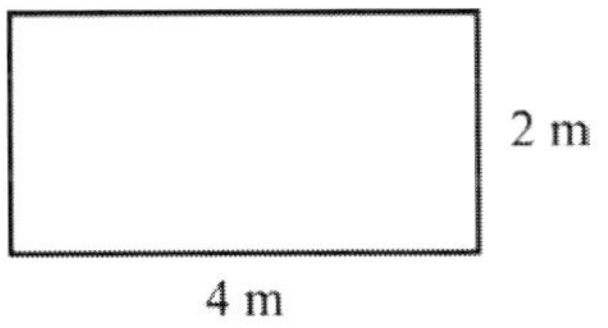

Use a formula to determine the perimeter of the garden.

Solution

Step 1
Develop a formula.

Since there are two lengths that are the same and two widths that are the same, you can use the formula $P = (2 \times l) + (2 \times w)$.

NOTES

Step 2
Apply the formula.

Substitute the given length (4 m) and width (2 m) into the formula.

$P = (2 \times 4) + (2 \times 2)$

$P = 8 + 4$

$P = 12$

The perimeter of Maria's garden is 12 m.

Time to Try 4

A carpenter builds a square wooden frame with 10 m long sides. Use a formula to find the perimeter of the frame.

PRACTICE EXERCISES

Use the following information to answer the next question.

On the geoboard shown, the space between two dots represents a 1 cm unit.

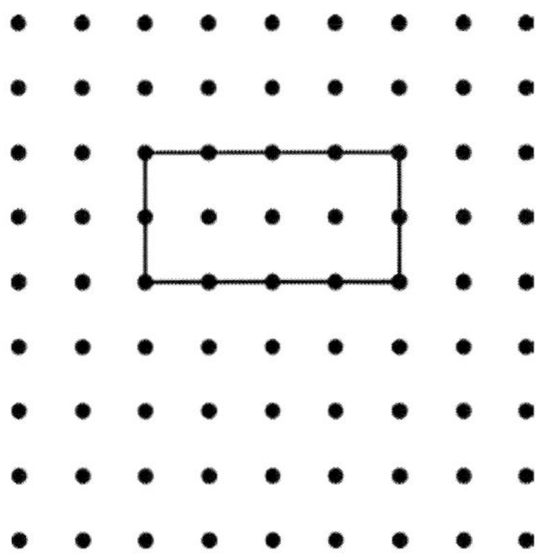

1. Explain how you can find the perimeter of the given rectangle.

Use the following information to answer the next question.

Jon measures the side lengths of a picture he sees in a book.

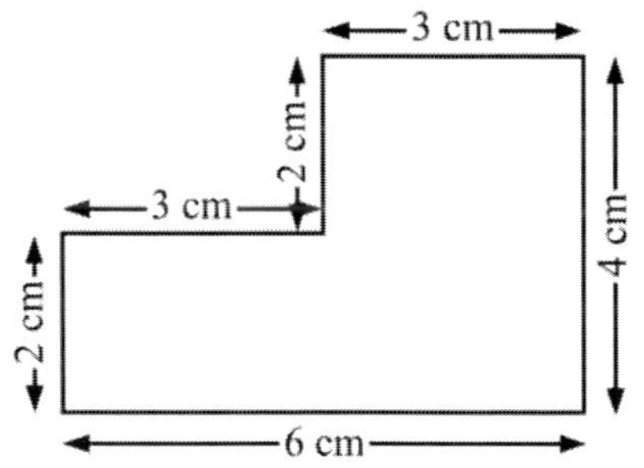

2. Use a formula to find the perimeter of the picture.

Use the following information to answer the next question.

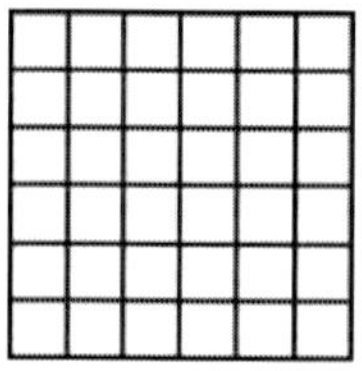

3. Explain how you can find the perimeter of the given shape.

Use the following information to answer the next question.

Jason's dad plans to build a fence around the new garden plot he dug.

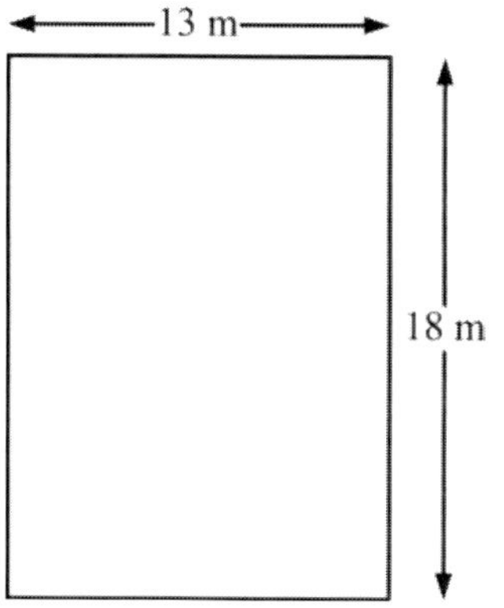

4. Use a formula to find how much fencing material Jason's dad needs to buy.

Lesson 2 DETERMINING THE AREA OF ANY RECTANGLE

NOTES

The area of any rectangle is the measurement of the space inside the rectangular shape. Area is expressed in square units (units^2), such as square centimetres (cm^2) and square metres (m^2).

USING MODELS TO EXPLAIN THE AREA OF RECTANGLES

The area of a rectangle is the number of square units needed to cover the surface of the rectangle without any overlapping of square units or spaces left between square units.

■ represents 1 square unit

Area can be found by counting square units, skip counting, repeated addition, or multiplication.

Example

This rectangle is covered in same-sized squares that each represent one square centimetre. The rectangle is covered with four rows of squares. There are eight squares in each row.

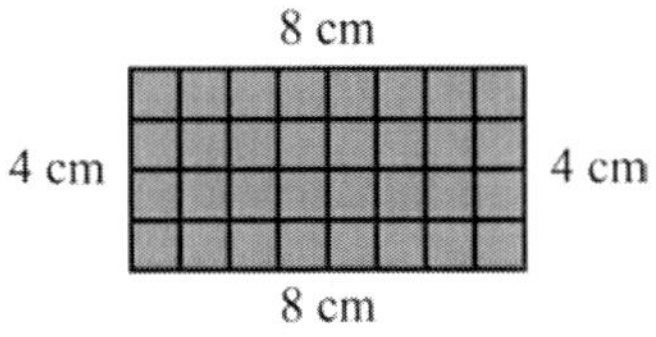

Explain how you can find the area of the rectangle.

Solution

One method to find the area is to count all the squares.
1, 2, 3,… 30, 31, 32

A quicker method is to count the number of squares that cover one length (8) and skip count up by the number (8) for the number of rows (4).
8, 16, 24, 32

Another method is to count the number of squares that cover one length (8) and repeatedly add that number (8) for the number of rows (4).
$8 + 8 + 8 + 8 = 32$

The quickest method is to multiply the number of squares in the length (8) by the number of squares in the width (4).
$8 \times 4 = 32$

Since there are a total of 32 squares that cover the surface of the rectangle, the area of the rectangle is 32 cm^2.

NOTES

Time to Try 1

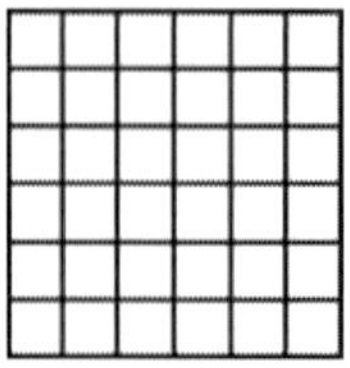

Explain one method that you can use to find the area of the given square.

GENERALIZING A RULE (FORMULA) FOR THE AREA OF A RECTANGLE

Since the edges of a rectangle are referred to as lengths and widths, you can let the letter l represent the length of the rectangle and the letter w represent the width of the rectangle.

Because the quickest method of finding the area of a rectangle is to use multiplication instead of repeated addition or skip counting, you can use a formula like $A = l \times w$.

SOLVING PROBLEMS INVOLVING THE AREA OF A RECTANGLE

When using a formula to determine area, substitute the numbers for the length and width of the rectangle into the formula. Be sure to include the correct unit of measure in your answer. Recall that area is always measured in square units (units^2).

NOTES

Example

A school principal plans to order some sand for the school sand pit. This diagram represents the length and width of the rectangular sand pit.

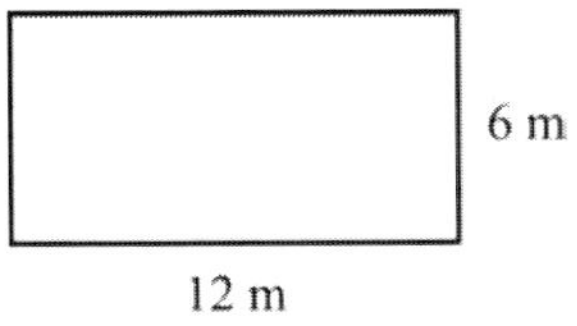

Use a formula to find the area of the sand pit.

Solution

To find the area, multiply the length of the sand pit by the width of the sand pit. Substitute the length (12 m) and the width (6 m) into the area formula.

$$\begin{aligned} A &= l \times w \\ &= 12 \times 6 \\ &= 72 \end{aligned}$$

The area of the rectangular sand pit is 72 m^2. The principal would need to order enough sand to cover 72 m^2.

Time to Try 2

Tom has a rectangular garden at the back of his house. He wants to buy enough peat moss to cover the entire garden.

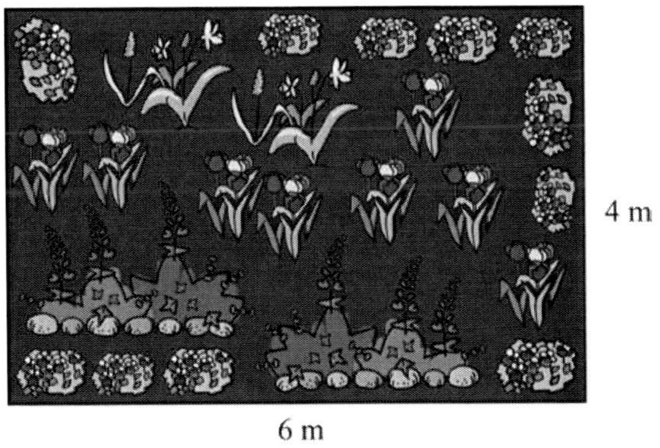

Use a formula to find the area he needs to cover.

PRACTICE EXERCISES

Use the following diagram to answer the next question.

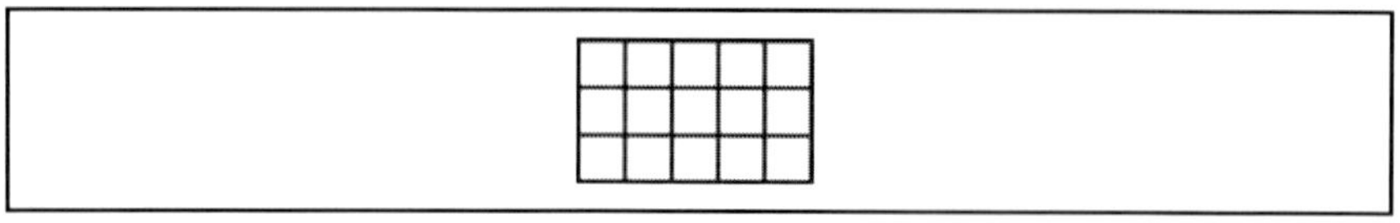

1. Explain two methods of finding the area of the given rectangle.

Use the following diagram to answer the next question.

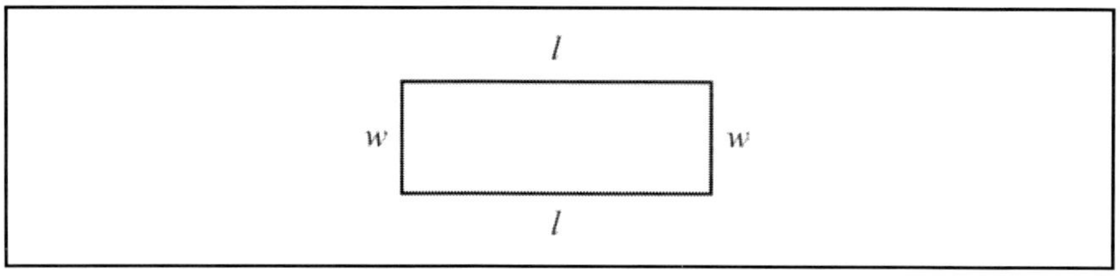

2. If the length of the given rectangle is 20 cm and the width is 10 cm, what is the area of the rectangle?

Use the following diagram to answer the next question.

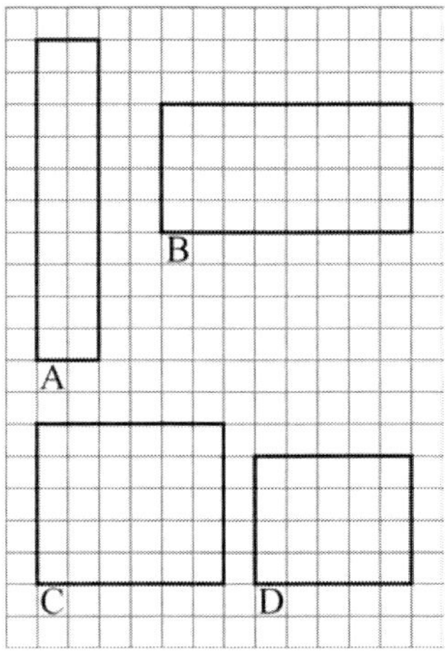

3. Which two rectangles have an area of 20 units^2?

4. A rectangular swimming pool measures 50 m by 30 m. Use a formula to find the area of the swimming pool.

Lesson 3 DETERMINING THE VOLUME OF ANY RIGHT RECTANGULAR PRISM

NOTES

Since a rectangular prism has a rectangular base, and all the edges of a rectangular prism meet at right angles (90° angles), all rectangular prisms are right rectangular prisms.

Volume is the amount of space that a rectangular prism holds. Volume is measured in cubic units (units^3), such as cubic centimetres (cm^3) and cubic metres (m^3).

USING MODELS TO EXPLAIN THE VOLUME OF RIGHT RECTANGULAR PRISMS

The volume of a right rectangular prism is the number of cubic units needed to fill the space of the prism without any overlapping of cubic units or spaces left between cubic units.

A cubic unit is a cube with equal side lengths of 1 unit on all six sides.

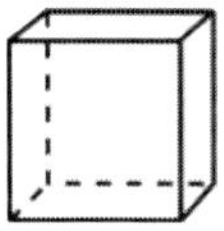
= 1 cubic unit

Volume can be determined by counting cubic units, skip counting, repeated addition, or multiplication.

Example

Jari uses cubic units to measure the volume of this right rectangular prism. He starts by filling the base of the prism with three rows of cubic units. Each row is made up of three cubic units.

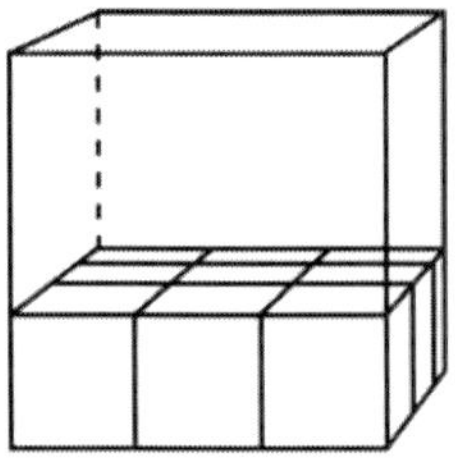

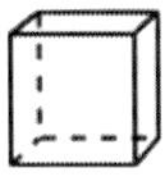
= 1 cubic unit (unit^3)

When Jari fills the rectangular prism to the top with cubic units, what will be the volume of the prism?

NOTES

Solution

Step 1
Find the number of layers of cubes needed.

Based on the size of the cubic units, two more layers of cubes will be needed to fill the rectangular prism. That means that the height of the prism will be 3 cubic units.

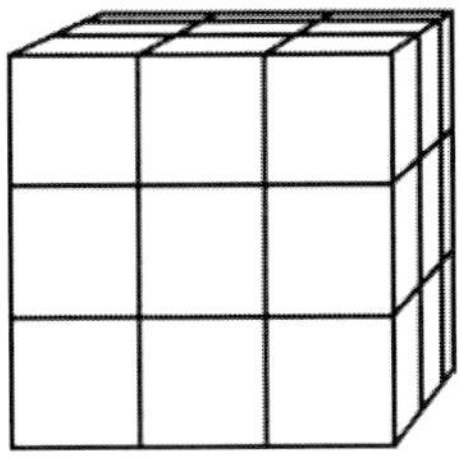

Step 2
Explain how you can find the total number of cubes used.

Start by counting the number of cubic units needed to cover the base.
1, 2, 3,… 9

Since there are three layers, skip count by 9 three times.
9, 18, 27

Another method is to add three 9s.
$9+9+9=27$

A quick method is to multiply the number of cubes in the base (the area of the base) by the number of cubes in the height.
$9\times 3=27$

Another quick method is to multiply the number of units in the length of the base (3) by the number of units in the width of the base (3) by the number of units in the height (3) of the prism.
$3\times 3\times 3=27$

The volume of Jari's right rectangular prism is 27 units^3.

NOTES

Time to Try 1

Mila measures the given rectangular prism by filling it with cm cubes.

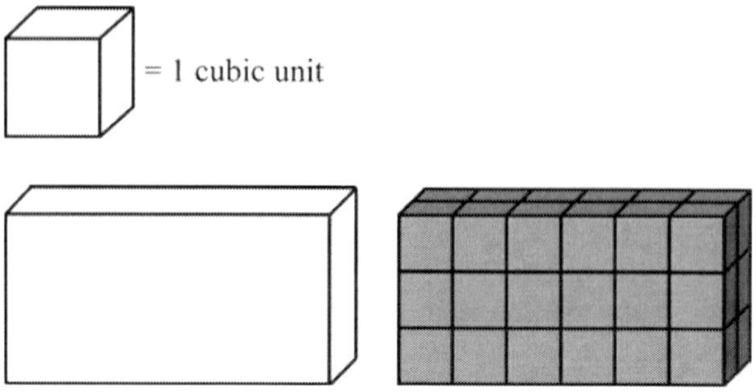

Explain one method of finding the volume of the right rectangular prism.

GENERALIZING A RULE (FORMULA) FOR THE VOLUME OF A RIGHT RECTANGULAR PRISM

A right rectangular prism has three dimensions: let l represent the number of cubic units in the length of the base; let w represent the number of cubic units in the width of the base; and let h represent the number of cubes in the height (number of layers) of the prism.

Since the quickest method for finding the volume of a rectangular prism is to use multiplication rather than repeated addition or skip counting, you can develop a formula like $V = l \times w \times h$.

Another method for finding the volume is to first find the area of the base, then multiply the area of the base by the height of the prism. To find the volume using this method, you can develop a formula for volume using the formula for area.

V = area of base × height of prism

$V = A \times h$

SOLVING PROBLEMS INVOLVING THE VOLUME OF A RIGHT RECTANGULAR PRISM

When using a formula to find volume, substitute the numbers for the length of the base, the width of the base, and the height of the prism into the formula. Be sure to include the correct unit of measure in your answer. Recall that volume is always measured in cubic units (units^3).

NOTES

Example

The following diagram represents the size of a storage shed that Mr. Barnes built.

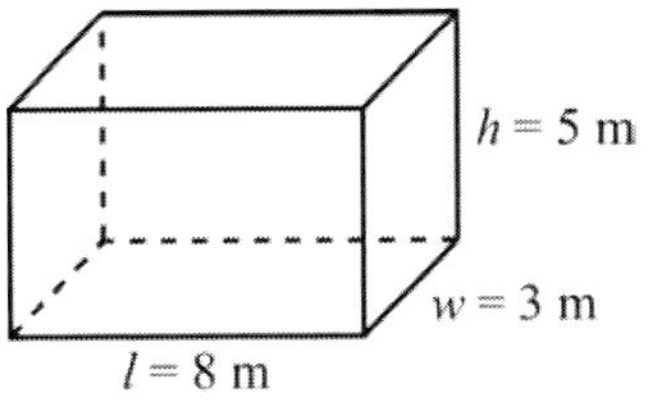

Use a formula to find the volume of the storage shed.

Solution

Since the length, width, and height measurements are given, you can use a formula like $V = l \times w \times h$.

Substitute the numbers for the length (8), the width (3), and the height (5) into the formula.

$$V = 8 \times 3 \times 5$$
$$= 120 \text{ m}^3$$

The volume of the storage shed is 120 m^3.

Time to Try 2

The area of the base of a right rectangular prism is 25 cm^2. The height of the prism is 6 cm. Use a formula to find the volume of the prism.

PRACTICE EXERCISES

Use the following diagram to answer the next question.

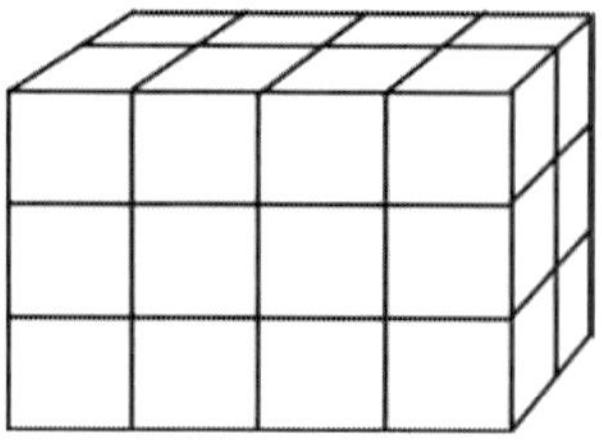

1. Explain one method for finding the volume of a right rectangular prism like the one shown.

Use the following information to answer the next question.

Lara has a box that she uses to store notes she receives from friends. The size of the box is shown in the given diagram.

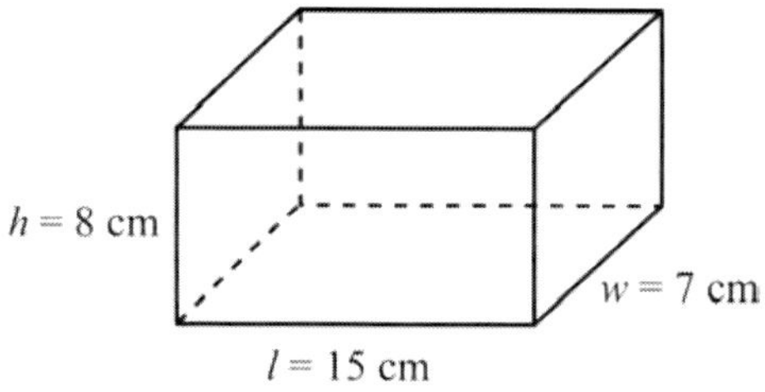

2. Use a formula to find the volume of the box.

Use the following diagram to answer the next question.

Ryder used 1 cm cubic units (1 cm^3) to measure the volume of this rectangular prism. Each layer has a length of 12 cubes and a width of 10 cubes.

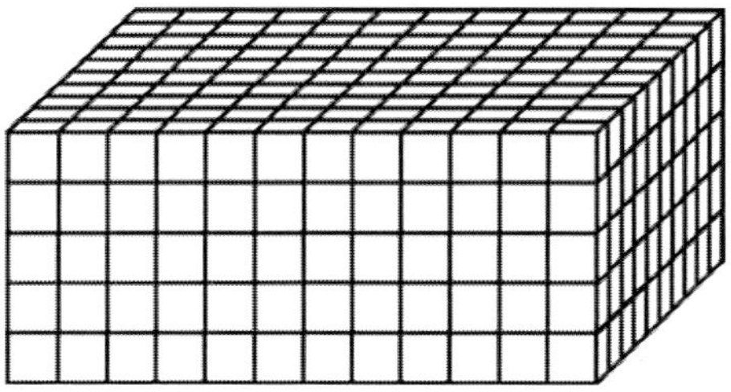

3. Use a formula to find the volume of the rectangular prism.

4. The length and width of the base of a pencil box is 5 cm and 6 cm, respectively. The height of the pencil box is 10 cm. Use a formula to find the volume of the prism.

REVIEW SUMMARY

- The perimeter of any polygon is the total distance around the shape of the polygon.
- Perimeter can be found by adding all the side lengths together.
- Some rules or formulas that can be developed for perimeter are as follows:
  - $P = s + s + s + s$
  - $P = l + w + l + w$
  - $P = (2 \times l) + (2 \times w)$
  - $P = 2(l + w)$
- The area of a rectangle is the measure of the space inside the rectangle.
- Area is expressed in square units, like cm^2 and m^2.
- Area can be found by multiplying the length of the rectangle by the width of the rectangle.
- A rule or formula that can be developed for area is $A = l \times w$.
- The volume of a right rectangle prism is the amount of space that the prism holds.
- Volume can be found by multiplying the length of the base by the width of the base by the height of the prism.
- Some rules or formulas that can be developed for volume are as follows:
  - $V = l \times w \times h$
  - $V = A \times h$

PRACTICE TEST

Use the following information to answer the next two questions.

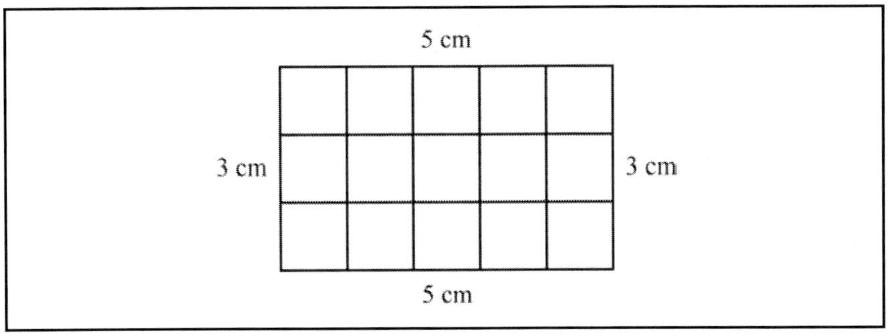

1. Explain how you can use the centimetre lengths to find the perimeter of the given rectangle.

2. Explain one method for using square units to find the area of the given rectangle.

Use the following diagrams to answer the next question.

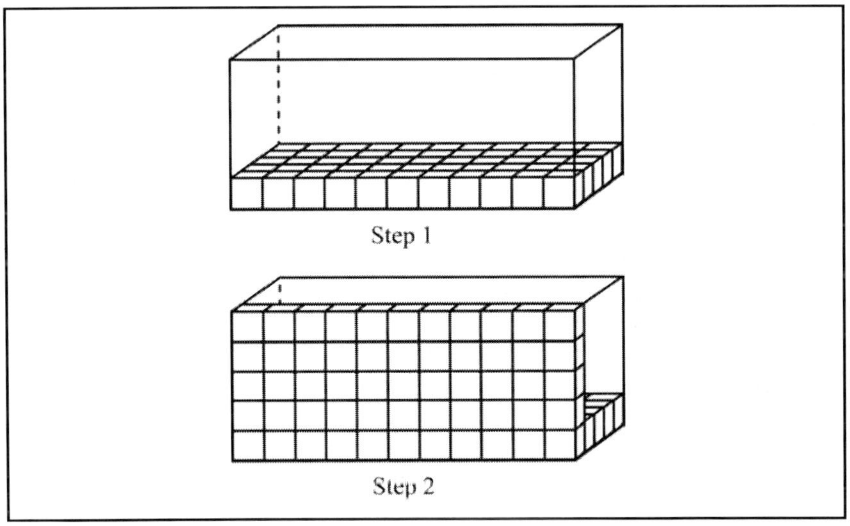

3. Explain how you can use cubic units to find the volume of the given rectangular prism.

Use the following diagram to answer the next two questions.

Brad's favourite game board looks like this:

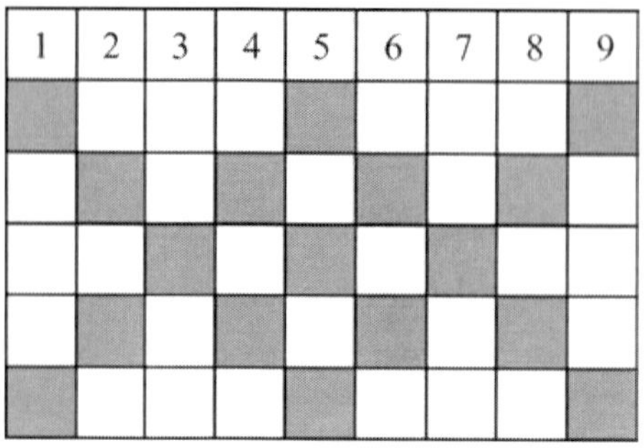

4. Use a formula to find the perimeter of the game board.

5. Use a formula to find the area of the game board.

Use the following diagram to answer the next question.

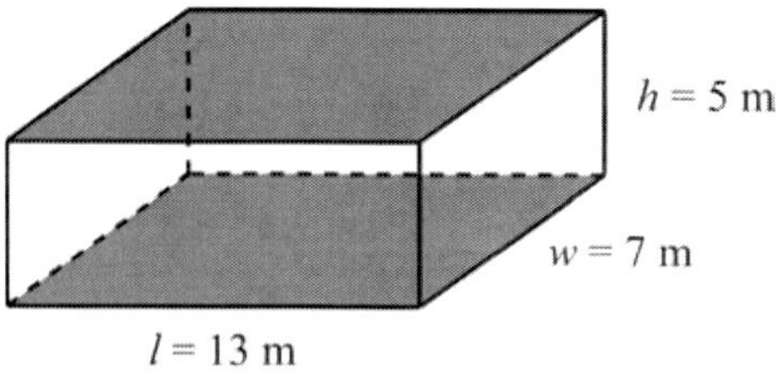

6. Use a formula to find the volume of the rectangular prism.

Use the following information to answer the next question.

Rob draws and labels this shape:

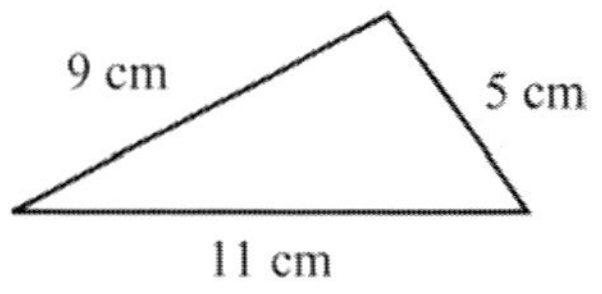

7. Use a formula to find the perimeter of the shape.

8. A rectangular park has a length of 40 m and a width of 30 m. What is the area of the park?

9. A rectangular water tank has a length of 5 m, a width of 4 m, and a height of 3 m. What is the volume of the tank?

NOTES

PROBLEM SOLVING

When you are finished this unit, you will be able to…

- Select an operation with which to solve a given problem
- Solve problems involving whole numbers and decimal numbers
- Estimate the solution to a given problem
- Determine the reasonableness of an estimated solution
- Use technology to solve problems involving large numbers

Lesson	Page	Completed on
1. Steps for Problem Solving	378	
2. Estimating Solutions	381	
3. Problem Solving with Large Numbers	385	
Review Summary	388	
Practice Test	389	
Answers and Solutions	at the back of the book	

PREREQUISITE SKILLS AND KNOWLEDGE

Prior to starting this unit, you should be able to…

- Select and use an estimation strategy (e.g., front-end rounding, compensation, compatible numbers) for a given problem
- Select and use mental strategies and number properties to solve basic multiplication and division facts
- Use mental strategies to solve multiplication problems involving whole numbers
- Multiply two-digit whole numbers by two-digit whole numbers
- Divide three-digit whole numbers by one-digit whole numbers
- Add and subtract decimal numbers to thousandths
- Multiply and divide decimal numbers by one-digit whole numbers

Lesson 1 STEPS FOR PROBLEM SOLVING

NOTES

It is important to pick the right operation with which to solve a problem. There are four main mathematical operators: addition (+), subtraction (–), multiplication (×), and division (÷).

A problem will often have keywords that can help you decide which operation to use.

Addition

Addition problems involve groups of different sizes being combined. Some keywords that point to addition include *in all*, *total*, *altogether*, *combined*, *increased*, *above*, and *gained.*

Subtraction

Subtraction problems involve groups of different sizes being taken away from an original, larger group. Some keywords that point to subtraction include *decreased*, *discount*, *fewer*, *left*, *difference*, *remains*, *loss*, and *owed.*

Multiplication

Multiplication problems involve groups of the same size being combined. Some keywords that point to multiplication include *product*, *of*, *times*, *double*, *triple*, *total*, *altogether*, and *in all.*

Division

Division problems involve a number being split into equal-sized groups. Some keywords that point to division include *each*, *split*, *divided*, *half*, *third*, *quarter*, *cut*, and *distributed.*

Be careful not to confuse addition problems with multiplication problems, or subtraction problems with division problems. Remember that multiplication and division problems always involve groups of equal size.

Read the question carefully and decide which operation is needed to solve the problem. Then, write an expression that can be used to solve the problem. After solving the problem, write a sentence to answer the question.

There is often more than one correct way to solve a problem.

NOTES

Example

Jeff has three cords. Each cord is 11.63 metres in length. How long are the three cords altogether?

Solution

Step 1
Determine the operation.

The three cords are of equal length. The keyword *altogether* means that multiplication can be used.

Step 2
Write an expression to solve the problem.

There are three cords. Each cord is 11.63 m long. The total length of the cords is given by 3 × 11.63 or 11.63 × 3.

This problem could also be solved by adding 11.63 + 11.63 + 11.63.

Step 3
Solve the problem.

$$\begin{array}{r} 11.63 \\ \times \quad 3 \\ \hline 34.89 \end{array}$$

The total length of the cords is 34.89 m.

Time to Try 1

Rachel baked 24 cookies. She gave away 10 cookies to her neighbour and 6 cookies to her best friend. How many cookies does Rachel have left?

PRACTICE EXERCISES

Use the following information to answer the next question.

Jane bought 6 boxes of crackers for a party. Each box had a mass of 7.48 grams.

1. What was the total mass of the 6 boxes?

Use the following information to answer the next question.

Lacey had $17.45 in her purse. She spent $4.55 on a magazine and $1.25 on a bottle of pop.

2. How much money does Lacey have left?

Use the following information to answer the next question.

Ms. Carlson bought 9.8 metres of cloth to make pillow covers. She cut the cloth into 7 equal lengths.

3. How long is each piece of cloth?

Use the following information to answer the next question.

Colton had 148 stamps in his collection. His grandma gave him 12 stamps to add to his collection. His brother gave him 7 stamps.

4. How many stamps does Colton now have in total?

Lesson 2 ESTIMATING SOLUTIONS

NOTES

In some situations, an exact answer is not required. Use estimation strategies such as front-end estimation or dropping the decimal digits to find an approximate solution.

Remember to first identify which operation will be required to solve the problem.

Example

Justin drove a total of 315.6 kilometres in 6 days, driving the same distance each day.

a) Applying the strategy of using the front-end number, estimate how many kilometres Justin drove each day.

Solution

Step 1
Determine the operation.

Justin drove an equal amount each day. The keyword *each* means that division can be used.

Step 2
Write an expression to solve the problem using estimation.

Use the front-end digit of the decimal number.
$315.6 \rightarrow 300$

The approximate distance Justin drove each day is given by $300 \div 6$.

Step 3
Solve the problem using estimation.

$$
\begin{array}{r}
50 \\
6\overline{)300} \\
\underline{30} \\
00 \\
\underline{0} \\
0
\end{array}
$$

Justin drove approximately 50 km each day.

NOTES

b) Determine if the estimate is reasonable by comparing it to the actual distance driven each day.

Solution

Step 1
Write an expression to solve the problem without estimation.

The distance Justin drove each day is given by $315.6 \div 6$.

Step 2
Solve the problem without estimation.

$$
\begin{array}{r}
52.6 \\
6\overline{)315.6} \\
\underline{30} \\
15 \\
\underline{12} \\
3\ 6 \\
\underline{3\ 6} \\
0
\end{array}
$$

Justin drove 52.6 km each day.

Step 3
Compare the estimated solution to the actual solution.

Subtract the estimated distance from the actual distance to find the difference between the two solutions.

$$
\begin{array}{r}
52.6 \\
\underline{-\ 50} \\
2.6
\end{array}
$$

There is only 2.6 km difference between the estimated solution and the actual solution, which in this case is not a significant amount.

Therefore, the estimate is reasonable.

When deciding if an estimate is reasonable, it is important to consider the particular situation. For example, a precise solution may be more important when money is involved.

NOTES

Time to Try 1

Evelyn saved up $23.00 of her allowance. She wants to use the money to decorate her room. Evelyn wants to buy a poster that costs $12.45 with tax and a figurine that costs $11.90 with tax.

a) Applying the strategy of dropping the decimal digits, estimate the total cost of both items.

b) Determine if the estimate is reasonable by comparing it to the actual cost.

PRACTICE EXERCISES

Use the following information to answer the next two questions.

Jill swam 9 laps. Each lap took 2.61 minutes to complete.

1. Applying the strategy of dropping the decimal digits, estimate the total time Jill took to complete the laps.

2. Determine if the estimate is reasonable by comparing it to the actual time it took Jill to complete the laps.

Use the following information to answer the next two questions.

Thomas has 218 building blocks in a basket. He uses 37 blocks to build a car and 103 blocks to build a house.

3. Applying the strategy of using the front-end number, estimate the number of blocks that are left in the basket.

4. Determine if the estimate is reasonable by comparing it to the actual number of blocks left in Thomas' basket.

Lesson 3 PROBLEM SOLVING WITH LARGE NUMBERS

NOTES

When problems involve very large numbers, it is often best to use technology, such as a calculator, to help solve the problem. Depending on the type of calculator used, you may have to press the operator button ([+], [−], [×], or [÷]) before entering the digits for the calculation.

Remember to read each question carefully and decide which operation will be necessary to solve the problem.

Example

The average distance from Earth to the moon is 384 392 km. A space shuttle leaves Earth and reaches the moon 5 days later. The shuttle travelled the same distance each day.

How far did the space shuttle travel in 1 day?

Solution

Step 1
Determine the operation.

Each day of the journey covered an equal distance. The keyword *each* means that division can be used.

Step 2
Write an expression to solve the problem.

The total distance is 384 392 km. The trip is divided into five equal parts.

The distance travelled in one day is given by 384 392 ÷ 5.

Step 3
Solve the problem.

Because the problem involves a very large number, use a calculator to perform the division.

[3] [8] [4] [3] [9] [2] [÷] [5] [=]

The shuttle travelled 76 878.4 km each day.

NOTES

Time to Try 1

Rebecca, Meghan, and David were competing as a team in a video game competition. Rebecca scored 35 295 218 points. Meghan scored 51 021 754 points. David scored 29 717 698 points.

What was the combined score of the three-person team?

PRACTICE EXERCISES

Use the following information to answer the next question.

Daniela is in charge of a fundraiser at her community centre. Her goal is to raise $10 000 for a national charity. Daniela's team collects $2 922.76 from a community garage sale, and $3 391.21 from a charity auction.

1. How much money does Daniela have left to collect before she reaches her goal?

Use the following information to answer the next question.

Kody is going on a road trip to Newfoundland. He decided to divide his trip into 12 days of travel. He will travel 538.54 km each day of the trip.

2. In total, how far will Kody travel to reach his destination?

Use the following information to answer the next question.

A chocolate factory produces 1 423 037 chocolate bars in one week. The factory owns 11 large trucks that can be used to transport the chocolate.

3. If the chocolate bars are split equally among the 11 trucks, how many bars will each truck need to carry?

REVIEW SUMMARY

- Look for keywords in a problem to help you identify which operation to use.
  - Addition keywords: *in all*, *total*, *altogether*, *combined*, *increased*, *above*, and *gained*
  - Subtraction keywords: *decreased*, *discount*, *fewer*, *left*, *difference*, *remains*, *loss*, and *owed*
  - Multiplication keywords: *product*, *of*, *times*, *double*, *triple*, *total*, *altogether*, and *in all*
  - Division keywords: *each*, *split*, *divided*, *half*, *third*, *quarter*, *cut*, and *distributed*
- Multiplication and division problems always involve groups of equal size.
- Use estimation strategies such as front-end estimation or dropping the decimal digits to find an approximate solution.
- When deciding if an estimate is reasonable, consider the particular situation at hand.
- When problems involve very large numbers, it is best to use technology, such as a calculator, to help solve the problem.

PRACTICE TEST

Use the following information to answer the next question.

Beth has a sticker book with 25 pages. Each page has 11 stickers on it.

1. What is the total number of stickers in Beth's sticker book?

Use the following information to answer the next question.

Nicholas has 823.5 g of birdseed. He fills the bird feeder in his front yard with 106.2 g of birdseed and the bird feeder in his backyard with 115.6 g of birdseed.

2. How much birdseed does Nicholas have left?

Use the following information to answer the next question.

Marissa is filling up the classroom fish tanks. She fills one fish tank with 1.42 L of water. She fills a second fish tank with 1.29 L of water, and a third fish tank with 1.61 L of water.

3. How much water did Marissa use in total?

Use the following information to answer the next question.

Dan has \$247.50 to spend on presents. He wants to split the money evenly among 5 family members.

4. How much money can Dan spend on each person's gift?

Use the following information to answer the next two questions.

Lorraine needs 45.2 yards of fabric to make slipcovers for her furniture. She has 25.8 yards of light-green fabric and 21.3 yards of dark-green fabric.

5. Applying the strategy of using the front-end digit, estimate the total amount of fabric Lorraine has with which to make her slipcover.

6. Determine if the estimate is reasonable by comparing it to the actual amount of fabric Lorraine has.

Use the following information to answer the next question.

The distance between Edmonton, Alberta, and Sydney, Australia, is 13 288.9 km. An airplane leaves Edmonton and travels 5 519.6 km toward Sydney.

7. How much farther does the airplane have left to travel before reaching Sydney?

S N A P

Student Notes and Problems

ANSWERS AND SOLUTIONS

CASTLE ROCK
RESEARCH CORP

PLACE VALUES

Lesson 1—Place Value and Large Numbers

TIME TO TRY ANSWERS AND SOLUTIONS

1.

Billions			Millions			Thousands			Ones		
H	T	O	H	T	O	H	T	O	H	T	O
	2	8	4	1	2	5	6	7	1	0	3

The digit 2 is in the ten billions place.
The value is 2 × 10 000 000 000, or
20 000 000 000.

The digit 8 is in the one billions place.
The value is 8 × 1 000 000 000, or
8 000 000 000.

The digit 4 is in the hundred millions place.
The value is 4 × 100 000 000, or
400 000 000.

The digit 1 is in the ten millions place.
The value is 1 × 10 000 000, or 10 000 000.

The digit 2 is in the one millions place.
The value is 2 × 1 000 000, or 2 000 000.

The digit 5 is in the hundred thousands place.
The value is 5 × 100 000, or 500 000.

The digit 6 is in the ten thousands place.
The value is 6 × 10 000, or 60 000.

The digit 7 is in the one thousands place.
The value is 7 × 1 000, or 7 000.

The digit 1 is in the hundreds place.
The value is 1 × 100, or 100.

The digit 0 is in the tens place. The value is 0. You still need to put the 0 in the chart to keep the other digits in the right place.

The digit 3 is in the ones place. The value is 3 × 1, or 3.

PRACTICE EXERCISES ANSWERS AND SOLUTIONS

1.

Millions			Thousands			Ones		
H	T	O	H	T	O	H	T	O
	2	3	4	0	3	7	9	6

The digit 2 is in the ten millions place.
The value is 2 × 10 000 000, or 20 000 000.

The digit 3 is in the one millions place.
The value is 3 × 1 000 000, or 3 000 000.

The digit 4 is in the hundred thousands place.
The value is 4 × 100 000, or 400 000.

The digit 0 is in the ten thousands place.
The value is 0. You still need to put the 0 in the chart to keep the digits in the right place.

The digit 3 is in the one thousands place.
The value is 3 × 1 000, or 3 000.

The digit 7 is in the hundreds place.
The value is 7 × 100, or 700.

The digit 9 is in the tens place. The value is 9 × 10, or 90.

The digit 6 is in the ones place. The value is 6 × 1, or 6.

3.

Billions			Millions			Thousands			Ones		
H	T	O	H	T	O	H	T	O	H	T	O
8	1	7	2	0	3	9	8	6	0	1	4

The digit 8 is in the hundred billions place.
The value is 8 × 100 000 000 000, or
800 000 000 000.

The digit 1 is in the ten billions place.
The value is 1 × 10 000 000 000, or
10 000 000 000.

The digit 7 is in the one billions place.
The value is 7 × 1 000 000 000, or
7 000 000 000.

The digit 2 is in the hundred millions place. The value is 2 × 100 000 000, or 200 000 000.

The digit 0 is in the ten millions place. The value is zero. You still need to put the 0 in the chart to keep the digits in the right place.

The digit 3 is in the one millions place. The value is 3 × 1 000 000, or 3 000 000.

The digit 9 is in the hundred thousands place. The value is 9 × 100 000, or 900 000.

The digit 8 is in the ten thousands place. The value is 8 × 10 000, or 80 000.

The digit 6 is in the one thousands place. The value is 6 × 1 000, or 6 000.

The digit 0 is in the hundreds place. The value is zero. You still need to put the 0 in the chart to keep the digits in the right place.

The digit 1 is in the tens place. The value is 1 × 10, or 10.

The digit 4 is in the ones place. The value is 4 × 1, or 4.

Lesson 2—Place Value and Decimal Numbers

TIME TO TRY ANSWERS AND SOLUTIONS

1.

Ones	**.**	**Parts of a Whole**				
O	**.**	**Tth**	**Hth**	**Thth**	**T Thth**	**H Thth**
4	.	9	0	1	2	8

The digit 4 is in the ones place. The value is 4 × 1, or 4.

The digit 9 is in the tenths place. The value is 9 × 0.1, or 0.9.

The digit 0 is in the hundredths place. The value is zero. You still need to put it in the chart to keep the other digits in the right place.

The digit 1 is in the thousandths place. The value is 1 × 0.001, or 0.001.

The digit 2 is in the ten-thousandths place. The value is 2 × 0.0001, or 0.0002.

The digit 8 is in the hundred-thousandths place. The value is 8 × 0.000 01, or 0.000 08.

PRACTICE EXERCISES ANSWERS AND SOLUTIONS

1.

Ones	**.**	**Parts of a Whole**			
O	**.**	**Tth**	**Hth**	**Thth**	**T Thth**
2	.	4	0	1	6

The digit 2 is in the ones place. The value is 2 × 1, or 2.

The digit 4 is in the tenths place. The value is 4 × 0.1, or 0.4.

The digit 0 is in the hundredths place. The value is zero. You still need to put the 0 in the chart to keep the digits in the right place.

The digit 1 is in the thousandths place. The value is 1 × 0.001, or 0.001.

The digit 6 is in the ten-thousandths place. The value is 6 × 0.0001, or 0.0006.

3.

Ones	**.**	**Parts of a Whole**				
O	**.**	**Tth**	**Hth**	**Thth**	**T Thth**	**H Thth**
8	.	6	3	2	4	1

The digit 8 is in the ones place. The value is 8 × 1, or 8.

The digit 6 is in the tenths place. The value is 6 × 0.1, or 0.6.

The digit 3 is in the hundredths place.
The value is 3 × 0.01, or 0.03.

The digit 2 is in the thousandths place.
The value is 2 × 0.001, or 0.002.

The digit 4 is in the ten-thousandths place.
The value is 4 × 0.0001, or 0.0004.

The digit 1 is in the hundred-thousandths place. The value is 1 × 0.000 01, or 0.000 01.

Lesson 3—Reading and Writing Numbers

TIME TO TRY ANSWERS AND SOLUTIONS

1. Work from left to right.

First, read the billions period. This is written in words and read as *fifty-one*, followed by the word *billion*.

Next, read the millions period. This is written in words and read as *one hundred twenty-nine*, followed by the word *million*.

Next, read the thousands period. This is written in words and read as *twenty-two*, followed by the word *thousand*.

Finally, read the ones period. This is written in words and read as *four hundred thirty-nine*.

The number 51 129 022 439 is read and written as *fifty-one billion, one hundred twenty-nine million, twenty-two thousand, four hundred thirty-nine*.

2. Work from left to right.

Start by writing the billions period in numeral form.
sixteen billion → 16

Leave a space, then write the millions period in numeral form.
one hundred forty five million → 145

Leave a space, then write the thousands period in numeral form.
sixty-two thousand → 062

Leave a space, then write the ones period in numeral form.
nine hundred two → 902

Written in numeral form, the number is 16 145 062 902.

3. **a)** Work from left to right.

First, read the whole numbers. These are written in words and read as *six hundred thirty-one*, followed by the word *and*.

Next, read the decimal numbers. These are written in words and read as *two*, followed by the place value *hundredths*.

The number 631.02 is read and written as *six hundred thirty-one and two hundredths*.

b) Work from left to right.

The whole number portion of the number is zero.

Read the decimal numbers, followed by the place value of the last digit.
The number 0.951 62 is read and written as *ninety-five thousand, one hundred sixty-two hundred-thousandths*.

4. **a)** Work from left to right

Start by writing the whole numbers in numeral form.
The whole number is zero → 0

Write a decimal point, followed by the decimal numbers in numeral form.
two hundred five thousandths → .205

Written in numeral form, the number is 0.205.

b) Work from left to right

Start by writing the whole numbers in numeral form.
seventy-three → 73

Write a decimal point, followed by the decimal numbers in numeral form.
five ten-thousandths → .0005

Written in numeral form, the number is 73.0005.

PRACTICE EXERCISES ANSWERS AND SOLUTIONS

1. First, read the billions period. This is written in words and read as *two hundred sixty-one*, followed by the word *billion*.

Next, read the millions period. This is written in words and read as *five*, followed by the word *million*.

Next, read the thousands period. This is written in words and read as *sixty-five*, followed by the word *thousand*.

Finally, read the ones period. This is written in words and read as *twenty-three*.

The number 261 005 065 023 is read and written as *two hundred sixty-one billion, five million, sixty-five thousand, twenty-three*.

3. The whole number portion of the number is zero.

Read the decimal numbers, followed by the place value of the last digit. The number 0.702 33 is read and written as *seventy thousand, two hundred thirty-three hundred-thousandths*.

5. Start by writing the whole numbers in numeral form.
five hundred sixty-seven → 567

Write a decimal point, followed by the decimal numbers in numeral form.
forty-five hundredths → .45

Written in numeral form, the number is 567.45.

Practice Test

ANSWERS AND SOLUTIONS

1.

Ones	.	Parts of a Whole				
O	**.**	**Tth**	**Hth**	**Thth**	**T Thth**	**H Thth**
6	.	5	9	7	1	3

The digit 6 is in the ones place. The value is 6 × 1, or 6.

The digit 5 is in the tenths place. The value is 5 × 0.1, or 0.5.

The digit 9 is in the hundredths place. The value is 9 × 0.01, or 0.09.

The digit 7 is in the thousandths place. The value is 7 × 0.001, or 0.007.

The digit 1 is in the ten-thousandths place. The value is 1 × 0.0001, or 0.0001.

The digit 3 is in the hundred-thousandths place. The value is 3 × 0.000 01, or 0.000 03.

3.

Billions			Millions			Thousands			Ones		
H	**T**	**O**	**H**	**T**	**O**	**H**	**T**	**O**	**H**	**T**	**O**
	6	2	4	0	3	9	7	2	1	8	5

The digit 6 is in the ten billions place. The value is 6 × 10 000 000 000, or 60 000 000 000.

The digit 2 is in the one billions place. The value is 2 × 1 000 000 000, or 2 000 000 000.

The digit 4 is in the hundred millions place. The value is 4 × 100 000 000, or 400 000 000.

The digit 0 is in the ten millions place. The value is zero. You still need to put the 0 in the chart to keep the digits in the right place.

The digit 3 is in the one millions place. The value is 3 × 1 000 000, or 3 000 000.

The digit 9 is in the hundred thousands place. The value is 9 × 100 000, or 900 000.

The digit 7 is in the ten thousands place. The value is 7 × 10 000, or 70 000.

The digit 2 is in the one thousands place. The value is 2 × 1 000, or 2 000.

The digit 1 is in the hundreds place. The value is 1 × 100, or 100.

The digit 8 is in the tens place. The value is 8 × 10, or 80.

The digit 5 is in the ones place. The value is 5 × 1, or 5.

5. First, read the thousands period. This is written in words and read as *six hundred thirty-five*, followed by the word *thousand.*

Next, read the ones period. This is written in words and read as *eight hundred fifteen.*

The number 635 815 is read and written as *six hundred thirty-five thousand, eight hundred fifteen.*

7. First, read the whole numbers. These are written in words and read as *twenty-three*, followed by the word *and.*

Next, read the decimal numbers. These are written in words and read as *eleven*, followed by the place value *hundredths.*

The number is read and written as *twenty-three and eleven hundredths.*

9. Start by writing the billions period in numeral form.
twenty billion → 20

Leave a space, then write the millions period in numeral form.
four hundred thirty-one million → 431

Leave a space, then write the thousands period in numeral form.
eighteen thousand → 018

Leave a space, then write the ones period in numeral form.
two hundred → 200

Written in numeral form, the number is 20 431 018 200.

11. Start by writing the whole numbers in numeral form.
six hundred sixty-eight → 668

Write a decimal point, followed by the decimal numbers in numeral form.
fourteen hundredths → .14

Written in numeral form, the number is 668.14

WORKING WITH DECIMALS

Lesson 1—Multiplication of Decimal Numbers by Multipliers from 0 to 9

TIME TO TRY ANSWERS AND SOLUTIONS

1. Calculate the product as a whole number.

$$\begin{array}{r} 5\,2213 \\ \times \quad 3 \\ \hline 15\,6639 \end{array}$$

There are four digits to the right of the decimal point in $5.\underline{2213}$. Therefore, there will be four digits to the right of the decimal point in the product.

$15\underline{6639}$

$$\begin{array}{r} 5.2213 \\ \times \quad 3 \\ \hline 15.6639 \end{array}$$

PRACTICE EXERCISES ANSWERS AND SOLUTIONS

1. Calculate the product as a whole number.

$$\begin{array}{r} 64\,33 \\ \times \quad 3 \\ \hline 192\,99 \end{array}$$

There are two digits to the right of the decimal point in $64.\underline{33}$. Therefore, there will be two digits to the right of the decimal point in the product.

$$\begin{array}{r} 64.33 \\ \times \quad 3 \\ \hline 192.99 \end{array}$$

3. Calculate the product as a whole number.

$$\begin{array}{r} 0\,852 \\ \times \quad 2 \\ \hline 1\,704 \end{array}$$

There are three digits to the right of the decimal point in $0.\underline{852}$. Therefore, there will be three digits to the right of the decimal point in the product.

$$\begin{array}{r} 0.852 \\ \times \quad 2 \\ \hline 1.704 \end{array}$$

5. Since there are three digits to the right of the decimal point in the decimal number, there must also be three digits to the right of the decimal point in the product.

$105.\underline{0}\,\underline{7}\,\underline{1} \times 3 = 315.\underline{2}\,\underline{1}\,\underline{3}$

$315213 \rightarrow 315.213$

Lesson 2—Dividing Decimals

TIME TO TRY ANSWERS AND SOLUTIONS

1. Calculate the quotient as a whole number.

$$\begin{array}{r} 2\,617 \\ 3\overline{)7\,851} \\ \underline{6} \\ 1\,8 \\ \underline{1\,8} \\ 05 \\ \underline{3} \\ 21 \\ \underline{21} \\ 0 \end{array}$$

There are three digits to the right of the decimal point in $7.\underline{851}$. Therefore, there will be three digits to the right of the decimal point in the quotient.

$2\underline{617}$

$$\begin{array}{r} 2.617 \\ 3\overline{)7.851} \end{array}$$

PRACTICE EXERCISES ANSWERS AND SOLUTIONS

1. Calculate the quotient as a whole number.

$$\begin{array}{r} 2\,51 \\ 6\overline{)15\,06} \\ \underline{12} \\ 3\,0 \\ \underline{3\,0} \\ 06 \\ \underline{6} \\ 0 \end{array}$$

There are two digits to the right of the decimal point in 15.<u>06</u>. Therefore, there will be two digits to the right of the decimal point in the quotient.

2.51

$$\begin{array}{r} 2.51 \\ 6\overline{)15.06} \end{array}$$

3. Calculate the quotient as a whole number.

$$\begin{array}{r} 25\,7 \\ 8\overline{)205\,6} \\ \underline{16} \\ 45 \\ \underline{40} \\ 5\,6 \\ \underline{5\,6} \\ 0 \end{array}$$

There is one digit to the right of the decimal point in 205.<u>6</u>. Therefore, there will be one digit to the right of the decimal point in the quotient.

25.7

$$\begin{array}{r} 25.7 \\ 8\overline{)205.6} \end{array}$$

5. Since there is one digit to the right of the decimal point in the decimal number, there must also be one digit to the right of the decimal point in the quotient.

$106.\underline{2} \div 9 = 11.\underline{8}$

Bob's solution has two digits to the right of the decimal point, so he did not put the decimal point in the correct place.

Lesson 3—Estimation Strategies for Multiplication and Division of Decimals

TIME TO TRY ANSWERS AND SOLUTIONS

1. Drop the decimal point and the digits to the right of the decimal point.

$36.64 \rightarrow 36$

Perform the division.

$$\begin{array}{r} 18 \\ 2\overline{)36} \\ \underline{2} \\ 16 \\ \underline{16} \\ 0 \end{array}$$

The actual quotient of $36.64 \div 2$ is greater than 12.

2. Use the front-end digit of the decimal number.

$252.36 \rightarrow 200$

Perform the multiplication.

$$\begin{array}{r} 200 \\ \underline{\times\ \ 4} \\ 800 \end{array}$$

The actual product of 252.36×4 is greater than 800.

3. Change the decimal number to a whole number that is compatible with division by 3.

One possible method is to increase 11.65 to 12, which is divisible by 3.
11.65 → 12

Perform the division.

$$\begin{array}{r} 4 \\ 3\overline{)12} \\ \underline{12} \\ 0 \end{array}$$

The actual quotient of 11.65 ÷ 3 is less than 4.

PRACTICE EXERCISES ANSWERS AND SOLUTIONS

1. Drop the decimal point and the digits to the right of the decimal point.
201.54 → 201

Perform the multiplication.

$$\begin{array}{r} 201 \\ \underline{\times \quad 6} \\ 1\,206 \end{array}$$

The actual product of 201.54 × 6 is greater than 1 206.

3. Use the front-end digit of the decimal number.
28.512 → 20

Perform the multiplication.

$$\begin{array}{r} 20 \\ \underline{\times \; 9} \\ 180 \end{array}$$

The actual product of 28.512 × 9 is greater than 180.

5. Change the decimal number to a whole number that is compatible with multiplication by 8.

One possible method is to decrease 41.56 to 40, which is easy to multiply.
41.56 → 40

Perform the multiplication.

$$\begin{array}{r} 40 \\ \underline{\times \; 8} \\ 320 \end{array}$$

The actual product of 41.56 × 8 is greater than 320.

Practice Test

ANSWERS AND SOLUTIONS

1. Calculate the product as a whole number.

$$\begin{array}{r} 43\,12 \\ \underline{\times \quad 3} \\ 129\,36 \end{array}$$

There are two digits to the right of the decimal point in 43.<u>12</u>. Therefore, there will be two digits to the right of the decimal point in the product.

$$\begin{array}{r} 43.12 \\ \underline{\times \quad 3} \\ 129.36 \end{array}$$

3. Calculate the quotient as a whole number.

$$\begin{array}{r} 16\,29 \\ 2\overline{)32\,58} \\ \underline{2} \\ 12 \\ \underline{12} \\ 0\,5 \\ \underline{4} \\ 18 \\ \underline{18} \\ 0 \end{array}$$

There are two digits to the right of the decimal point in 32.<u>58</u>. Therefore, there will be two digits to the right of the decimal point in the quotient.

$$\begin{array}{r} 16.29 \\ 2\overline{)32.58} \end{array}$$

5. Drop the decimal point and the digits to the right of the decimal point.
$403.92 \rightarrow 403$

Perform the multiplication.

$$\begin{array}{r} 403 \\ \underline{\times \quad 7} \\ 2\,821 \end{array}$$

The actual product of 403.92×7 is greater than 2 821.

7. Change the decimal number to a whole number that is compatible with division by 6.

One possible method is to decrease 73.014 to 72, which is divisible by 6.
$73.014 \rightarrow 72$

Perform the division.

$$\begin{array}{r} 12 \\ 6\overline{)72} \\ \underline{6} \\ 12 \\ \underline{12} \\ 0 \end{array}$$

The actual quotient of $73.014 \div 6$ is greater than 12.

9. Since there is one digit to the right of the decimal point in the decimal number, there must also be one digit to the right of the decimal point in the quotient.
$29.\underline{4} \div 2 = 14.\underline{7}$

Alesha's solution has two digits to the right of the decimal point, so she did not put the decimal point in the correct place.

INTEGERS

Lesson 1—Understanding Integers

TIME TO TRY ANSWERS AND SOLUTIONS

1. Since the lake is frozen, its temperature must be below zero.

Therefore, the temperature of the lake would be represented by a negative integer.

2. Since the elevation is above sea level, the integer will be positive.

The elevation is represented by the integer +2 697 m.

3. Jackie's score is a positive integer. Therefore, her score was one above par.

PRACTICE EXERCISES ANSWERS AND SOLUTIONS

1. A tall mountain would have an elevation above sea level.

Therefore, the elevation would be represented by a positive integer.

3. Since the temperature is below zero, the integer will be negative.

The temperature is represented by the integer –10 C.

5. Since the score is above par, the integer will be positive.

The score is represented by the integer +4.

Lesson 2—Integers on a Number Line

TIME TO TRY ANSWERS AND SOLUTIONS

1. The integer +6 is positive. Positive integers get larger as you move to the right on a number line.

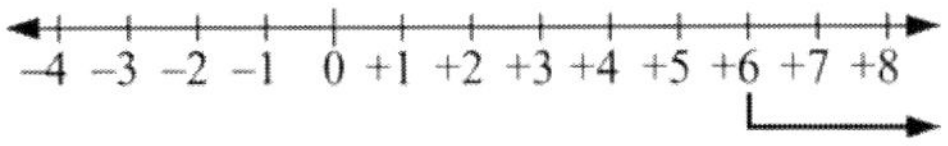

The integers +7 and +8 are larger than +6.

2. The integer –6 is negative. Negative integers get smaller as you move to the left on a number line.

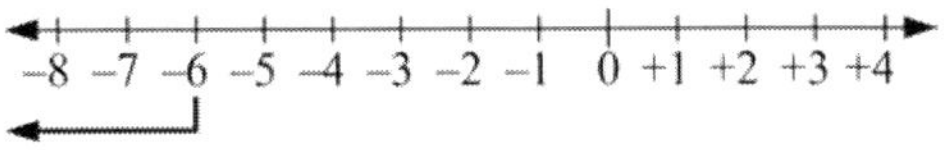

The integers –7 and –8 are smaller than –6.

3. Draw a number line, and place the integers on it.

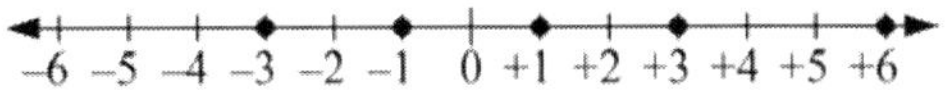

Of the given integers, –3 is the smallest because it is farthest to the left of 0.

The integer –1 is larger than –3 because it is closer to 0.

The integer +1 is larger than –1 because it is to the right of 0.
The integer +3 is larger than +1 because it is farther to the right of 0.

Of the given integers, +6 is the largest because it is farthest to the right of 0.

The integers in order from smallest to largest are –3, –1, +1, +3, and +6.

PRACTICE EXERCISES ANSWERS AND SOLUTIONS

1.

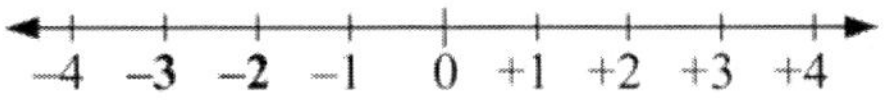

3. Draw a number line, and place the given integers on it.

Of the given integers, –7 is the smallest because it is farthest to the left of 0.

The integer –1 is larger than –7 because it is closer to 0.

The integer +1 is larger than –1 because it is to the right of 0.

The integer +3 is larger than +1 because it is farther to the right of 0.

Of the given integers, +5 is the largest because it is farthest to the right of 0.

The integers in order from smallest to largest are –7, –1, +1, +3, and +5.

5. Draw a number line, and place the given integers on it.

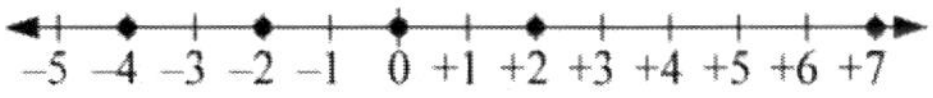

Of the given integers, –4 is the smallest because it is farthest to the left of 0.

The integer –2 is larger than –4 because it is closer to 0.

The integer 0 is next. It is not positive or negative.

The integer +2 is larger than 0 because it is to the right of 0.

Of the given integers, +7 is the largest because it is farthest to the right of 0.

The integers in order from smallest to largest are –4, –2, 0, +2, and +7.

Lesson 3—Comparing Integers

TIME TO TRY
ANSWERS AND SOLUTIONS

1. a) When comparing two negative integers, the integer that is farther to the left of zero is the smaller number.

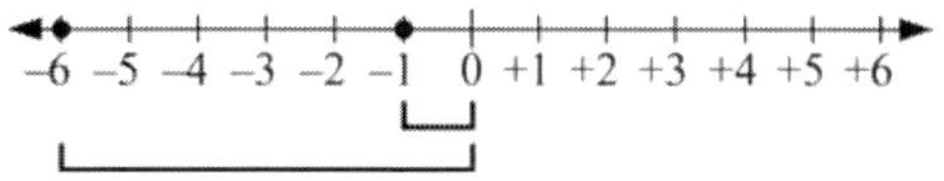

In this case, –1 is closer to zero than –6, so –1 is greater than –6.
–1 > –6

b) When comparing a negative integer with a positive integer, the negative integer is always less than the positive integer.

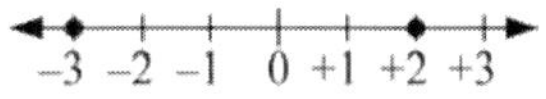

In this case, –3 is negative and +2 is positive, so –3 is less than +2.
–3 < +2

c) When comparing two positive integers, the integer that is farther to the right of zero is the greater number.

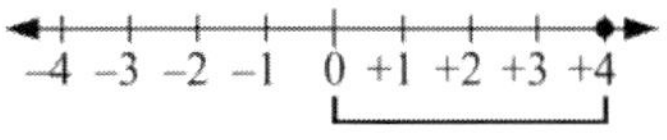

+4 and +4 are the same distance from zero because those numbers are equal.
+4 = +4

d) Negative integers always have a value less than 0.

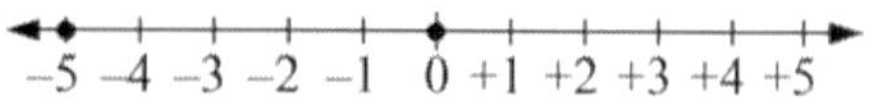

Therefore, –5 is less than 0.
–5 < 0

PRACTICE EXERCISES
ANSWERS AND SOLUTIONS

1. When comparing a negative integer with a positive integer, the negative integer is always less than the positive integer.

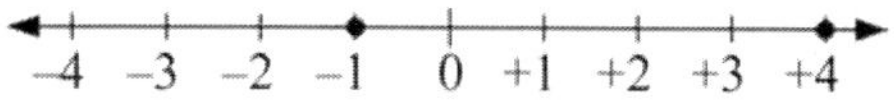

In this case, –1 is negative and +4 is positive, so –1 is less than +4.
–1 < +4

3. When comparing two negative integers, the integer that is closer to zero is the larger number.

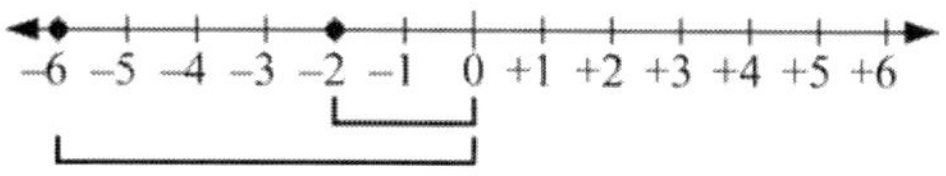

In this case, –2 is closer to zero than –6, so –2 is greater than –6.
–2 > –6

5. When comparing a positive integer with a negative integer, the positive integer is always greater than the negative integer.

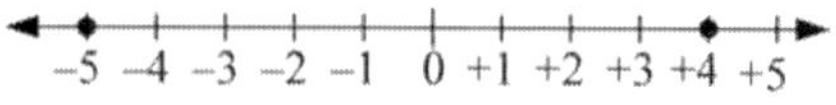

In this case, +4 is positive and –5 is negative, so +4 is greater than –5.
+4 > –5

Practice Test

ANSWERS AND SOLUTIONS

1.

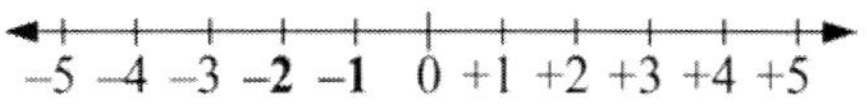

3. Since the elevation is below sea level, the integer will be negative.

The elevation is represented by the integer –140 m.

5. Draw a number line, and place the integers on it.

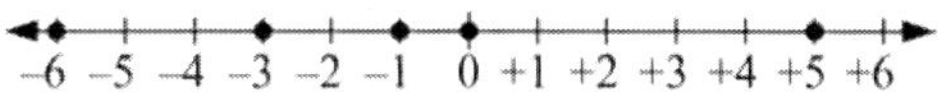

Of the given integers, –6 is the smallest because it is farthest to the left of 0.

The integer –3 is larger than –6 because it is closer to 0.

The integer –1 is larger than –3 because it is closer to 0.

The 0 is next. It is not positive or negative.

Of the given integers, +5 is the largest because it is the only number to the right of 0.

The integers in order from smallest to largest are –6, –3, –1, 0, and +5.

7. When comparing two positive integers, the integer that is closer to zero is the smaller number.

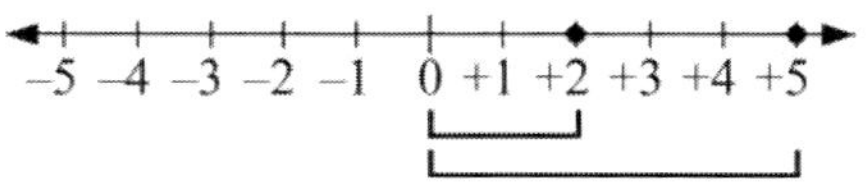

In this case, +2 is closer to zero than +5, so +2 is smaller than +5.
+2 < +5

9. When comparing a positive integer with a negative integer, the positive integer is always greater than the negative integer.

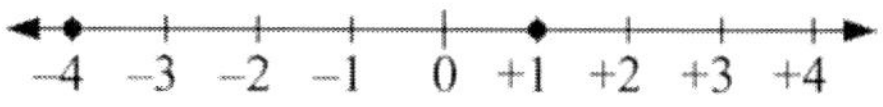

In this case, +1 is positive and –4 is negative, so +1 is greater than –4.
+1 > –4

11. When comparing a negative integer with a positive integer, the negative integer is always less than the positive integer.

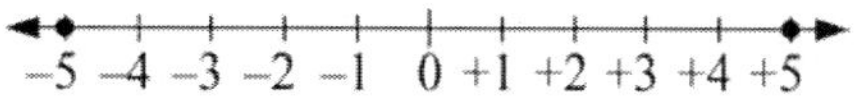

In this case, –5 is negative and +5 is positive, so –5 is less than +5.
–5 < +5

MULTIPLES AND FACTORS

Lesson 1—Multiples

TIME TO TRY ANSWERS AND SOLUTIONS

1. To find five multiples, start with one 20 and add four more: $20 + 20 + 20 + 20 + 20$.

Repeated Addition	Multiples
	20
$20 + 20 = 40$	40
$40 + 20 = 60$	60
$60 + 20 = 80$	80
$80 + 20 = 100$	100

The first five multiples of 20 are 20, 40, 60, 80, and 100.

2. To find four multiples, start with 4 and skip count by 4s for three more counts.

[4], 5, 6, 7, [8], 9, 10, 11, [12], 13, 14, 15, [16]

The first four multiples of 4 are 4, 8, 12, and 16.

3. Multiply 6 by 1, then by 2, and so on for four counts.

Multiplication	Multiples
$6 \times 1 = 6$	6
$6 \times 2 = 12$	12
$6 \times 3 = 18$	18
$6 \times 4 = 24$	24

The first four multiples of the number 6 are 6, 12, 18, and 24.

4. **Step 1**
List the multiples of both numbers.

The multiples of 6 are 6, 12, 18, 24, 30, 36,…
The multiples of 9 are 9, 18, 27, 36,…

Step 2
Identify the multiples that are common to both lists.
6, 12, [18], 24, 30, [36]

9, [18], 27, [36]

The first two common multiples of the numbers 6 and 9 are 18 and 36.

5. **Step 1**
Identify the common multiples of the two numbers.
10, 20, [30], 40, 50, [60],…

15, [30], 45, [60], 75,…

The common multiples of 10 and 15 are 30 and 60.

Step 2
Identify the lowest common multiple (LCM).

Since $30 < 60$, 30 is the lowest common multiple of 10 and 15 because it has the least value.

6. This type of problem can easily be solved by determining the lowest common multiple of the two given numbers.

Multiples of 15: 15, 30, 45, [60],…
Multiples of 20: 20, 40, [60],…

Since the first multiple that is common to both 15 and 20 is 60, the two timers will beep together 60 minutes after they both start.

PRACTICE EXERCISES ANSWERS AND SOLUTIONS

1. **Step 1**
Identify the first four multiples of 21.

The multiples of 21 are 21, 42, 63, and 84.

Step 2
Explain the strategy you used to identify them.

Multiplication is a strategy that works well when using large numbers.

To find the first four multiples, start by multiplying 21 by 1, then by 2, by 3, and by 4. The products are the multiples.
$21 \times 1 = 21$
$21 \times 2 = 42$
$21 \times 3 = 63$
$21 \times 4 = 84$

3. **Step 1**
Identify the multiples of 4 and 12, up to 40.

The multiples of 4 are 4, 8, 12, 16, 20, 24, 28, 32, 36, 40.

The multiples of 12 are 12, 24, 36.

Step 2
Identify the numbers common to both lists.
4, $\boxed{8}$, 12, $\boxed{16}$, 20, $\boxed{24}$, 28, $\boxed{32}$, 36, $\boxed{40}$

$\boxed{8}$, $\boxed{16}$, $\boxed{24}$, $\boxed{32}$, $\boxed{40}$

Up to 40, the common multiples of 4 and 8 are 8, 16, 24, 32, 40.

5. **Step 1**
Determine the lowest common multiple of 5 and 7.

(Bob) Multiples of 5: 5, 10, 15, 20, 25, 30, $\boxed{35}$
(Brett) Multiples of 7: 7, 14, 21, 28, $\boxed{35}$

The lowest common multiple of 5 and 7 is 35.

After 35 minutes of jogging at the same speeds, Bob and Brett will be at the starting point at the same time.

Bob will be on his 7th lap and Brett will be on his 5th lap.

Lesson 2—Prime and Composite Numbers

TIME TO TRY ANSWERS AND SOLUTIONS

1. **Step 1**
Write a different one-digit prime number.

Any of these numbers works: 3, 5, or 7.

Step 2
Explain why the number is a prime number.

These numbers are all prime numbers because they have exactly two factors: 1 and the number itself.
$1 \times 3 = 3$
$1 \times 5 = 5$
$1 \times 7 = 7$

2. **Step 1**
Write a one-digit composite number.

Any of these numbers works: 4, 6, 8, or 9.

Step 2
Explain why the number is a composite number.

All these numbers are composite numbers because they all have more than two factors.

$1\times4=4$ $2\times2=4$	$1\times6=6$ $2\times3=6$	$1\times8=8$ $2\times4=8$	$1\times9=9$ $3\times3=9$

3. A prime number is a whole number that has exactly two factors: 1 and the number itself. A composite number is a whole number with two or more factors.

Therefore, the number 0 is neither prime nor composite because it does not have any factors.

4. **Step 1**
Identify the even composite numbers.

Since all even numbers except 2 are composite numbers, 8, 10, 12, 14, and 44 are all composite numbers. They all have more than two factors.

Step 2
Sort the odd numbers.

There are four numbers left: 19, 25, 27, and 43.

Of these numbers, 25 and 27 are composite because they have more than two factors.

$1\times5=25$ $5\times5=25$	$1\times7=27$ $3\times9=27$

Therefore, 19 and 43 are the prime numbers because they each only have two factors: 1 and themselves.
$1\times19=19$
$1\times43=43$

The numbers should be sorted like this:

Prime Numbers	Composite Numbers
19, 43	8, 10, 12, 14, 25, 27, 44

PRACTICE EXERCISES ANSWERS AND SOLUTIONS

1. The number 49 has more than two factors:
$1\times49=49$
$7\times7=49$

Therefore, 49 is a composite number.

3. A prime number is a whole number that has exactly two factors: 1 and the number itself. A composite number is a whole number with two or more factors.

The number 1 is neither prime nor composite because it has only one factor: 1.

5. **Step 1**
Identify the composite numbers.

14, 28, 36, 57, and 62 are composite numbers. They all have more than two factors.

Step 2
Sort the prime numbers.

There are four prime numbers: 7, 17, 23, and 41.

All these numbers are prime numbers because they each have only two factors: 1 and themselves.

The numbers should be sorted like this:

Prime Numbers	Composite Numbers
7, 17, 23, 41	14, 28, 36, 57, 62

Lesson 3—Factors

TIME TO TRY ANSWERS AND SOLUTIONS

1. **Step 1**
Draw or model arrays for the number 24.

Think of all the ways to arrange 24 same-sized squares to make a rectangle.

The following arrays show all the numbers that result in a product of 24 when multiplied by each other.

$1 \times 24 = 24$

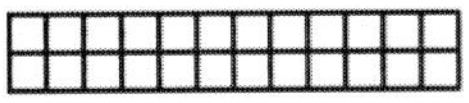

$2 \times 12 = 24$

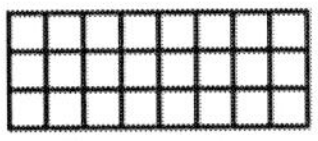

$3 \times 8 = 24$

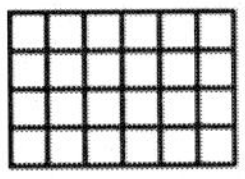

$4 \times 6 = 24$

Step 2
Identify all the factors of the number 24.

The number 24 has eight factors: 1, 2, 3, 4, 6, 8, 12, and 24.

2. **Step 1**
Think of all the factors of 30.
$1 \times 30 = 30$
$2 \times 15 = 30$
$3 \times 10 = 30$
$5 \times 6 = 30$

Step 2
List the factors.

The number 30 has eight factors: 1, 2, 3, 5, 6, 10, 15, 30

3. Yes, 3 is a factor of 99 because 99 can be divided evenly by 3.
$99 \div 3 = 33$

4. **Step 1**
Identify the factors of 28 and 32.

The factors of 28 are 1, 2, 4, 7, 14, 28.
The factors of 32 are 1, 2, 4, 8, 16, 32.

Step 2
Identify the common factors.

The common factors are 1, 2, 4.

Step 3
Enter the numbers in a Venn diagram.

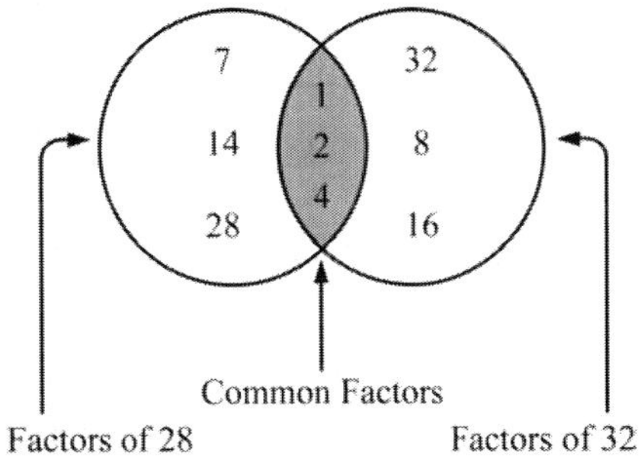

5. **Step 1**
Identify the factors of 15.
$1 \times 5 = 15$
$3 \times 5 = 15$

The factors are 1, 3, 5, 15.

Step 2
Identify the factors of 20.
$1 \times 20 = 20$
$2 \times 10 = 20$
$4 \times 5 = 20$

The factors are 1, 2, 4, 5, 10, 20.

Step 3
Identify the common factors.
$\boxed{1}$, 3, $\boxed{5}$, 15

$\boxed{1}$, 2, 4, $\boxed{5}$, 10, 20

The common factors are 1 and 5.

Step 4
Identify the greatest common factor.

Since $5 > 1$, the greatest common factor of 15 and 20 is 5.

6. **Step 1**
Identify the common factors.

Factors of 32: $\boxed{1}$, $\boxed{2}$, $\boxed{4}$, $\boxed{8}$, 16, 32
Factors of 40: $\boxed{1}$, $\boxed{2}$, $\boxed{4}$, 5, $\boxed{8}$, 10, 20, 40

Step 2
Identify the greatest common factor.

The greatest common factor of 32 and 40 is 8.

Therefore, the greatest number of loot bags that Brenna can make is 8.

Each loot bag will have 4 candies ($32 \div 8 = 4$) and 5 pencils ($40 \div 8 = 5$).

PRACTICE EXERCISES ANSWERS AND SOLUTIONS

1. **Step 1**
Think of all the ways to arrange 16 same-sized squares into a rectangle.

The following arrays show three sets of numbers that result in a product of 16 when multiplied by each other.

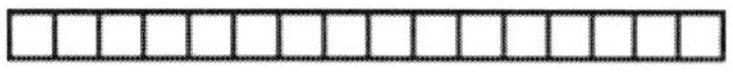

$1 \times 16 = 16$

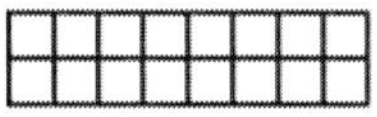

$2 \times 8 = 16$

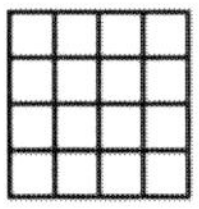

$4 \times 4 = 16$

Step 2
Identify all the factors of the number 16.

The number 16 has five factors: 1, 2, 4, 8, 16.

3. Yes, 4 is a factor of 140 because 140 can be divided evenly by 4.
$140 \div 4 = 35$

5. **Step 1**
Identify the factors of 20.
$1 \times 20 = 20$
$2 \times 10 = 20$
$4 \times 5 = 20$

The factors are 1, 2, 4, 5, 10, 20.

Step 2
Identify the factors of 35.
$1 \times 35 = 35$
$5 \times 7 = 35$

The factors are 1, 5, 7, 35.

Step 3
Identify the common factors.
$\boxed{1}$, 2, 4, $\boxed{5}$, 10, 20

$\boxed{1}$, $\boxed{5}$, 7, 35

Step 4
Identify the greatest common factor.

Since $5 > 1$, the greatest common factor of 20 and 35 is 5.

7. **Step 1**
Determine the number of quarters in each group.

Divide the total number of quarters by the number of groups.
$25 \div 5 = 5$

There will be 5 quarters in each of the five groups.

Step 2
Determine the number of dimes in each group.

Divide the total number of dimes by the number of groups.
$35 \div 5 = 7$

There will be 7 dimes in each of the five groups.

When five groups of coins are made, there will be five quarters and seven dimes in each group.

Lesson 4—Prime Factorization

TIME TO TRY ANSWERS AND SOLUTIONS

1. Following is one possible factor tree for the number 36.

Step 1
Start drawing the factor tree.

Begin with the number 36 at the top of the tree. Draw two branches that will each lead to a factor of the number 36.

Two factors of 36 are 3 and 12 ($3 \times 12 = 36$). Place these factors in the factor tree.

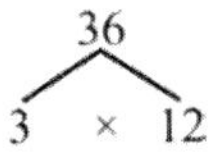

Step 2
Rewrite the composite number 12.

The number 12 can be rewritten as 3×4. Draw two branches out of the number 12 and place the factors 3 and 4 in the factor tree.

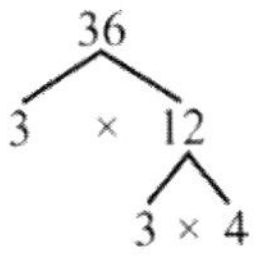

Step 3
Change the composite number 4 into prime factors.

The number 4 can be rewritten as 2×2. Draw two branches out of the number 4 and place the factors 2 and 2 in the factor tree.

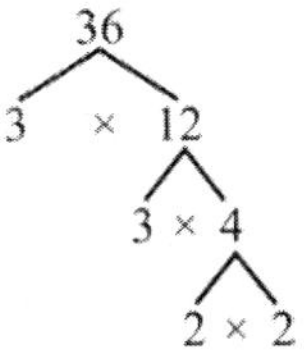

The numbers at the bottoms of the branches are all prime numbers, so the factorization is complete.

The prime factorization of 36 is $3 \times 3 \times 2 \times 2$.

Regardless of which factor tree used, the prime factors will be the same, but they may be expressed in a different order.

2. **Step 1**
Divide the number 45 by the smallest prime number (3) that divides evenly.

$$\begin{array}{r} 15 \\ 3\overline{)45} \end{array}$$

Step 2
Divide 15 by the smallest prime number (3) that divides evenly.

$$\begin{array}{r} 5 \\ 3\overline{)15} \\ 3\overline{)45} \end{array}$$

Since 5 is a prime number, the factorization is complete. The prime factorization of 45 is $3 \times 3 \times 5$.

PRACTICE EXERCISES ANSWERS AND SOLUTIONS

1. **Step 1**
Draw a factor tree.
Following is one possible factor tree.

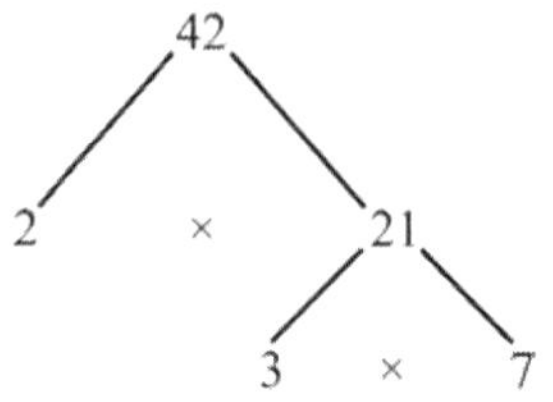

Step 2
Identify the prime factors of 42.
The prime factorization of 42 is $2 \times 3 \times 7$.

3. **Step 1**
Identify the prime factors.

The prime factorization of 81 is $3 \times 3 \times 3 \times 3$.

Step 2
Explain your strategy.

Using repeated division by the smallest prime number, the process would look like this:

$$\begin{array}{r} 3 \\ 3\overline{)9} \\ 3\overline{)27} \\ 3\overline{)81} \end{array}$$

Practice Test

ANSWERS AND SOLUTIONS

1. To find five multiples, start with 13 and add four more 13s. $(13+13+13+13+13)$

Repeated Addition	Multiples
	13
$13+13=26$	26
$26+13=39$	39
$39+13=52$	52
$52+13=65$	65

The first five multiples of 13 are 13, 26, 39, 52, and 65.

3. Multiply 15 by 1, then by 2, and so on for five counts.

Multiplication	Multiples
$15\times1=15$	15
$15\times2=30$	30
$15\times3=45$	45
$15\times4=60$	60
$15\times5=75$	75

The first five multiples of the number 15 are 15, 30, 45, 60, and 75.

5. **Step 1**
Identify common multiples of the two numbers.
3, [6], 9, [12], 15, [18], 21,...

[6], [12], [18], 24, 30,...

The common multiples of 3 and 6 are 6, 12, 18,...

Step 2
Identify the lowest common multiple (LCM).

Since $6 < 12 < 18$, 6 is the lowest common multiple of 3 and 6 because it has the least value.

7. The number 7 is a prime number because it has only two factors: 1 and itself.
$7 \times 1 = 7$
$1 \times 7 = 7$

It can only be divided equally by 1 or itself.
$7 \div 1 = 7$
$7 \div 7 = 1$

9. Step 1
Identify the even composite numbers.

Since all even numbers except 2 are composite numbers, 12, 22, 28, and 46 are all composite numbers. They all have more than two factors.

Step 2
Sort the odd numbers.

There are five odd numbers: 3, 5, 17, 31, and 39.

Of these numbers, 39 is a composite number because it has more than two factors.
$1 \times 39 = 39$
$3 \times 13 = 39$

The prime numbers are 3, 5, 7, and 31 because they each only have two factors: 1 and themselves.
$1 \times 3 = 3$
$1 \times 5 = 5$
$1 \times 17 = 17$
$1 \times 31 = 31$

The numbers should be sorted like this:

Prime Numbers	Composite Numbers
3, 5, 17, 31	12, 22, 28, 39, 46

11. Step 1
Think of all the factors that result in 36.

Start with 1 and work your way up.
$1 \times 36 = 36$
$2 \times 18 = 36$
$3 \times 12 = 36$
$4 \times 9 = 39$
$6 \times 6 = 36$

Step 2
List the factors.

The number 36 has nine factors: 1, 2, 3, 4, 6, 9, 12, 18, 36.

Note that if the number is repeated, like 6 (6×6), it is only listed once.

13. Step 1
List the factors of the number 30.
1, 2, 3, 5, 6, 10, 15, 30

Step 2
List the factors of the number 36.
1, 2, 3, 4, 6, 9, 12, 18, 36

Step 3
Compare the two sets of factors and identify the common factors.
$\boxed{1}$, $\boxed{2}$, $\boxed{3}$, 5, $\boxed{6}$, 10, 15, 30

$\boxed{1}$, $\boxed{2}$, $\boxed{3}$, 4, $\boxed{6}$, 9, 12, 18, 36

The common factors are 1, 2, 3, and 6.

15. Step 1
Determine the common factors of 48 and 42.

Factors of 48: $\boxed{1}$, $\boxed{2}$, $\boxed{3}$, 4, $\boxed{6}$, 8, 12, 16, 24, 48
Factors of 42: $\boxed{1}$, $\boxed{2}$, $\boxed{3}$, $\boxed{6}$, 7, 14, 21, 42

The common factors of 48 and 42 are 1, 2, 3, and 6.

Step 2
Identify the greatest common factor.

Since $6 > 3 > 2 > 1$, the greatest common factor is 6.

Therefore, the greatest number of groups that Mrs. Palmer can make is 6 groups.

17. The repeated division by prime factors is as follows:

$$\begin{array}{r} 11 \\ 3\overline{)33} \\ 2\overline{)66} \end{array}$$

The prime factorization of 66 can be expressed as $2 \times 3 \times 11$.

IMPROPER FRACTIONS AND MIXED NUMBERS

Lesson 1—Improper Fractions and Mixed Numbers

TIME TO TRY
ANSWERS AND SOLUTIONS

1. The denominator (8) shows that each shape used must be divided into eight equal parts.

The numerator (9) shows that nine of the parts must be shaded (one whole and one part of the other shape).

The model can look like this:

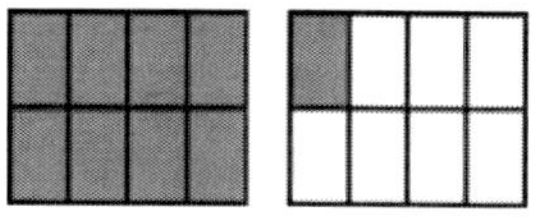

2. Step 1
Determine the denominator.

Since the strips are divided into 6 equal parts, the denominator of the improper fraction is 6.

Step 2
Determine the numerator.

Count all of the parts that are shaded (from both strips). Since there are 9 shaded parts, the numerator is 9.

The symbolic form of the given improper fraction model is $\frac{9}{6}$.

3. The denominator (3) shows that each shape used must be divided into 3 equal parts.

The whole number (4) shows that there are four whole shapes that are all shaded.

The numerator (2) shows that two of the three parts of the fifth shape must be shaded.

The model can look like this:

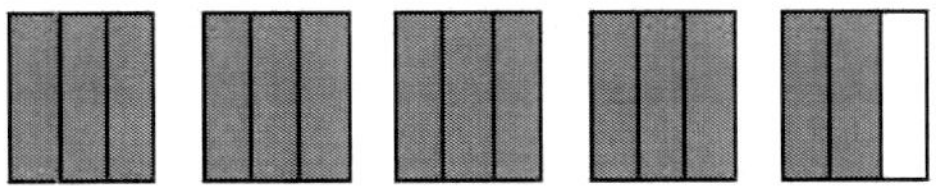

4. **Step 1**
Determine the whole number.

Count the number of shapes that are completely shaded. There are 2 completely shaded shapes, so 2 is the whole number part of the mixed number.

Step 2
Determine the numerator.

Count the number of shaded parts in the partially shaded shape. Since six parts are shaded, the numerator of the proper fraction is 6.

Step 3
Determine the denominator.

Count the number of equal parts of the shapes. There are ten equal parts in each shape, so the denominator is 10.

The symbolic form of the given mixed number model is $2\frac{6}{10}$.

PRACTICE EXERCISES ANSWERS AND SOLUTIONS

1. The denominator (5) shows that each shape used needs to be divided into five equal parts.

The numerator (8) shows that eight of the parts must be shaded (one whole and three parts of the other shape).

The model can look like this.

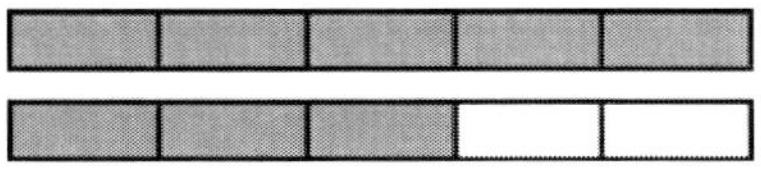

3. The model form of the improper fraction could look like this:

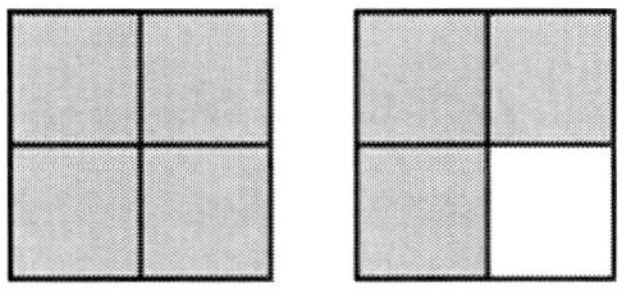

There are two same-sized squares divided into four equal parts. One square with all four parts shaded represents one whole. One square with three parts shaded represents three out of four.

5. The denominator (6) shows that the shape used must be divided into 6 equal parts.

The whole number (2) shows that there are two whole shapes that are all shaded.

The numerator (2) shows that two of the six parts of the third shape must be shaded.

The model can look like this:

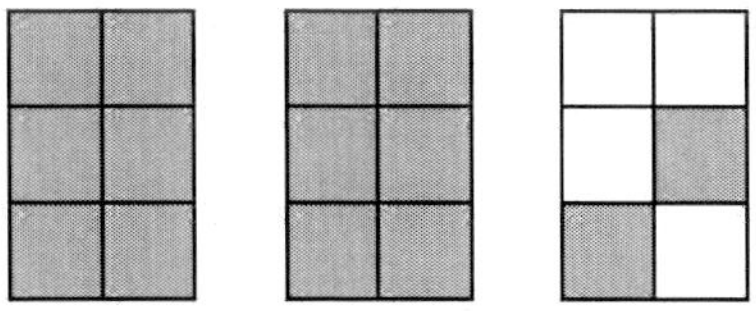

Note that the shaded parts in the partially shaded shape do not need to be side by side.

Lesson 2—Expressing Improper Fractions as Mixed Numbers and Mixed Numbers as Improper Fractions

TIME TO TRY ANSWERS AND SOLUTIONS

1. **Step 1**
Divide the numerator by the denominator to get a whole number and a remainder.

$$\begin{array}{r} 2R5 \\ 6\overline{)17} \\ \underline{12} \\ 5 \end{array}$$

Step 2
Write the mixed number.

- Write the whole number 2 on the left.
- Write the remainder 5 as the numerator of the fraction.
- Keep the same denominator (6).

The improper fraction $\frac{17}{6}$ is expressed as the mixed number $2\frac{5}{6}$.

2. **Step 1**
Determine the numerator.

Multiply the whole number (3) by the denominator (4).
$3 \times 4 = 12$.

Add the numerator (2) to the product.
$12 + 2 = 14$

The number 14 becomes the numerator of the improper fraction.

Step 2
Write the improper fraction.

Recall that the denominator (4) stays the same in both the mixed number and the related improper fraction.

The mixed number $3\frac{2}{4}$ is expressed as the improper fraction $\frac{14}{4}$.

PRACTICE EXERCISES ANSWERS AND SOLUTIONS

1. Multiply the whole number by the denominator and add the numerator to the product.
$1 \times 2 = 2$
$2 + 1 = 3$

The numerator of the improper fraction is 3.

Keep the same denominator of 2.

The mixed number $1\frac{1}{2}$ is expressed as the improper fraction $\frac{3}{2}$.

3. Multiply the whole number by the denominator and add the numerator to the product.
$12 \times 10 = 120$
$120 + 9 = 129$

The numerator of the improper fraction is 129. Keep the same denominator of 10.

The mixed number $12\frac{9}{10}$ is expressed as the improper fraction $\frac{129}{10}$.

5. Step 1
Determine who is correct.

Matt is correct.

Step 2
Justify your answer.

Matt is correct because he wrote the whole number 1 on the left of the fraction, the remainder 4 as the numerator of the mixed number, and kept the same denominator of 5.

Sophia is incorrect because she wrote the whole number as the numerator and the remainder as the whole number.

Lesson 3—Ordering Fractions on a Number Line

TIME TO TRY ANSWERS AND SOLUTIONS

1. Since the denominators are the same, order the numerators from least to greatest.
$3 < 5 < 10$
   - Start at the 0 and count to the third tick to place $\frac{3}{10}$.
   - Count to the fifth tick to place $\frac{5}{10}$.
   - Count to the tenth tick to place $\frac{10}{10}$, which is equivalent to the whole number 1.

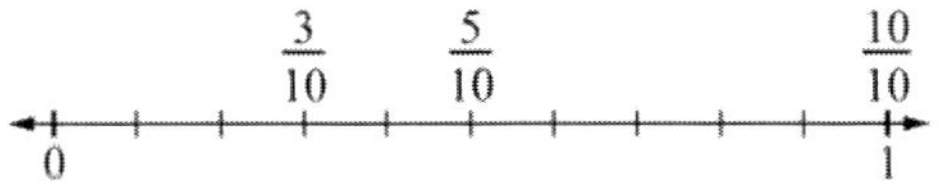

2. Step 1
Convert the improper fraction to a mixed number.
$\frac{12}{8} = 12 \div 8 = 1\frac{4}{8}$

Step 2
Make equivalent fractions with the same denominator.

Since the number line is marked in fourths, use 4 as the common denominator.
$2\frac{1}{4} = 2\frac{1}{4}$

$1\frac{4}{8} = \frac{4 \div 2}{8 \div 2} = 1\frac{2}{4}$

Step 3
Place the fractions on the number line.
- Start at 1 and count to the second tick to place $1\frac{2}{4}$ $(1\frac{4}{8} = \frac{12}{8})$.
- Start at the 2 and count to the first tick to place $2\frac{1}{4}$.

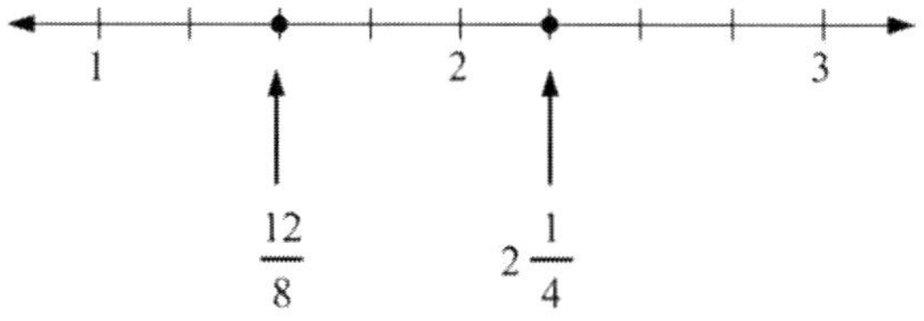

PRACTICE EXERCISES ANSWERS AND SOLUTIONS

1. Step 1
Since the number line is marked in tenths, make equivalent fractions with the denominator of 10.
$\frac{8}{10} = \frac{8}{10}$

$\frac{3}{5} = \frac{3 \times 2}{5 \times 2} = \frac{6}{10}$

Step 2
Place the fractions on the number line.

- Start at the 0 and count to the sixth tick to place $\frac{6}{10}\left(\frac{3}{5}\right)$.
- Count to the eighth tick to place $\frac{8}{10}$.

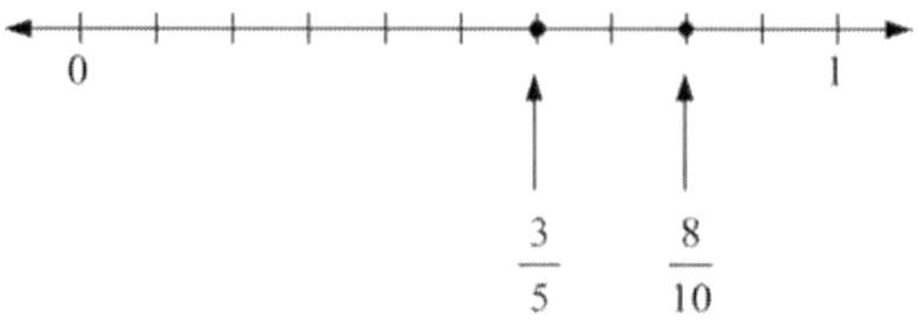

3. **Step 1**
Express the improper fraction as a mixed number.

$\frac{11}{5}=11\div 5=2\frac{1}{5}$

Step 2
Make equivalent fractions.

Since the number line is marked in tenths, use 10 as the common denominator.

$2\frac{1}{5}=\frac{1\times 2}{5\times 2}=2\frac{2}{10}$

$2\frac{2}{5}=\frac{2\times 2}{5\times 2}=2\frac{4}{10}$

$2\frac{8}{10}=2\frac{8}{10}$

Step 3
Place the fractions on the number line.

- Start at 2 and count to the second tick to place $2\frac{2}{10}\left(\frac{11}{5}\right)$.
- Count to the fourth tick to place $2\frac{4}{10}\left(2\frac{2}{5}\right)$.
- Count to the eighth tick to place $2\frac{8}{10}$.

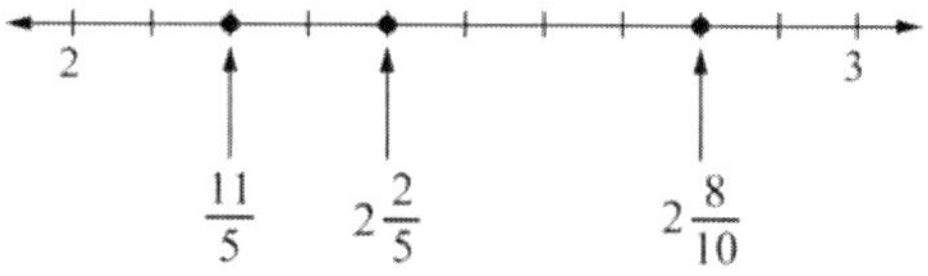

Practice Test

ANSWERS AND SOLUTIONS

1. An improper fraction is a fraction that represents a number that is equal to or greater than one. As a result, in an improper fraction, the numerator is always equal to or greater than the denominator.

This set of circles shows how the improper fraction $\frac{17}{10}$ represents a number greater than one whole. It represents 10 shaded parts of one whole plus seven shaded parts of the next circle.

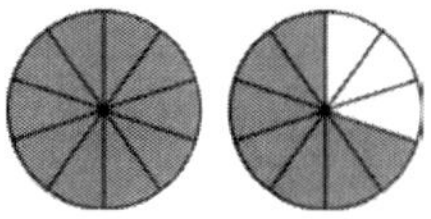

3. The denominator (3) shows that the shape used must be divided into 3 equal parts.

The whole number (3) shows that there are three whole shapes that are all shaded.

The numerator (1) shows that one of the three parts of the fourth shape must be shaded.

The model can look like this:

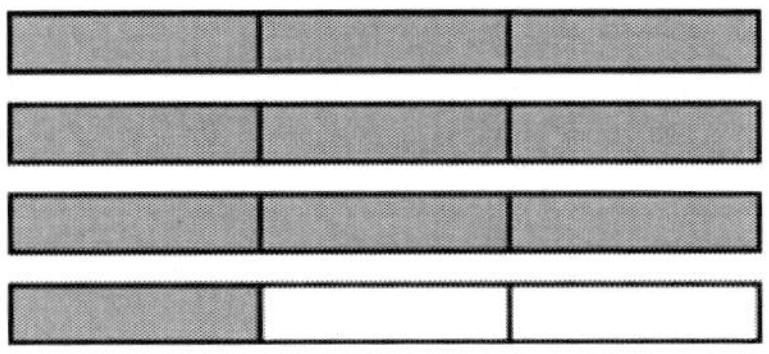

5. **Step 1**
Determine the whole number.

There are five completely shaded shapes, so 5 is the whole number part of the mixed number.

Step 2
Determine the numerator.

Count the number of shaded parts in the partially shaded shape. Since three parts are shaded, the numerator of the proper fraction is 3.

Step 3
Determine the denominator.

Count the number of equal parts of the shapes. There are five equal parts in each shape, so the denominator is 5. The symbolic form of the given mixed number model is $5\frac{3}{5}$.

7. **Step 1**
Determine the numerator.

Multiply the whole number (17) by the denominator (4).
$17 \times 4 = 68$.

Add the numerator (1) to the product.
$68 + 1 = 69$

The number 69 becomes the numerator of the improper fraction.

Step 2
Write the improper fraction.

Recall that the denominator (4) stays the same in both the mixed number and the related improper fraction.

The mixed number $17\frac{1}{4}$ is expressed as the improper fraction $\frac{69}{4}$.

9. **Step 1**
Make equivalent fractions.

Since the number line is marked in tenths, use the common denominator of 10.

$$\frac{1}{5} = \frac{1 \times 2}{5 \times 2} = \frac{2}{10}$$

$$\frac{1}{2} = \frac{1 \times 5}{2 \times 5} = \frac{5}{10}$$

For $\frac{6}{15}$, first simplify the fraction and then make an equivalent fraction with a denominator of 10.

$$\frac{6}{15} = \frac{6 \div 3}{15 \div 3} = \frac{2}{5}$$

$$\frac{2}{5} = \frac{2 \times 2}{5 \times 2} = \frac{4}{10}$$

$$\frac{9}{10} = \frac{9}{10}$$

Step 2
Place the fractions on the number line.

- Start at 0 and count to the second tick to place $\frac{2}{10}\left(\frac{1}{5}\right)$.

- Count to the fourth tick to place $\frac{4}{10}\left(\frac{6}{15}\right)$.

- Count to the fifth tick to place $\frac{5}{10}\left(\frac{1}{2}\right)$.

- Count to the ninth tick to place $\frac{9}{10}$.

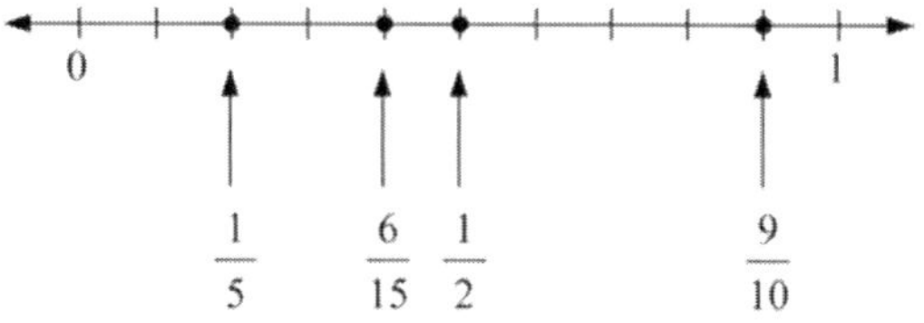

RATIOS

Lesson 1—Using Ratio to Compare Two Quantities

TIME TO TRY ANSWERS AND SOLUTIONS

1. The first part of the comparison is the number of white parts, so 4 is the number written first. Write the word *to* to the right of 4.

The second part of the ratio is the number of shaded parts, so 2 is the number written to the right of the word *to*.

The ratio 4 to 2 expresses the comparison of white parts to shaded parts.

2. Start by drawing a square grid that has a length of four units and a width of four units. Draw the grid so that it has a total of 16 square units.

Since the first part of the comparison represents the number of shaded squares, shade four squares.

Since the number to the right of the colon represents the number of white squares, leave the remaining 12 squares white.

The pictorial or concrete model could look like this.

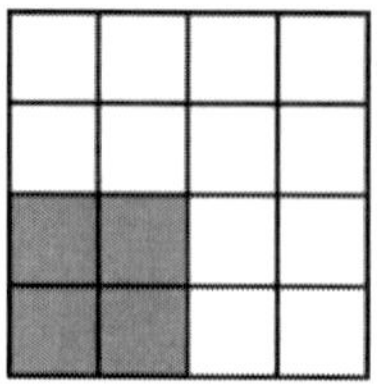

3. **Step 1**
Determine the first number in the ratio.

Since the first quantity being compared is the number of white circles, the first number in the ratio is 4.

Step 2
Determine the second number.

Since the second quantity being compared is the number of shaded circles, the second number in the ratio is 5.

Step 3
Write the ratio in three ways.

The ratio comparing white circles to shaded circles is expressed as $4:5$, $\frac{4}{5}$, or 4 to 5.

4. **Step 1**
Determine the numerator.

The first part of the comparison is the number of wins, so 7 is the numerator of the ratio fraction.

Step 2
Determine the denominator.

The number of losses is the second part of the comparison, so 3 is the denominator of the ratio fraction.

The ratio fraction that compares wins with losses is $\frac{7}{3}$.

PRACTICE EXERCISES ANSWERS AND SOLUTIONS

1. The first part of the comparison is the number of grey triangles, so 5 is the numerator of the fraction.

The second part of the ratio is the number of black triangles, so 4 is the denominator of the fraction.

The ratio fraction $\frac{5}{4}$ expresses the comparison of grey triangles to black triangles.

3. **Step 1**
Determine the first number in the ratio.

Since the first quantity being compared is the number of As, the first number in the ratio is 5.

Step 2
Determine the second number.

Since the second quantity being compared is the number of Bs, the second number in the ratio is 3.

Step 3
Write the ratio in three ways.

The ratio comparing As with Bs is expressed as $5:3$, $\frac{5}{3}$, or 5 to 3.

5. **Step 1**
Determine the first number in the ratio.

Since the first number of the legend is the number of centimetres, the first number in the ratio is 1.

Step 2
Determine the second number in the ratio.

Since the second quantity being compared is the number of kilometres, the second number in the ratio is 10.

The ratio on the legend using a colon is expressed as $1:10$.

Step 3
Explain what the ratio means.

Since 1 centimetre (first number) is being compared with 10 kilometres (second number), the ratio $1:10$ means that the length of 1 cm on the map compares with (is equal to) 10 km in real life.

This means that if two towns on the map are 1 cm apart, in real life, the towns are actually 10 km apart.

Lesson 2—Part/Part and Part/Whole Ratios

TIME TO TRY ANSWERS AND SOLUTIONS

1. The ratio $\frac{8}{4}$ is a part/part ratio. The number of squares is being compared with the number of circles.

2. The ratio 16 to 53 is a part/whole ratio. The number of shaded squares is being compared with the total number of squares.

PRACTICE EXERCISES ANSWERS AND SOLUTIONS

1. The ratio $5:20$ is a part/whole ratio. The number of squares is being compared with the total number of shapes.

3. You are asked to write a part/whole ratio.

There are 4 triangles, so 4 becomes the numerator of the fraction. There are 20 shapes in all, so 20 becomes the denominator of the fraction.

The ratio fraction that compares the number of triangles with the total number of shapes is $\frac{4}{20}$.

5. The ratio $5:12$ is a part/whole ratio. The number of crows is being compared with the total number of birds.

7. You are asked to write a whole/part ratio.

The whole is 12 because there are 12 birds in all. The part is 7 because there are 7 sparrows.

Since you are comparing the whole with the part, the 12 comes first and the 7 comes after it.

The ratio you write can be expressed as 12 to 7, $\frac{12}{7}$, or $12:7$.

Lesson 3—Solving Problems Involving Ratios

TIME TO TRY ANSWERS AND SOLUTIONS

1. **Step 1**
Identify what is being compared.

The number of boys (10) is being compared with the total number of students.

Step 2
Calculate the total number of students.

Add the number of girls and the number of boys.
$15+10=25$

Step 3
Write the part/whole ratio.

The ratio of boys to the total number of students can be expressed as 10 to 25, $\frac{10}{25}$, or $10:25$.

PRACTICE EXERCISES ANSWERS AND SOLUTIONS

1. **Step 1**
Identify what is being compared.

The number of girls is being compared with the total number of students (30).

Step 2
Calculate the number of girls.

Subtract the number of boys from the total number of students.
$30-18=12$

Step 3
Write the whole/part ratio.

The ratio of students to the number of girls can be expressed as 30 to 12, $\frac{30}{12}$, or $30:12$.

3. **Step 1**
Explain the part/part ratio.

For every three boys on the team, there is one girl.

Step 2
Calculate the number of boys.

Since there are five girls on the team, multiply 5 by 3 to determine the number of boys on the team.
$5\times3=15$

There are 5 girls and 15 boys on the soccer team.

Practice Test

ANSWERS AND SOLUTIONS

1. In this part/part ratio, the 5 hats are being compared with the 3 caps. Therefore, the 5 is placed to the left of the colon and the 3 is placed to the right of the colon.

The ratio of hats to caps is $5:3$.

3. **Step 1**
Interpret the fraction ratio.

The numerator of the fraction (3) represents the first part of the comparison, which is the number of rectangles.

The denominator (2) represents the second part of the comparison, which is the number of triangles.

Step 2
Draw the ratio.

Draw three rectangles and two triangles.

The pictorial or concrete model could look like this.

5. The ratio $\frac{4}{3}$ is a part/part ratio. The number of triangles (4) is being compared with the number of rectangles (3).

7. You are asked to write a part/whole ratio.

The part is 4 because there are 4 triangles.

The whole is 18 because there are 18 shapes in all.

Since you are comparing the part with the whole, the 4 comes first and the 18 comes after it.

The ratio you write can be expressed as $4:18$, $\frac{4}{18}$, and 4 to 18.

9. **Step 1**
Identify what is being compared.

The number of boys is being compared with the total number of students (32).

Step 2
Calculate the number of boys.

Subtract the number of girls from the total number of students.
$32-19=13$

Step 3
Write the part/whole ratio.

The ratio of boys to the total number of students can be expressed as 13 to 32, $\frac{13}{32}$, or $13:32$.

PERCENTAGE

Lesson 1—Understanding Percentage

TIME TO TRY ANSWERS AND SOLUTIONS

1. Since a hundredths grid is made up of 100 squares and percentage is a ratio out of 100, shade 9 of the 100 squares to show that 9% of the grid is shaded.

The pictorial representation could look like this.

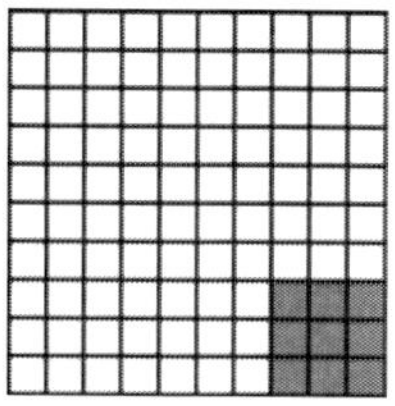

2. In the number 65, there are 6 tens and 5 ones. The model that illustrates 65% should be built with 6 ten rods and 5 units.

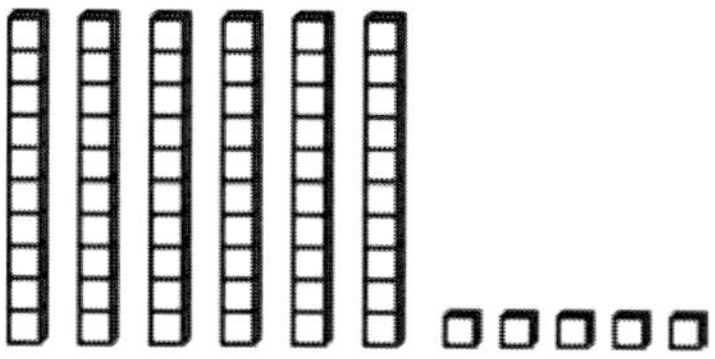

3. **Step 1**
Count the ten rods by 10s.

There are seven ten rods: 10, 20, 30,…60, 70

Step 2
Count the units by 1s.

Continue counting from 70: 71, 71, 73.

The model illustrates 73%.

4. One-half is the same as $\frac{50}{100}$ (fifty cents out of one hundred cents). Jason spent about 50% of his money.

PRACTICE EXERCISES ANSWERS AND SOLUTIONS

1. Percentage is a ratio out of 100, so 25% can be explained as $\frac{25}{100}$ or $25:100$, where 25 is the number of parts out of a total of 100 parts.

3. In the number 38, there are 3 tens and 8 ones. The model that illustrates 38% should be built with 3 ten rods and 8 units.

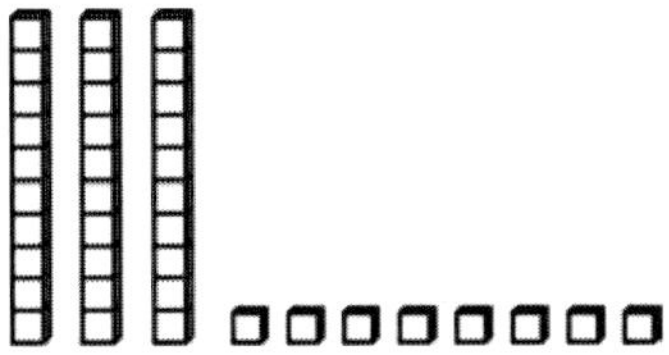

5. Count the number of squares that are not shaded.
10, 20, 30,...70, 80, 81

Since 81 of the 100 squares are not shaded, 81% of the grid is not shaded.

7. Recall that percentage represents a ratio out of 100. Since 60 out of 100 students $\left(\frac{60}{100}\right)$ are boys, then 60% of the students are boys.

Lesson 2—Expressing a Percentage as a Fraction and a Decimal

TIME TO TRY ANSWERS AND SOLUTIONS

1. 33% means 33 parts out of 100 parts.
The denominator of the related fraction is 100. The digits in the percentage (33) become the numerator.
$33\% = \frac{33}{100}$

2. 89% means 89 parts out of 100 parts.
Since there are two digits in the percentage, write the two digits to the right of the decimal point in the decimal number.
$89\% = 0.89$

PRACTICE EXERCISES ANSWERS AND SOLUTIONS

1. 30% means 30 parts out of 100 parts.
The denominator of the related fraction is 100. The digits of the percentage (30) become the numerator.
$30\% = \frac{30}{100}$

3. Recall that 9% means nine parts out of 100 parts, as shown on this hundredths grid.

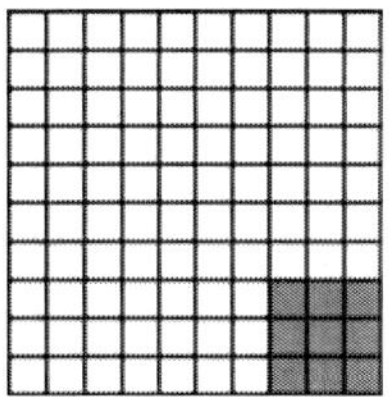

Since the related decimal number must have two digits to the right of the decimal point, place a 0 to the left of the 9 when writing the decimal number. The 0 will serve as a place holder for the tenths position.
$9\% = 0.09$

5. **Step 1**
Express 1% as a fraction.

Since percentage is a ratio out of 100, the denominator is 100. The digit in the percentage (1) becomes the numerator.
$1\% = \frac{1}{100}$

Step 2
Express 1% as a decimal.

Since there is only one digit in the percentage, place a 0 to the left of the 1 when you write the decimal number. The 0 will act as a place holder for the tenths place.
$1\% = 0.01$

Lesson 3—Solving Problems Involving Percentage

TIME TO TRY ANSWERS AND SOLUTIONS

1. **Step 1**
Determine the percentage of blond-, black-, and brown-haired students.

Add the percentages of students with blond, black, and brown hair.
$23 + 12 + 54 = 89$

Out of 100 students, 89% have blond, black, or brown hair.

Step 2
Determine the percentage of red-haired students. Since there are 100 students represented, the percentages need to add up to 100.

Subtract 89% from 100% to determine the percentage of students with red hair.
$100 - 89 = 11$

Out of the 100 students, 11% have red hair.

PRACTICE EXERCISES ANSWERS AND SOLUTIONS

1. **Step 1**
Calculate the total number of apples.
$34 + 42 + 24 = 100$

Since there were 100 apples in total, and percentage is a ratio out of 100, 34% of the apples were green, 42% of the apples were red, and 24% of the apples were yellow.

Step 2
Determine the percentage of red and yellow apples.
$42 + 24 = 66$

Out of the 100 apples, 66% were either red or yellow.

3. **Step 1**
Calculate the amount of money put into the bank.
$\$100 - \$55 = \$45$

Of the $100, Jason put $45 into his bank account.

Step 2
Determine the percentage of money put into the bank.

Since percentage is a ratio out of 100, Jason put 45% of the money he earned into his bank account.

Practice Test

ANSWERS AND SOLUTIONS

1. To illustrate 30%, shade 30 of the 100 squares.

The grid could look like this.

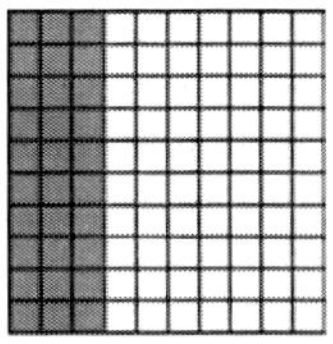

3. Since percentage is a ratio out of 100, and 11 of the 100 squares are shaded, 11% of the squares are shaded.

5. Since there are 100 seats and each person must buy a ticket to get a seat, there must be 100 tickets for sale.

If 83 of the 100 tickets were sold, then 83 of the 100 seats will be filled.

Recall that percentage represents a ratio out of 100. If 83 of the 100 seats are filled, then 83% of the seats are filled.

Therefore, if 83 tickets are sold, 83% of the seats will be filled.

7. **Step 1**
Express 5% as a fraction.

Since percent is a ratio out of 100, the denominator is 100. The digit in the percent (5) becomes the numerator.

$5\% = \frac{5}{100}$

Step 2
Express 5% as a decimal.

Since there is only one digit in the percentage, place a 0 to the left of the 5 when writing the decimal number. The 0 will act as a place holder for the tenths place.
$5\% = 0.05$

PATTERNS AND RELATIONS

Lesson 1—Using Tables and Graphs to Represent Patterns and Relationships

TIME TO TRY
ANSWERS AND SOLUTIONS

1. **Step 1**
Write the term numbers.

Since there are four terms in the number pattern, write the numbers 1, 2, 3, and 4 under the term number column.

Term Number	Term
1	
2	
3	
4	

Step 2
Write the terms.

Write the four terms in the term column, in the order they appear.

Term Number	Term
1	21
2	30
3	39
4	48

2. **Step 1**
Write the ordered pairs.

The ordered pairs that show the locations of the points are (1, 10), (2, 20), (3, 30), (4, 40).

Step 2
Create the table of values.

Write the first numbers of the ordered pairs in the Term Number column.

Write the second numbers of the ordered pairs in the Term column, in the order they appear.

Term Number	Term
1	10
2	20
3	30
4	40

3. **Step 1**
Translate the pattern to a table of values.

Term Number	Term
1	7
2	5
3	3
4	1

Step 2
Write the term numbers and terms as ordered pairs.

The ordered pairs are (1, 7), (2, 5), (3, 3), (4, 1)

Step 3
Plot the ordered pairs on a graph.

The graph could look like this.

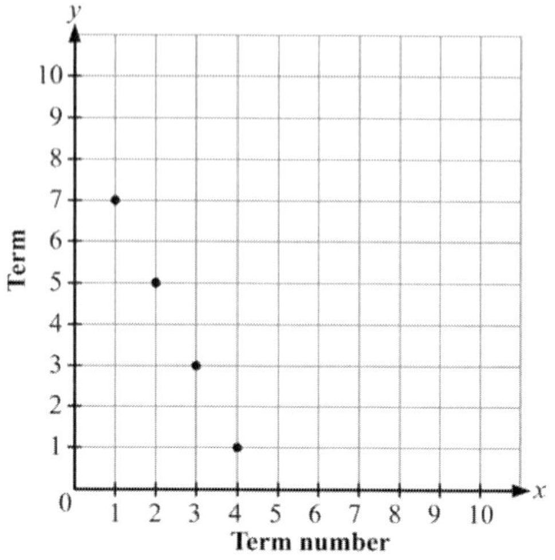

4. **Step 1**
Determine how the term numbers change.

The term numbers (time in minutes) start at 0 and increase by 1 each time.

Step 2
Determine how the terms change.

The corresponding terms (distances in cm) start at 0 and increase by 10 each time.

Step 3
Determine the relationship between term numbers and corresponding terms.

The relationship between the term numbers and their corresponding terms (time and its corresponding distance) is that the tortoise travels 10 cm every minute.

PRACTICE EXERCISES ANSWERS AND SOLUTIONS

1. **Step 1**
Write the term numbers.

Since there are five terms in the pattern of numbers, write the numbers 1, 2, 3, 4, and 5 on the left side of the table of values.

Term Number	Term
1	
2	
3	
4	
5	

Step 2
Write the terms.

Write the five terms in the term column, in the order they appear.

Term Number	Term
1	2
2	3
3	5
4	8
5	12

3. **Step 1**
Translate the pattern to a table of values.

Term Number	Term
1	2
2	7
3	12

Step 2
Write the term numbers and terms as ordered pairs.

The ordered pairs are (1, 2), (2, 7), (3, 12).

Step 3
Plot the ordered pairs on a graph.

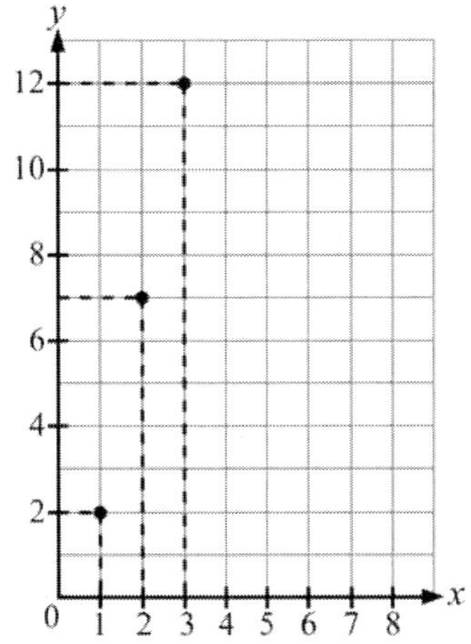

The graph could look like this.

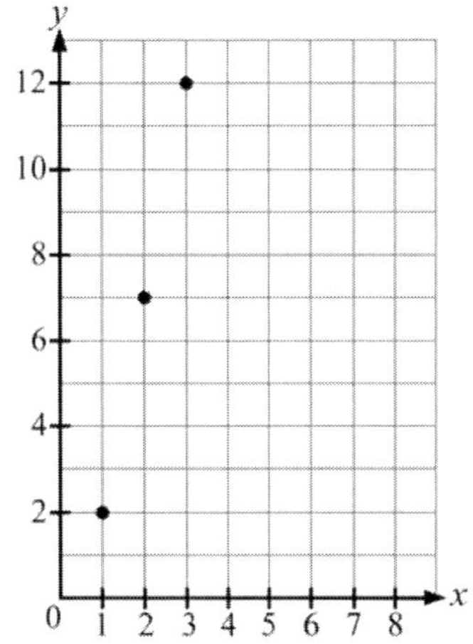

5. **Step 1**
Determine how the term numbers change.

The term numbers (number of pens) start at 1 and increase by 1 each time.

Step 2
Determine how the terms change.

The corresponding terms (number of pencils) start at 4 and increase by 4 each time.

Step 3
Determine the relationship between pens and pencils.

The relationship between number of pens and the corresponding number of pencils is that the cost of one pen is equal to the cost of four pencils.

For every pen purchased, four pencils could be purchased for the same cost.

Lesson 2—Using Relationships Within Tables to Solve Problems

TIME TO TRY ANSWERS AND SOLUTIONS

1. **Step 1**
Determine how the numbers change from term to term.

The pattern starts with 3. Each term is multiplied by 2 to get the next term.
$3 \times 2 = 6$
$6 \times 2 = 12$
$12 \times 2 = 24$

Step 2
Describe the relationship.

The relationship between any two consecutive terms in the pattern is that each successive term is two times greater than the previous term.

The relationship can be expressed as the rule $n \times 2$, where n represents any given term.

2. **Step 1**
Determine how each term number changes to its corresponding term.
$1 + 7 = 8$
$2 + 7 = 9$
$3 + 7 = 10$
$4 + 7 = 11$

Step 2
Describe the relationship.

The relationship between any term number and its corresponding term is that the term is seven more than the corresponding term number.

The relationship can be expressed as the rule $k + 7$, where k represents any given term number.

3. **Step 1**
Determine the relationship.
$87 - 5 = 82$
$82 - 5 = 77$
$77 - 5 = 72$

Each term is 5 less than the previous term.

Step 2
Make a prediction.

To get from term number 4 to 7, subtract 5 for three more counts. A good prediction is 57.

Step 3
Verify the prediction.

Term Number	Term
5	$72 - 5 = 67$
6	$67 - 5 = 62$
7	$62 - 5 = 57$

4. Apply the pattern rule of $+6$.

Fill in the terms up to the number 26.

Term Number	Term
1	2
2	8
3	14
4	20
5	26

The fifth term has a value of 26.

5. **Step 1**
Determine the relationship.

Only 12 and 6 are consecutive numbers, which means that 12 is divided by 2 to get 6.

This shows that the pattern rule is to divide each term by 2 to get the next term. The rule can be expressed as $\div 2$ or $n \div 2$.

Step 2
Start with 96 and apply the pattern rule.

Term Number	Term
1	96
2	$96 \div 2 = 48$
3	$48 \div 2 = 24$
4	$24 \div 2 = 12$
5	6
6	$6 \div 2 = 3$

The 2nd, 3rd, and 6th terms, respectively, are 48, 24, and 3.

6. **Step 1**
Determine how the numbers change from term to term.
$16 - 2 = 14$
$14 - 2 = 12$
$12 - 2 = 10$
$10 - 2 = 8$

The pattern rule can be described as $n - 2$.

Step 2
Apply the rule.

- Start with a row of 16 stars.
- The second row will have two fewer stars (14).
- The third row will have two fewer stars (12).
- The fourth row will have two fewer stars (10).
- The fifth row will have two fewer stars (8).

The concrete or pictorial representation can look like this.

Row 1 ☆☆☆☆☆☆☆☆☆☆☆☆☆☆☆☆

Row 2 ☆☆☆☆☆☆☆☆☆☆☆☆☆☆

Row 3 ☆☆☆☆☆☆☆☆☆☆☆☆

Row 4 ☆☆☆☆☆☆☆☆☆☆

Row 5 ☆☆☆☆☆☆☆☆

7. **Step 1**
Determine the relationship.

Each term number is divided by 3 to result in the next term.
$324 \div 3 = 108$
$108 \div 3 = 36$
$36 \div 3 = 12$
$12 \div 3 = 4$

Step 2
Identify the error.

The fifth term is incorrect. It does not follow the pattern rule of $n \div 3$.
$12 \div 3 = 4$, not 3.

8. **Step 1**
Identify the relationship.

Since four people can sit on each side of a square table, 16 people can be seated at each table. The pattern is to start with 16 and add 16 to each term to get the next term.

Step 2
Create a table of values.

Start with 16, and apply the rule of +16 up to the number 112.

Term Number (Number of tables)	**Term** (Number of people)
1	16
2	$16+16=32$
3	$32+16=48$
4	$48+16=64$
5	$64+16=80$
6	$80+16=96$
7	$96+16=112$

Mrs. Smith will need seven tables to seat 112 people.

PRACTICE EXERCISES ANSWERS AND SOLUTIONS

1. **Step 1**
Determine how the numbers change from term to term.

The pattern starts with 2 and each term is multiplied by 4 to get the next term.

$2\times4=8$
$8\times4=32$
$32\times4=128$

Step 2
Describe the relationship.

The relationship between any two consecutive terms in the pattern is that each successive term is 4 times greater than the previous term.

The relationship can be expressed as the rule $\times4$ or $n\times4$, where n represents any given term.

3. **Step 1**
Determine the relationship.

$1+11=12$
$12+11=23$
$23+11=34$

Each term is 11 more than the previous term.

Step 2
Make a prediction.

To get from term number 4 to 7, add 11 for three more counts. A good prediction is 67.

Step 3
Verify the prediction.

Term Number	**Term**
5	$34+11=45$
6	$45+11=56$
7	$56+11=67$

5. **Step 1**
Determine the relationship.

Only 4 and 2 are consecutive numbers, which means that 4 is divided by 2 to get 2.

This shows that the pattern rule is to divide each term by 2 to get the next term.

The rule can be expressed as $\div 2$ or $n\div2$ where n represents any term.

Step 2
Start with 64 and apply the pattern rule of $\div 2$.

Term Number	**Term**
1	64
2	$64\div2=32$
3	$32\div2=16$
4	$16\div2=8$

Step 3
Verify the pattern rule for terms 5 and 6.

5	$8 \div 2 = 4$
6	$4 \div 2 = 2$

The missing terms are 32, 16, and 8.

7. **Step 1**
Determine the relationship.

Each term number is 5 more than the previous term.
$87 + 5 = 92$
$92 + 5 = 97$
$97 + 5 = 102$
$102 + 5 = 107$

Step 2
Identify the error.

The fourth term is incorrect. It does not follow the pattern rule of $n + 5$.
97 + 5 = 102, not 112

Since the fourth term has an error, the fifth term also has an error, because 5 was added to the previous term, which was incorrect.

Practice Test

ANSWERS AND SOLUTIONS

1. **Step 1**
Write the term numbers.

Since there are five terms in the pattern of numbers, write the numbers 1, 2, 3, 4, and 5 on the left side of the table of values.

Term Number	Term
1	
2	
3	
4	
5	

Step 2
Write the terms.

Write the five terms in the term column, in the order they appear.

Term Number	Term
1	1
2	2
3	4
4	8
5	16

3. **Step 1**
Translate the pattern to a table of values.

Term Number	Term
1	0
2	1
3	2
4	3

Step 2
Write the term numbers and terms as ordered pairs.

The ordered pairs are (1, 0), (2, 1), (3, 2), (4, 3).

Step 3
Plot the ordered pairs on a graph.

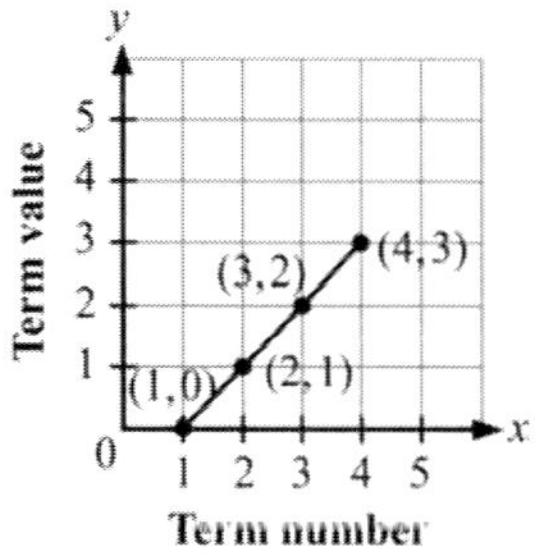

The graph can look like this.

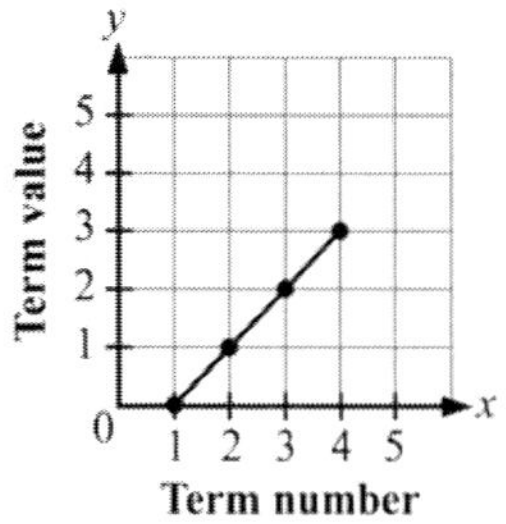

5. Step 1
Determine how the numbers change from term to term.

The pattern starts with 625 and each term is divided by 5.
$625 \div 5 = 125$
$125 \div 5 = 25$
$25 \div 5 = 5$

Step 2
Describe the relationship.

The relationship between any two consecutive terms in the pattern is that each term is divided by five to result in the next term.

The relationship can be expressed as the rule $\div 5$ or $n \div 5$, where n represents any given term.

7. Step 1
Determine the relationship.
$13 + 8 = 21$
$21 + 8 = 29$
$29 + 8 = 37$

Each term is 8 more than the previous term.

Step 2
Make a prediction.

To get from term number 4 to 7, add 8 for three more counts. A good prediction is 61.

Step 3
Verify the prediction.

Term Number	Term
5	$37 + 8 = 45$
6	$45 + 8 = 53$
7	$53 + 8 = 61$

9. Step 1
Determine the relationship.

Only 4 and 11 are consecutive numbers, which means that 7 has been added to 4 to result in a sum of 11.

This shows that the pattern rule is to add 7 to each term to result in the next term. The rule can be expressed as $+7$ or $n + 7$.

Step 2
Start with 11 and apply the pattern rule.

Term Number	Term
1	4
2	11
3	$11 + 7 = 18$
4	$18 + 7 = 25$
5	$25 + 7 = 32$
6	$32 + 7 = 39$

Step 3
Verify the pattern rule for term 6.

The missing terms are 18, 25, and 32.

11. Step 1
Determine the relationship.

Each term number is multiplied by 7 to result in the corresponding term.
$3 \times 7 = 21$
$4 \times 7 = 28$
$5 \times 7 = 35$
$6 \times 7 = 42$
$7 \times 7 = 49$

Step 2
Identify the error.

The fourth term is incorrect. It does not follow the pattern rule of $n \times 7$.
$4 \times 7 = 28$, not 32.

ANGLES

Lesson 1—Identifying and Classifying Angles According to Their Measures

TIME TO TRY ANSWERS AND SOLUTIONS

1. The first angle is a reflex angle because its measure is greater than 180° and less than 360°.

The second angle is a straight angle because its measure is exactly 180°.

PRACTICE EXERCISES ANSWERS AND SOLUTIONS

1. The top angle is an example of an obtuse angle because its measure of 103° is greater than 90° and less than 180°.
$90 < 103 < 180$

The bottom angle is an example of an reflex angle because its measure of 217° is greater than 180° and less than 360°.
$180 < 217 < 360$

3. The angle on the left is an example of an acute angle because its measure of 47° is less than 90°.
$47 < 90$

The angle on the right is an example of a right angle because its measure is 90°. Recall that a square corner in an angle always indicates a right angle.

Lesson 2—Estimating and Determining Angle Measures in Degrees

TIME TO TRY ANSWERS AND SOLUTIONS

1. The given angle is greater than a 90°angle, but is less than a 180° angle. However, the angle is almost a straight line, so it is very close to a 180° angle.

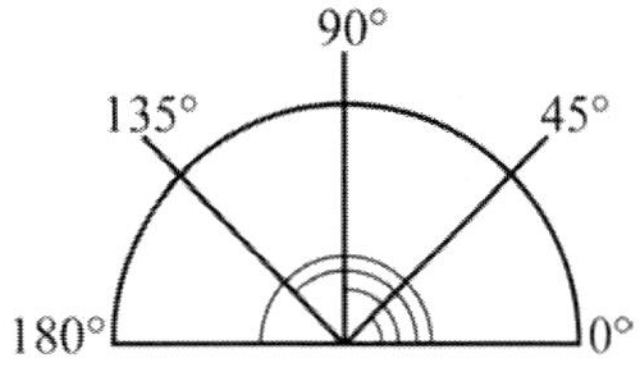

A good estimate of the angle measure is about 170°, which is very close to 180°.

2. **Step 1**
Place the midpoint of the protractor (centre) on the vertex of the angle.

Step 2
Line up the midpoint and the 0° line of the protractor with the bottom ray of the angle.

Step 3
Start from 0° on the inner scale, and count the number of degrees to where the other ray lines up.

The measure of the given angle is 135°.

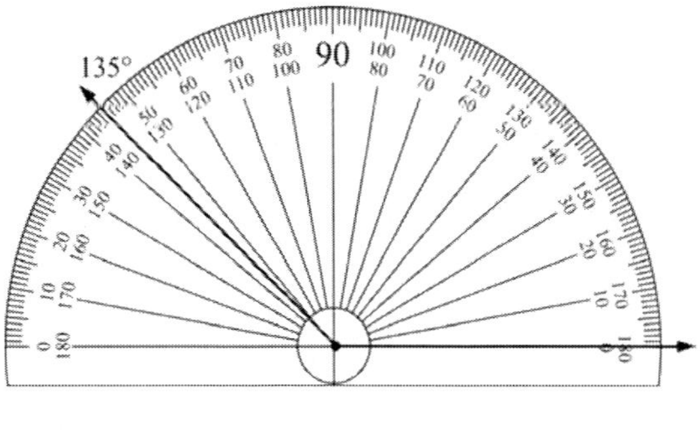

3. **Step 1**
Use a protractor to measure the smaller angle. The smaller angle has a measure of 70°.

Step 2
Subtract 70° from 360°, which is the measure of a full turn.
$360° - 70° = 290°$

The measure of the reflex angle is 290°.

PRACTICE EXERCISES ANSWERS AND SOLUTIONS

1. The given angle is greater than a right angle and less than a straight angle.

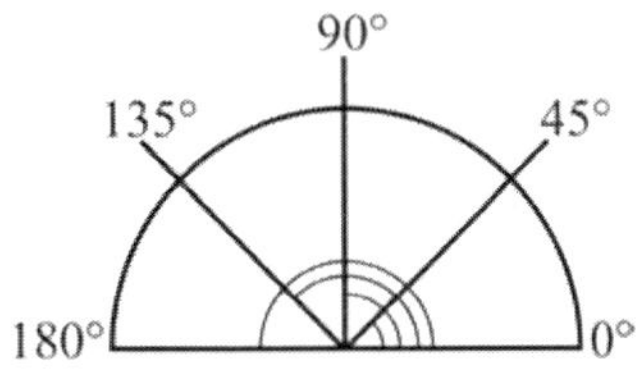

Since the given angle is almost halfway between 90° and 135°, but a little closer to 90°, a good estimate would be about 110°.

3. **Step 1**
First method

Start at the 0° line on the inner scale on the right side of the protractor, and count the degrees by 10s until you reach ray *AB*.
10, 20, 30, …, 120, 130

Step 2
Second method

Add the measures of the two smaller angles. Angle *DBC* is a right angle, so it is 90°. Angle *ABD* has a given measure of 40°.
$90° + 40° = 130°$

The measure of angle *ABC* is 130°.

5. **Step 1**
Position the protractor.

Place the centre of the protractor on the vertex, and line up the zero with the first ray of the angle.

Step 2
Read the scale.

Count by 10s from 0 to where the second ray intersects the curved edge of the protractor.
10, 20, …, 70, 80, 90

The measure of the given angle is 90°.

Lesson 3—Drawing Angles With and Without a Protractor

TIME TO TRY ANSWERS AND SOLUTIONS

1. Recall that a 180° angle looks like a straight line. Be sure to label the midpoint where the two rays meet.

2. **Step 1**
Draw a ray.

Step 2
Place the protractor on the ray so that the midpoint of the protractor lines up with the endpoint of the ray.

Start at the 0 on the inner scale, as shown on this protractor, and count up to 77°. Make a tick at 77°.

Step 3
Remove the protractor.

Using a ruler, join the tick at 77° to the endpoint of the original ray.

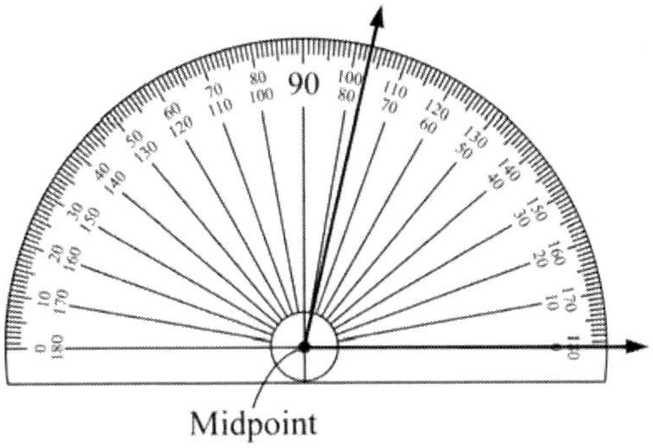

3. **Step 1**
Calculate the size of the smaller (inner) angle.
$360° - 300° = 60°$

Step 2
Draw the 60° angle.

Label the outside rotation as 300°.

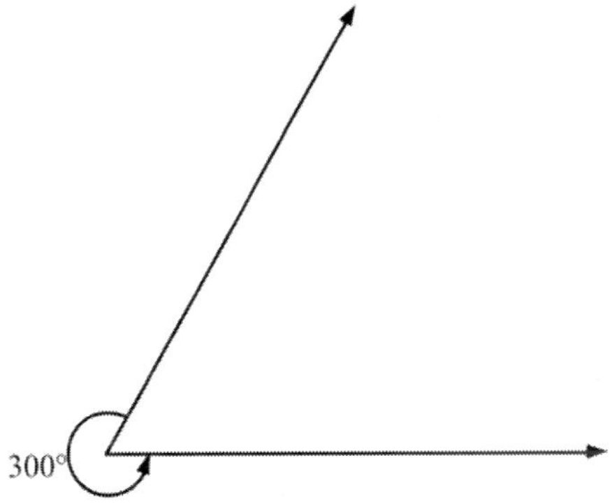

PRACTICE EXERCISES ANSWERS AND SOLUTIONS

1. To construct a 30° angle, you need to count from 0 to 30° on your protractor.

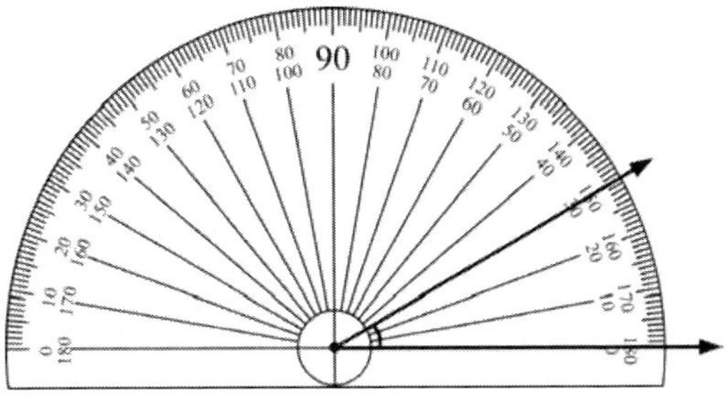

3. Recall that the reference angle of 90° forms a square corner where the two rays meet at a vertex. The rays are perpendicular to each other.

Three orientations of the 90° angles you sketch could look like these:

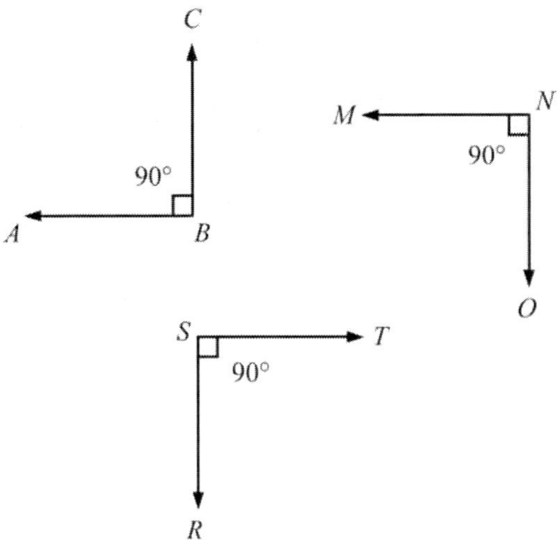

5. **Step 1**
Draw a ray pointing to the left.

Step 2
Place your protractor on the ray so the midpoint of the protractor lines up with the vertex.

Step 3
Start with the 0° on the left side of the protractor and count to 63°, marking the degree with a tick.

Join the tick and the vertex of the angle. Label the angle.

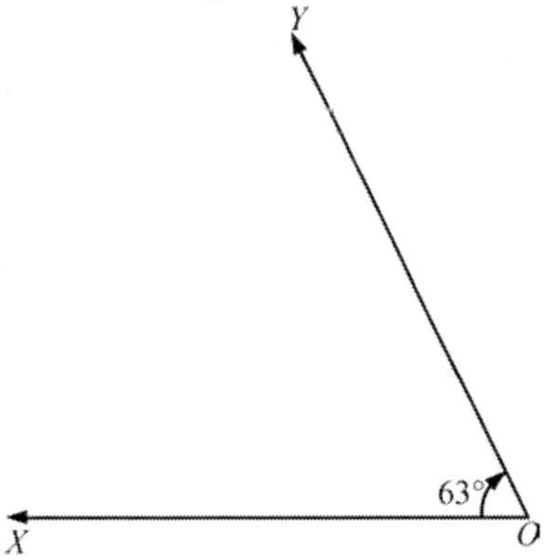

Lesson 4—The Sum of Interior Angles

TIME TO TRY ANSWERS AND SOLUTIONS

1. Recall that a square corner marks a right angle with a measure of 90°.

Step 1
Calculate the sum of the given angles.
$90° + 45° = 135°$

Step 2
Subtract 135° from 180°.
$180° - 135° = 45°$

The missing angle measure is 45°.

2. **Step 1**
Calculate the sum of the three given angles.
$125° + 118° + 56° = 299°$

Step 2
Calculate the difference between the sum and 360°.
$360° - 299° = 61°$

The missing angle measure is 61°.

Check your work by adding the four angle measures.
$125° + 118° + 56° + 61° = 360°$

PRACTICE EXERCISES ANSWERS AND SOLUTIONS

1. Since the interior angles of all triangles add up to 180°, add the two given angle measures, and subtract the sum from 180°.
$57° + 48° = 105°$
$180° - 105° = 75°$

The measure of the angle marked with an x is 75°.

3. **Step 1**
Explain how to find the missing measures.

Since square corners are used to show right angles that have a measure of 90°, you know that both angles have a measure of 90°.

Step 2
Justify your answer.

Add the four measures.

If the sum is 360°, you know that the angle measures are correct.
$110° + 70° + 90° + 90° = 360°$

5. **Step 1**
Calculate the sum of the given angles.
$125° + 118° + 56° = 299°$

Step 2
Calculate the difference between the sum and 360°.
$360° - 299° = 61°$

The missing angle measure is 61°.

Practice Test

ANSWERS AND SOLUTIONS

1. Angle *A* is an acute angle because it is less than a right angle of 90°.

   Angle *B* is a right angle because the two perpendicular rays meet at a square corner, forming a 90° angle.

   Angles *C* and *D* are both obtuse angles because they are both greater than 90° and less than 180°.

3. Compare the size of the angle to the benchmark angles of 45° and 90°.

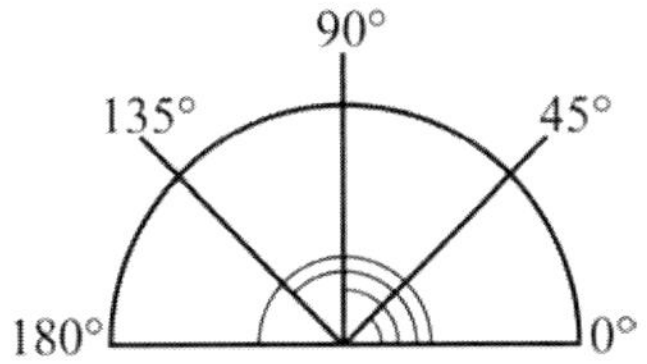

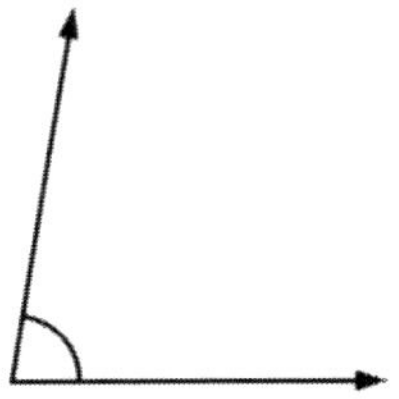

   The angle is between 90° and 45°, but it is closer to 90° than 45°.

   A good estimate would be about 80°.

5. **Step 1**
   Determine the measure of angle *ABC*.

   Angle *ABC* has a measure of 100°.

   Step 2
   Explain how you arrived at your answer.

   Since the angle is facing left, start from the 0 on the left side of the protractor (the outer scale) and count by 10s until you reach the other ray of the angle.
   10, 20, 30,…, 80, 90, 100

7. **Step 1**
   Measure the smaller (inside) angle.

   The smaller (inside) angle has a measure of 60°.

   Step 2
   Calculate the measure of the reflex angle.
   $360° - 60° = 300°$

   The given reflex angle measures 300°.

9. The angle you draw could look like this:

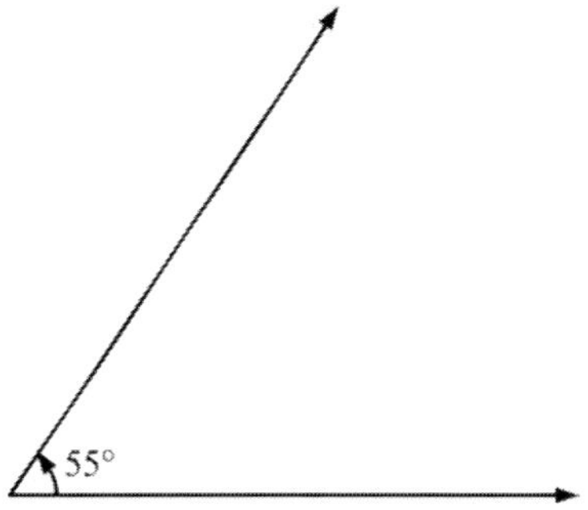

   To draw an angle in this orientation, follow these steps:
   1. Draw a ray pointing to the right.
   2. Place the protractor on the ray, and mark the point that shows 55°.
   3. Join the point to the vertex of the ray with a straight line.
   4. Mark the angle with a small arc, and label the measurement.

11. **Step 1**
Calculate the measure of the smaller (inside) angle.
$360° - 245° = 115°$

Step 2
Draw an angle measuring 115°.

Step 3
Mark the outside rotation with an arc, and label the measurement.

The angle you draw could look like this:

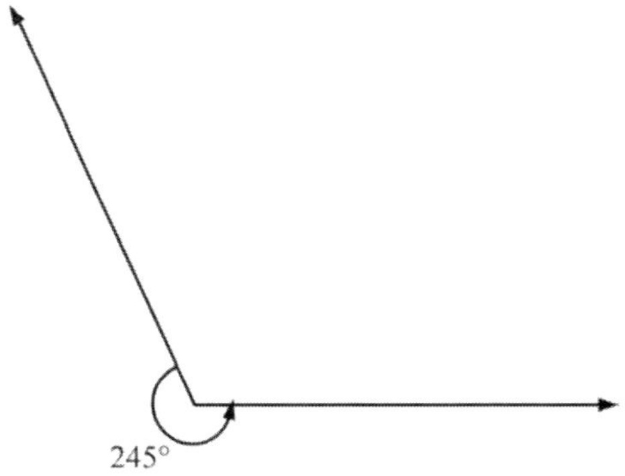

13. Since all the angles in a quadrilateral add up to 360°, to find the missing interior angle of the quadrilateral, add the three given angles, and subtract the total from 360°.
$79° + 63° + 117° = 259°$
$360° - 259° = 101°$

The missing interior angle of the given quadrilateral is 101°.

TWO-DIMENSIONAL SHAPES

Lesson 1—Comparing Triangles

TIME TO TRY ANSWERS AND SOLUTIONS

1. This triangle is an isosceles triangle because two side lengths are the same (6 cm) and one side length is **different** (10 cm).

2. This is an obtuse triangle because one of the angles (120°) is an obtuse angle. Recall that an obtuse angle is greater than 90° and less than 180°.

3. This is an isosceles triangle because it has two equal side **lengths**, as indicated by the ticks on two sides.

The triangle can also be classified as an acute triangle because all three angles are less than 90°.

4. The **triangles** in group 1 all have right angles, as shown by the square corners.

None of the triangles in group 2 have right angles.

The sorting rule used was *right angles or no right angles.*

PRACTICE EXERCISES ANSWERS AND SOLUTIONS

1. This is **an** acute triangle because all three angles are less than 90°.

3. The triangle can be classified as an equilateral triangle because all three sides are equal in length.

   The triangle can also be classified as an acute triangle because each of the angles in an **equilateral** triangle have a measure of 60° (less than 90°).

5. **Step 1**
   Identify the triangles that belong to the group labelled "all equal side lengths."

   Triangle C is an equilateral triangle, so it is the only triangle of the set that belongs to this group.

   Step 2
   Identify the triangles that belong to the group labelled "not all equal side lengths."

   Triangles A and B belong to this group because neither of them have all equal sides.

Lesson 2—Drawing and Replicating Triangles

TIME TO TRY ANSWERS AND SOLUTIONS

1. **Step 1**
   Draw a base line of 7 cm.

   Step 2
   At one endpoint of the line segment, measure a 60° angle using a protractor.

   At the other endpoint of the line segment, measure a second 60° angle using a protractor.

   Extend the rays of the angles drawn until they intersect. This will form the third 60° angle.

   Step 3
   Label the base line and the three 60° angles.

   The triangle will look like this:

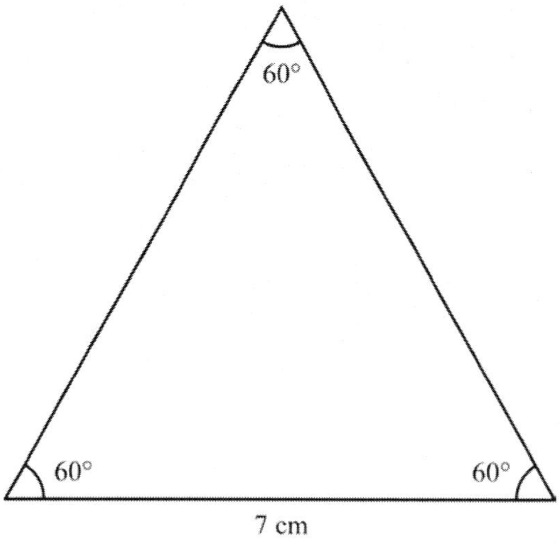

2. Replicate the triangle in a different orientation. Show that the two are congruent by labelling the replicated triangle the same way that the original triangle is labelled.

   This triangle is an example of one different orientation.

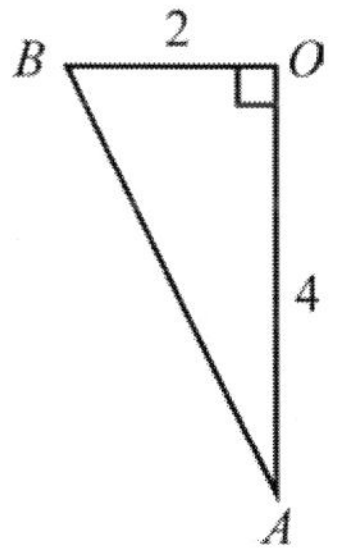

PRACTICE EXERCISES ANSWERS AND SOLUTIONS

1. **Step 1**
Draw the base line of 7 cm.

Step 2
Use a protractor to draw an angle of 30° at one end of the line. Draw an angle of 60° at the other end of the line.

Extend the rays of the two drawn angles until they intersect. This will make the third angle.

Step 3
Label the base line and the two angles you measured.

The triangle could look like this.

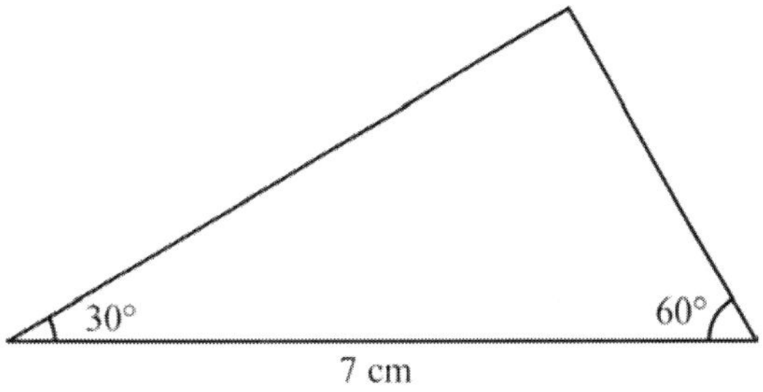

3. These triangles are three examples of different orientations of the given triangle:

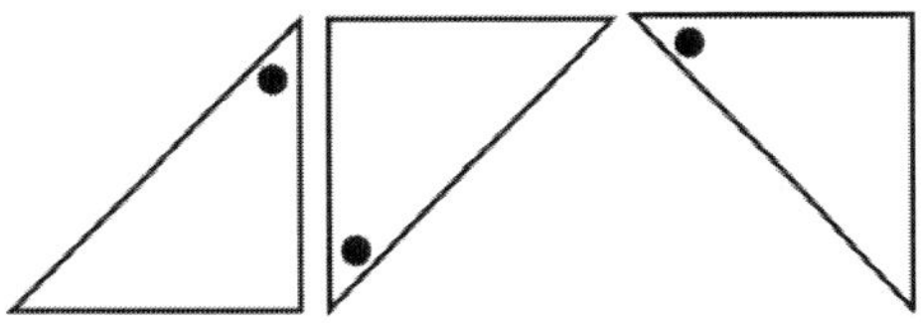

Lesson 3—Regular and Irregular Polygons

TIME TO TRY ANSWERS AND SOLUTIONS

1. **Step 1**
Sort the shapes using the sorting rule.

Regular polygons	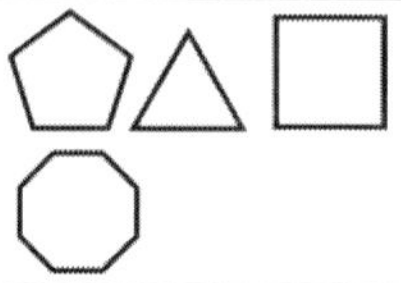
Irregular polygons	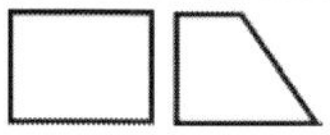

Step 2
Explain the rule.

Shapes that have all congruent sides (same lengths) and all congruent angles (same measure of degrees) are to be placed in the group labelled "regular polygons."

Shapes that do not have all congruent sides and angles are to be placed in the group labelled "irregular polygons."

PRACTICE EXERCISES ANSWERS AND SOLUTIONS

1. **Step 1**
Identify the type of shape used.

Carla used a regular polygon to create the design.

Step 2
Justify your answer.

Carla used a regular hexagon that had six same-sized side lengths and six same-measure angles.

3. **Step 1**
Sort the shapes.

All equal angles	Frames P, Q, R
Not all equal angles	Frame S

Step 2
Justify the sorting.

Every angle in the shape must be the same size to belong to the group labelled "all equal angles."

Frame P is a square and frame Q is a rectangle, so they both have four right angles. Frame R is a regular pentagon, so it has five equal angles.

To belong to the group labelled "not all equal angles," the shapes must have some angles that are not equal.

Frame S has some equal angles, but all four angles are not equal.

5. **Step 1**
Identify the sorting rule.

The sorting rule can be expressed in different ways:

1. Regular polygons or irregular polygons
2. All equal side lengths or not all equal side lengths
3. All equal angle measures or not all equal angle measures
4. All equal side lengths and angle measures or not all equal side lengths and angle measures

Step 2
Explain the sorting rule.

All the shapes placed in group A are regular polygons because each shape has all equal side lengths and all equal angle measures.

All the shapes placed in group B are irregular polygons because none of the shapes have all equal side lengths and all equal angle measures.

Lesson 4—Congruency in Regular Polygons

TIME TO TRY ANSWERS AND SOLUTIONS

1. **Step 1**
Identify the congruent triangle.

Triangle *D* is congruent to triangle *A*.

Step 2
Explain how you can demonstrate the congruency of the two triangles.

One way to demonstrate the congruency is to trace triangle *A* onto a piece of paper and then cut the shape out. Place the cut triangle over each given triangle to see which one matches all the side lengths and angles. Triangles *D* and *A* match in side lengths and angle measures.

PRACTICE EXERCISES ANSWERS AND SOLUTIONS

1. **Step 1**
Identify the shape used.

Gino used shape C to make the set of congruent triangles.

Step 2
Explain your answer.

All the shapes in the set Gino made are equilateral triangles. Shape C is the only triangle shown that is an equilateral triangle and also has the same-sized sides as the three drawn shapes in the set.

3. **Step 1**
Determine if the shapes are congruent.

Yes, the two squares are congruent.

Step 2
Use measurement to justify your answer.

Count the number of square units for the length and width of each square.

The width of each square is three units, and the length of each square is three units. Therefore, square A and square B are congruent.

Practice Test

ANSWERS AND SOLUTIONS

1. The given triangle is classified as an acute triangle because the three angle measures are all less than 90°.

3. Based on its interior angles, the given triangle can be classified as a right triangle because it has one right angle (90°).

   Based on its side lengths, the triangle can be classified as a scalene triangle because the three sides are all different lengths.

5. **Step 1**
Draw a base line of 6 cm.

Step 2
Draw the angles.

Use a protractor to draw a 60° angle at one end of the line.

Draw another 60° angle at the other end of the line.

The intersection of the lines forms the third angle.

Step 3
Label the base line and the two given angles.

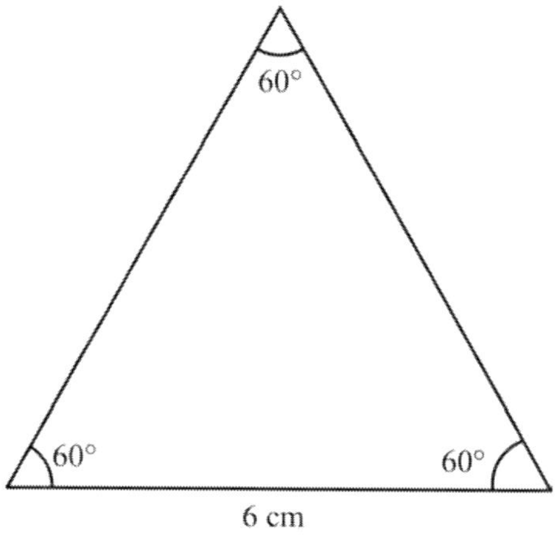

7. **Step 1**
Sort the shapes.

Polygon	
Non-polygon	

Step 2
Explain the sorting rule.

To belong to the polygon group, all the shapes must be closed figures made out of straight lines that intersect at the vertices.

The circle is a closed figure, but it belongs to the non-polygon group because it is not made of straight lines.

9. **Step 1**
Determine if all the shapes are congruent.

No, all the shapes Eric drew are not congruent. Only two of the shapes are congruent.

Step 2
Justify your answer.

All four shapes are regular pentagons, and all four pentagons have equal-sized angles. However, they are not all congruent because their side lengths are not all the same length. Only the two middle pentagons are congruent.

TRANSFORMATIONS

Lesson 1—Successive and Combined Transformations

TIME TO TRY
ANSWERS AND SOLUTIONS

1. **Step 1**
Identify the first transformation.

The first transformation is a slide. The figure slides three units to the right. The slide can also be expressed as R3.

Step 2
Identify the second transformation.

The second transformation is also a slide. The figure slides vertically upward four units. The slide can also be expressed as U4.

Step 3
Describe the successive transformations.

To result in the image shown, the given shape slid to the right for three units and then slid up for four units. The transformations can also be expressed as R3, U4.

Step 4
Determine congruency through measurement.

Each vertex in the shape in position 1 is three squares from the next vertex. Each vertex in the image in position 2 is three squares from the next vertex.

All the angles in the figure in position 1 are right angles, and all the angles in the image in position 2 are right angles.

Since all the side lengths are the same and all the angles have the same measure, the image in position 2 is congruent to the figure in position 1.

2. **Step 1**
Identify the first transformation.

The first transformation is a reflection.

The figure in position *A* is two units above the flip line. It is flipped so the image is now two units below the flip line, as shown in position *B*.

Step 2
Identify the second transformation.

The figure in position *B* is rotated 90° clockwise then slid three units right, as shown in position *C*.

Step 3
Describe the combined transformations.

The shape in position *A* was reflected and then rotated and slid to result in the image seen in position *C*.

3.

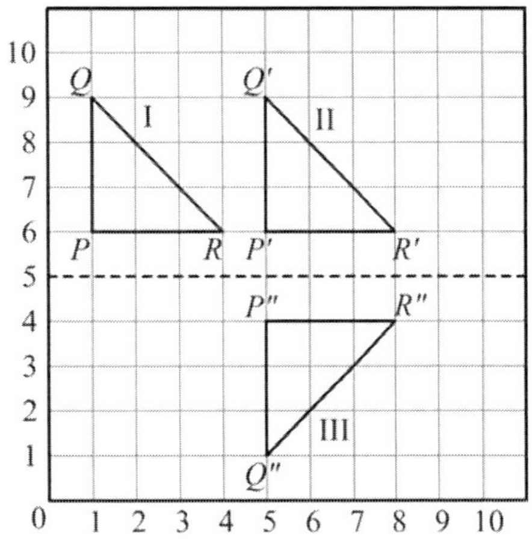

The image in position II shows the first translation, which is a slide to the right of four units (R4).

The image in position III shows the image after the second transformation, which is a flip across the line of reflection.

PRACTICE EXERCISES ANSWERS AND SOLUTIONS

1. When describing slides, describe the horizontal slide first and then the vertical slide.

Step 1
Identify the first transformation.

The first transformation is a slide of three units to the right (R3).

Step 2
Identify the second transformation.

After the slide to the right, the shape slid two units downward (D2).

Step 3
Describe the successive transformations.

Shape *PQRS* slid to the right for three units and then slid down for two units (R3, D2) to result in image *P'Q'R'S'*.

3.

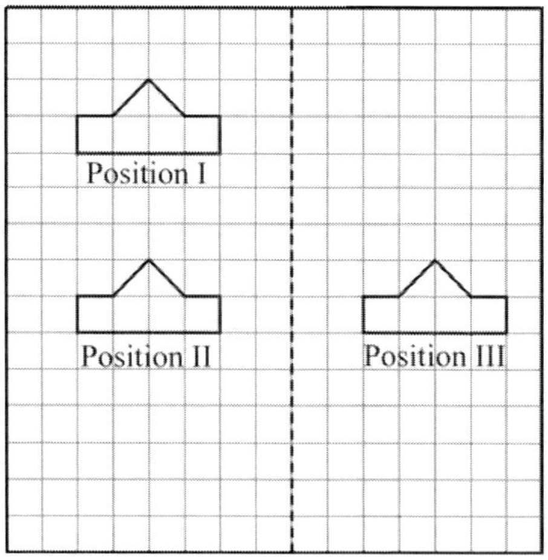

The image in position II shows the first translation, a slide downward of five units (D5).

The image in position III shows the image after the second transformation, a flip across the line of reflection.

5. **Step 1**
Identify the first transformation.

The first transformation is a slide of six units to the right (R6).

Step 2
Identify the second transformation.

After the slide to the right, the shape slid five units downward (D5).

Step 3
Describe the successive transformations.

Shape *ABCD* slid to the right for six units and then slid down for five units (R6, D5) to result in image $A'B'C'D'$.

7. **Step 1**
Draw and label the first transformation.

Figure *ABCD* is reflected across the flip line, so the new image is located at $A'(9,5)$, $B'(7,8)$, $C'(6,8)$, and $D'(6,5)$.

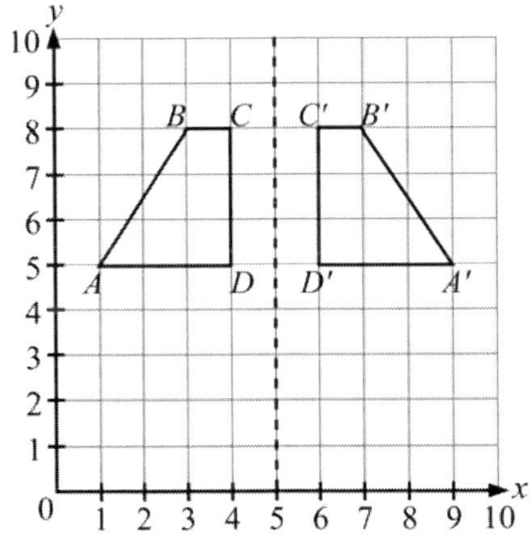

Step 2
Draw and label the second transformation.
Image $A'B'C'D'$ is translated down three units, so the new figure is located at $A''(9,2)$, $B''(7,5)$, $C''(6,5)$, and $D''(6,2)$.

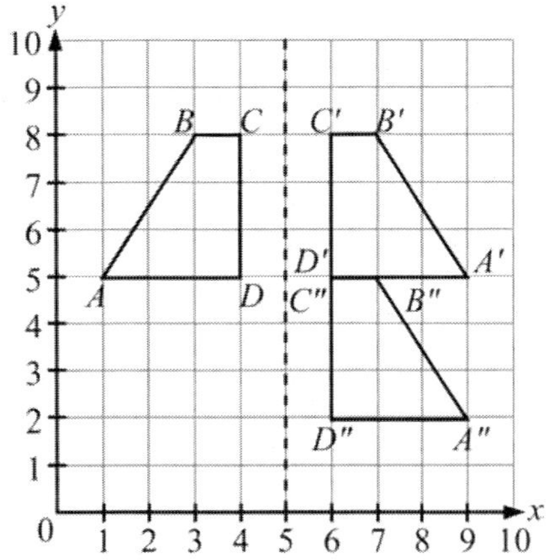

Lesson 2—Designs Created from Successive Transformations

TIME TO TRY ANSWERS AND SOLUTIONS

1. **Step 1**
Start the design.

Use the original shape for the first shape of the design.

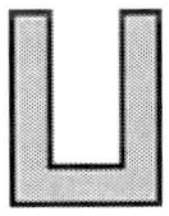

Step 2
Continue the design.

Slide the original shape diagonally upward (↗) to create the next position.

Slide the second shape diagonally downward (↘) to create the third position.

Your design should be similar to the one shown here.

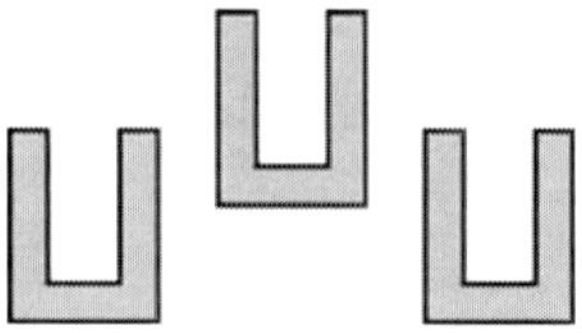

2. **Step 1**
Identify the basic shape.

The basic shape used was a set of five concentric 90° angles.

Step 2
Describe the transformations used.

The basic shape was translated five times to the right to create the given design.

PRACTICE EXERCISES ANSWERS AND SOLUTIONS

1. The original shape used is a set of three concentric angles.

3. The original image was translated to the right three times to create the design.

5. **Step 1**
Identify the shape used to create the design.

The design is created from five congruent hexagons.

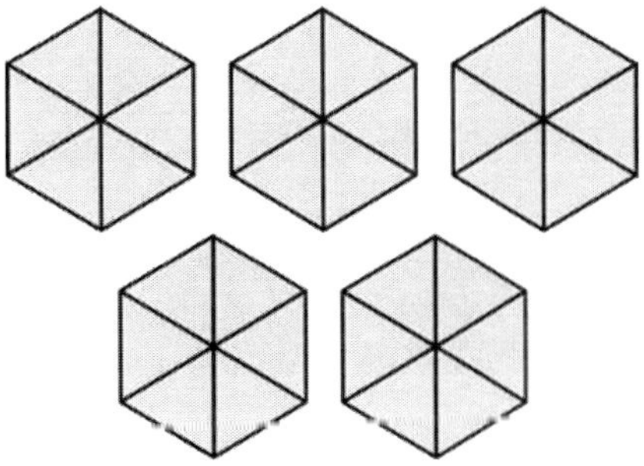

Step 2
Identify the shape used to create each hexagon.

Each hexagonal shape is made from rotated triangles.

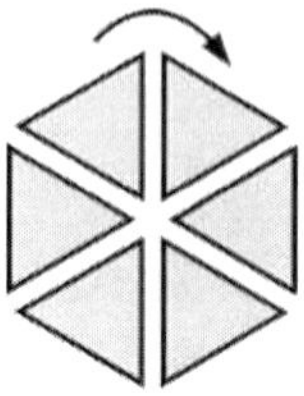

Therefore, the original starting shape of the design is a triangle.

7. The transformations used to create the design are repeated reflections.

The first shape is reflected to create the second shape of the design. The second shape is then reflected to create the third shape of the design, and so on.

Each time the shape is flipped to the right (→), a reflection, or opposite image, of the previous shape will result.

Practice Test

ANSWERS AND SOLUTIONS

1. The transformations used to create the design are repeated reflections.

The original shape is reflected to create the second shape of the design. The second shape is then reflected to create the third shape of the design, and so on.

Each time the shape is flipped to the right (→), a reflection, or opposite image, of the previous shape will result.

3. The design could have been created from either successive translations or successive reflections.

Since the design in the square has vertical and horizontal symmetry, translations and reflections will appear identical.

5. **Step 1**
Identify the first transformation.

The first transformation is a slide to the right of six units (R6).

Step 2
Identify the second transformation.

The second transformation is a slide downward of two units (D2).

This is the image shown in position II.

Step 3
Identify the third transformation.

The image shown in position II is flipped across the line of reflection to result in the image shown in position III.

Step 4
Describe the combined transformations.

To move from position I to position III, the shape slid to the right for six units, slid downward for two units, and then flipped over the line of reflection.

7. The transformations used to create the design are repeated rotations and reflections.

The original shape is the capital letter Q. The Q is rotated 90° clockwise and reflected to create the second image. The second image is rotated 90° clockwise to create the third image. These two successive transformations are repeated to create the given design.

9. There are actually three possible combinations of transformations that could move shape R into the position of image S.

Step 1
Describe one possible combination.

The first transformation could be a reflection. If it is a reflection, then the second transformation must be a translation.

The combined transformations would be a reflection and a translation.

Step 2
Describe another possible combination.

The first transformation could be a 180° clockwise rotation. If the first transformation is a rotation, then the next transformation must be a reflection.

The combined transformations would be a rotation and a reflection.

Step 3
Describe another possible combination.

The first transformation could be a translation to the left $(\leftarrow)$. If the first transformation is a translation to the left, the next transformation must be a reflection.

The combined transformations would be a translation and a reflection.

FIRST QUADRANT OF A CARTESIAN PLANE

Lesson 1—Identifying Points and Distances Between Points

TIME TO TRY ANSWERS AND SOLUTIONS

1. Recall that when you write an ordered pair, write the horizontal coordinate (x-axis) first and the vertical coordinate (y-axis) second.

The coordinates for the three given points are as follows:
Point A: (1, 3)
Point B: (4, 7)
Point C: (7, 1)

2. The ordered pairs (1, 2), (2, 3), and (3, 1) were used to produce the triangle.

Point A is 1 unit to the right of the origin and 2 units above it. The ordered pair is (1, 2).

Point B is 2 units to the right of the origin and 3 units above it. The ordered pair is (2, 3).

Point C is 3 units to the right of the origin and 1 unit above it. The ordered pair is (3, 1).

3. **Step 1**
Identify the interval used.

The interval is 1, so you will count by 1s.

Step 2
Start with the horizontal distance.

Count the number of spaces between the two points.

Point C is 9 spaces to the right of point A. You can also describe this as 9R.

Step 3
Describe the vertical distance.

Count the number of spaces between the two points.

Point C is 4 spaces up from point A. You can also describe this as 4U.

Step 4
Put the two descriptions together.

The distance between point A and point C is 9 to the right (horizontal) and 4 up (vertical), or 9R and 4U.

PRACTICE EXERCISES ANSWERS AND SOLUTIONS

1.

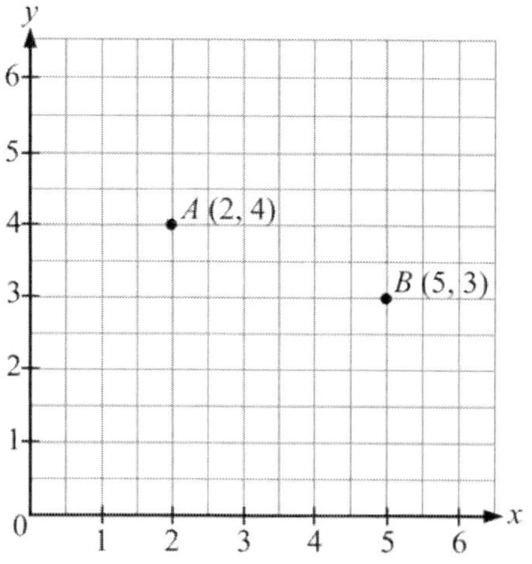

The ordered pair for point A is (2, 4).
The ordered pair for point B is (5, 3).

3. Point A is 3 spaces to the right of point C. Since the interval is 1, count the spaces by 1s.

The distance between the two points is 3 units, or 3L.

5. The distance between points C and B is 8 spaces to the right and 6 spaces down, or 8R, 6D.

Lesson 2—Plotting Points in the First Quadrant

TIME TO TRY ANSWERS AND SOLUTIONS

1. **Step 1**

Plot the ordered pair (1, 7).

- Start at the origin.
- Move to the right along the x-axis until you get to 1 (the first number in the ordered pair).
- Move up along the y-axis until you get to 7 (the second number in the ordered pair).
- Plot the point where these two lines intersect.

Step 2

Plot the ordered pair (3, 1).

- Start at the origin.
- Move to the right along the x-axis until you get to 3 (the first number in the ordered pair).
- Move up along the y-axis until you get to 1 (the second number in the ordered pair).
- Plot the point where these two lines intersect.

Step 3

Plot the ordered pair (5, 3).

- Start at the origin.
- Move to the right along the x-axis until you get to 5 (the first number in the ordered pair).
- Move up along the y-axis until you get to 3 (the second number in the ordered pair).
- Plot the point where these two lines intersect.

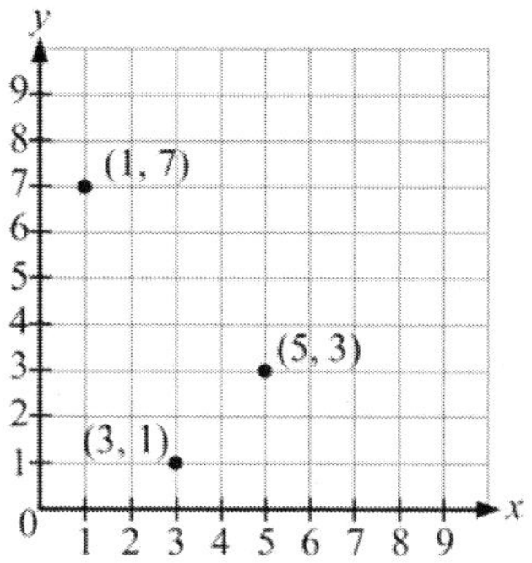

2. **Step 1**

Plot the ordered pair (2, 8).

- Start at the origin.
- Move to the right along the x-axis until you get to 2 (the first number in the ordered pair).
- Move up along the y-axis until you get to 8 (the second number in the ordered pair).
- Plot the point where these two lines intersect.

Step 2

Plot the ordered pair (3, 5).

- Start at the origin.
- Move to the right along the x-axis until you get to the line that represents 3 (the first number in the ordered pair). The 3 is the line between the numbered lines for 2 and 4.
- Move up along the y-axis until you get to the line that represents 5 (the second number in the ordered pair). The 5 is the line between the numbered lines for 4 and 6.
- Plot the point where these two lines intersect.

Step 3
Plot the ordered pair (8, 3).

- Start at the origin.
- Move to the right along the x-axis until you get to 8 (the first number in the ordered pair).
- Move up along the y-axis until you get to the line that represents 3 (the second number in the ordered pair). The 3 is the line between the numbered lines for 2 and 4.
- Plot the point where these two lines intersect.

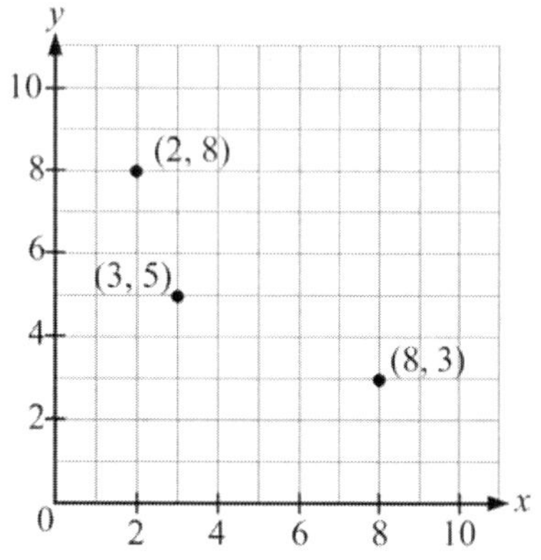

3. The shape created is a parallelogram, as shown.

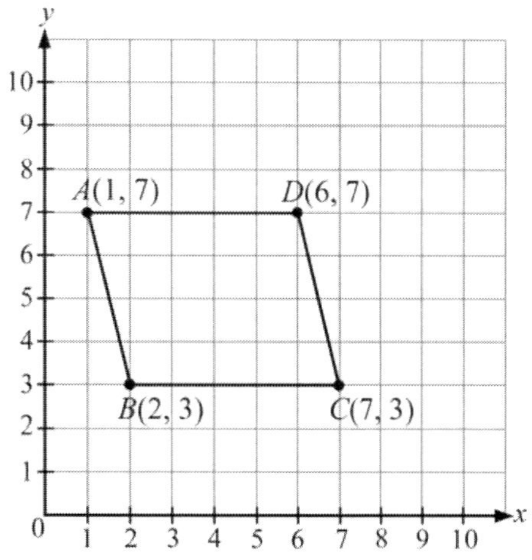

PRACTICE EXERCISES ANSWERS AND SOLUTIONS

1. The plotted points (1, 3), (3, 5), (6, 6), and (8, 0) should be shown as follows:

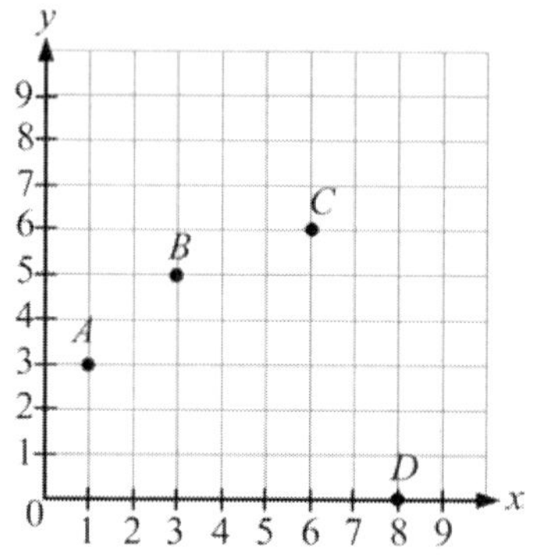

3. The plotted points $A(5, 5)$, $B(10, 12)$, and $C(25, 25)$ should be shown as follows:

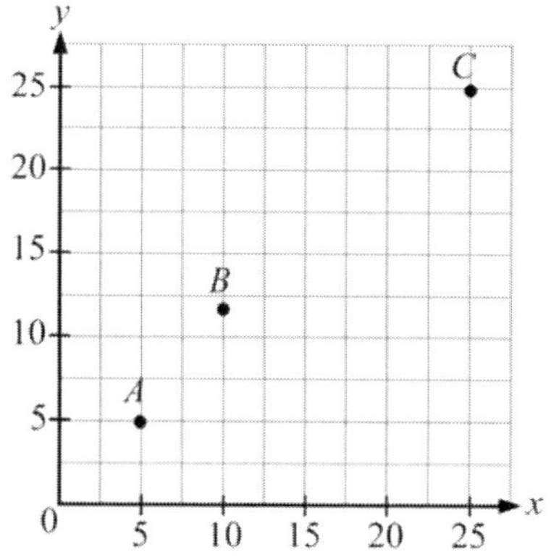

Note that the halfway line between 10 and 15 represents 12.5. Therefore, the point representing 12 should be just below the line.

5. The shape created when the points are connected is a triangle, as shown.

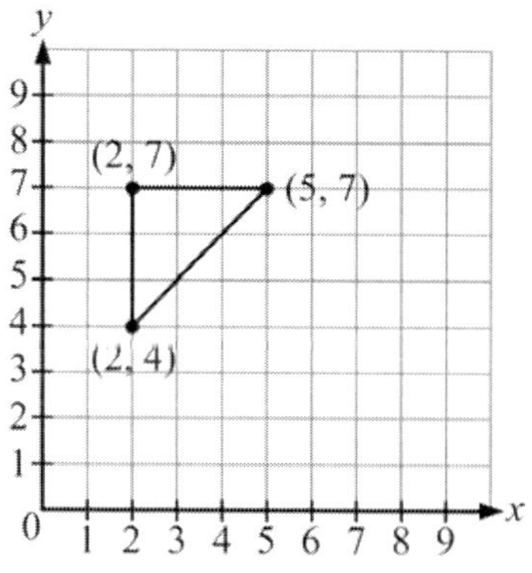

Lesson 3—Single Transformations in the First Quadrant

TIME TO TRY ANSWERS AND SOLUTIONS

1. **Step 1**
Identify the coordinates of the shape labelled Figure 1.

Starting at the top left vertex of the triangle and moving in a counter-clockwise direction, the coordinates of the triangle are (1, 9), (1, 5), and (5, 5).

Step 2
Describe the positional change in vertices.
Figure 1 $\rightarrow$ Figure 2
(1, 9) $\rightarrow$ (6, 9)
(1, 5) $\rightarrow$ (6, 5)
(5, 5) $\rightarrow$ (10, 5)

2. **Step 1**
Draw the translated image.

When you slide the triangle up three squares diagonally ($\nearrow$), the image you draw should look like this:

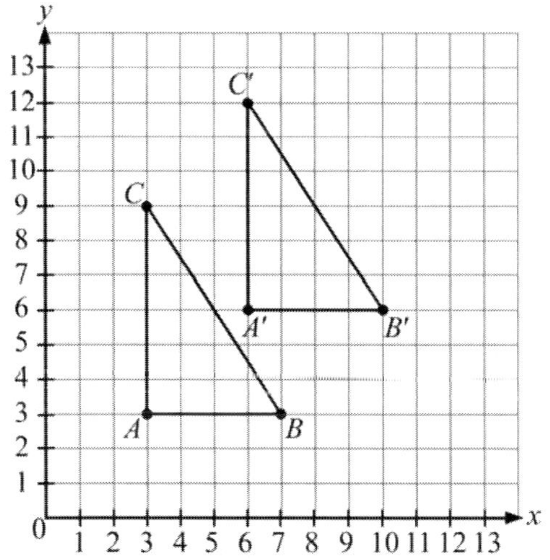

Step 2
Identify the vertices of the translated image, labelled $A'B'C'$.

The coordinates of the vertices of the transformed shape are $A'(6, 6)$, $B'(10, 6)$, and $C'(6, 12)$.

PRACTICE EXERCISES ANSWERS AND SOLUTIONS

1. **Step 1**
Identify the coordinates of the shape in position 1.

Starting at the top left corner of the square and moving in a counterclockwise direction, the coordinates of the square are (1, 4), (1, 1), (4, 1), and (4, 4).

Step 2
Describe the positional change in vertices.
Position 1 $\rightarrow$ Position 2
(1, 4) $\rightarrow$ (1, 10)
(1, 1) $\rightarrow$ (1, 7)
(4, 1) $\rightarrow$ (4, 7)
(4, 4) $\rightarrow$ (4, 10)

3. **Step 1**
Identify the coordinates of triangle *ABC* in position 1.

The coordinates of the triangle are *A*(5, 7), *B*(3, 9), and *C*(3, 7).

Step 2
Describe the positional change in vertices.
Position 1 $\rightarrow$ Position 2
$A(5, 7) \rightarrow A(5, 7)$
$B(3, 9) \rightarrow B(7, 5)$
$C(3, 7) \rightarrow C(7, 7)$

Note that vertex *A* stays the same because it is the point of the 180° rotation that transformed the shape into its image.

5. **Step 1**
Draw the reflected image.

The reflected image should look like this:

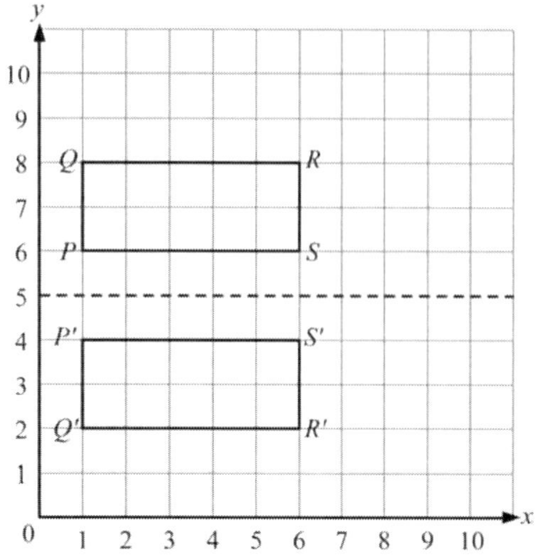

Step 2
Identify the vertices of the reflected image labelled $P'Q'R'S'$.

The coordinates of the vertices of the reflected image are $P'(1, 4)$, $Q'(1, 2)$, $R'(6, 2)$, and $S'(6, 4)$.

Practice Test

ANSWERS AND SOLUTIONS

1. Recall that the first number in an ordered pair refers to the x-axis and the second number refers to the y-axis. Point T is located at (2, 5).

3. Recall that the first number in an ordered pair refers to the x-axis. Point R is located at the origin of the x-axis, so the first number of the ordered pair is 0.

The ordered pair that describes the location of point R is (0, 1).

5. **Step 1**
Determine the interval used.

The interval used is 5.

Step 2
Determine the distance.

Since point A is located on the line marked 15 on the x-axis and point B is located on the line marked 35 on the x-axis, start with 15 and count the spaces by 5s until you get to 35.
$\boxed{15}$...5,10,15...$\boxed{35}$

Another way is to subtract the two numbers.
$35 - 15 = 20$

The distance between point A and point B is 20 units.

7. The plotted points for (10, 0), (30, 15), and (50, 40) look like this:

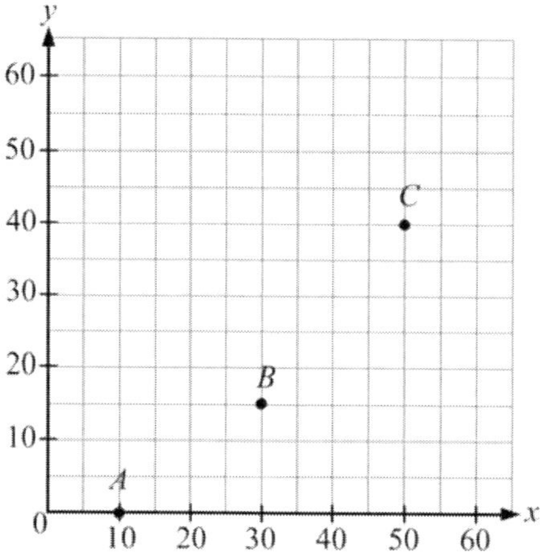

9. The ordered pairs (3, 3), (3, 7), and (6, 3) were used to produce the triangle.

- Point A is 3 units to the right of the origin and 3 units above it. The ordered pair is (3, 3).
- Point B is 3 units to the right of the origin and 7 units above it. The ordered pair is (3, 7).
- Point C is 6 units to the right of the origin and 3 units above it. The ordered pair is (6, 3).

11. Step 1
Draw the reflected image.

The reflected image should look like this:

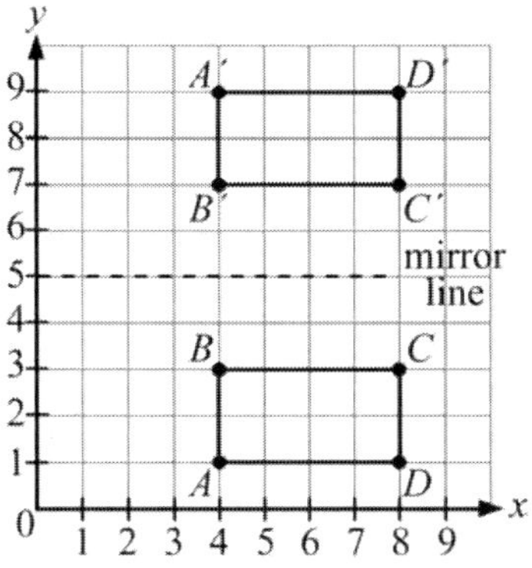

Step 2
Identify the vertices of the reflected image labelled $A'B'C'D'$.

The coordinates of the vertices of the reflected image are $A'(4, 9)$, $B'(4, 7)$, $C'(8, 7)$, and $D'(8, 9)$.

EXPRESSIONS AND EQUATIONS

Lesson 1—Order of Operations

TIME TO TRY ANSWERS AND SOLUTIONS

1. Since there are no brackets and there is no multiplication or division, go directly to step 3: Add and subtract in order from left to right.
$20 - 8 = 12$
$12 + 3 = 15$

$20 - 8 + 3 = 15$

2. **Step 1**
Determine who is correct.

Ben is correct.

Step 2
Justify your answer.

Ben followed the order of operations. He did the operation in the brackets first: $(2 + 3 = 5)$. He then divided 50 by the sum: $50 \div 5 = 10$.

Cora is incorrect because she did not follow the order of operations. She first divided $50 \div 2 = 25$. She then added 3 to the quotient: $25 + 3 = 28$.

3. $15 \div (16 \div 2 - 3) = 3$

First operation	Division inside brackets: $16 \div 2 = 8$	$15 \div (8 - 3)$
Second operation	Subtraction inside brackets: $8 - 3 = 5$	$15 \div 5$
Third operation	Division: $15 \div 5 = 3$	$15 \div (16 \div 2 - 3) = 3$

4. **Step 1**
Identify the type of calculator used.

Darin used a scientific calculator.

Step 2
Justify your answer.

Since Darin typed in the expression just as it was presented ($10+2\times5=$) and got the correct answer, he used a scientific calculator. Scientific calculators are programmed to automatically follow the order of operations.

If the calculator had not been a scientific calculator, typing the expression as is would result in the incorrect answer of 60.

PRACTICE EXERCISES ANSWERS AND SOLUTIONS

1. **Step 1**
Determine who is correct.

Jesse is correct.

Step 2
Justify your answer.

Jesse followed the order of operations. He multiplied $12\times2=24$ first and then subtracted the product from 26: $26-24=2$.

Becky is incorrect because she did not follow the order of operations. She subtracted $26-12=14$ first and then multiplied the product by 2.

3. $(2\times5)\times3\div15=2$

First operation	Multiplication inside brackets: $2\times5=10$	$10\times3\div15$
Second operation	Multiplication: $10\times3=30$	$30\div15$
Third operation	Division: $30\div15=2$	$(2\times5)\times3\div15=2$

5. $(5\times2)\div(4+1)=2$

First operation	Multiplication inside brackets: $5\times2=10$	$10\div(4+1)$
Second operation	Addition inside brackets: $4+1=5$	$10\div5$
Third operation	Division: $10\div5=2$	$(5\times2)\div(4+1)=2$

7. $9+3\times4-6=15$

First operation	Multiplication: $3\times4=12$	$9+12-6$
Second operation	Addition: $9+12=21$	$21-6$
Third operation	Subtraction: $21-6=15$	$9+3\times4-6=15$

Lesson 2—Generalizations Arising from Number Relationships

TIME TO TRY ANSWERS AND SOLUTIONS

1. **Step 1**
Write a formula.

Using the variables l for length, w for width, and P for perimeter, one formula that explains how to determine the perimeter of the rectangle is $P=l+w+l+w$.

Step 2
Explain the formula.

To use this perimeter formula, substitute the length (5 cm) and the width (3 cm) of the rectangle into the formula and then add the four dimensions to determine the distance all around the rectangle.
$P = l + w + l + w$
$P = 5 + 3 + 5 + 3$
$P = 16$ cm

2. **Step 1**
Write an equation.

This equation shows that the two numbers can be multiplied in any order: $a \times b = b \times a$. The variable a represents one number and the variable b represents a different number.

Step 2
Justify your answer with an example.

Let a represent the number 8 and b represent the number 6.

Since $a \times b = b \times a$, then
$8 \times 6 = 6 \times 8$
$48 = 48$

3. **Step 1**
Determine the relationship.

Each term number is 8 more than the previous term number.

Step 2
Write an expression using the variable x.

The relationship between the term numbers (the pattern rule) can be expressed as $x + 8$, where the variable x represents any term number.

4. **Step 1**
Determine how each term changes.

To get to the next term, each term is multiplied by 2 and then 1 is added to the product.

Step 2
Represent the pattern rule as an expression.

The pattern rule can be represented by the expression $2x + 1$.

PRACTICE EXERCISES ANSWERS AND SOLUTIONS

1. **Step 1**
Write a formula.

One formula that can be used is $A = s \times s$.

Step 2
Explain the formula.

The variable s in the formula represents the length of one side of the square. Since all the sides in a square are equal, substitute the length of the side into the formula, and then multiply the number by itself.

3. The value of the variable m is 83.

The commutative property of addition states that numbers can be added in any order; therefore, $83 + 19 = 19 + 83$.

5. **Step 1**
Determine the relationship.

Each number in the In column is doubled and then one is added to the doubled number.

Step 2
Write an expression using the variable k.

The relationship between the In and Out numbers can be expressed as $(k \times 2) + 1$ or $2k + 1$, where the variable k represents any number in the In column.

7. **Step 1**
Determine the relationship.

Each term number of the sequence is multiplied by 3 and then 1 is subtracted from the product.

Step 2
Write an expression.

The pattern rule can be expressed as $(h \times 3) - 1$ or $3h - 1$, where h represents any number in the pattern.

Lesson 3—Expressing Problems as Equations

TIME TO TRY ANSWERS AND SOLUTIONS

1. **Step 1**
Identify the unknown.

The unknown is the number of packages of socks the coach needs to buy. This number will be represented by the variable *m*.

Step 2
Express the problem as an equation.

- The expression that shows the number of boys and girls is $7 + 8$.
- The expression that shows the number of pairs of sock needed is $3 \times m$.
- Since each boy and girl will get a pair of socks, the two expressions are equal to each other.

The problem expressed as an equation is $7 + 8 = 3 \times m$.

Step 3
Solve the problem symbolically.

$$7 + 8 = 3 \times m$$
$$15 = 3 \times m$$
$$15 \div 3 = m$$
$$5 = m$$

Since $m = 5$, the coach needs five packages of socks so that each boy and girl on the team can get one pair of socks.

Step 4
Solve the problem concretely.

Use connecting cubes to help you understand and solve the problem.

Connect three cubes to represent one package of 3 pairs of socks.

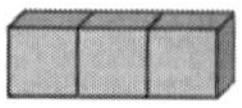

Make groups of cubes until you reach 15 cubes to represent one pair of socks for each person.

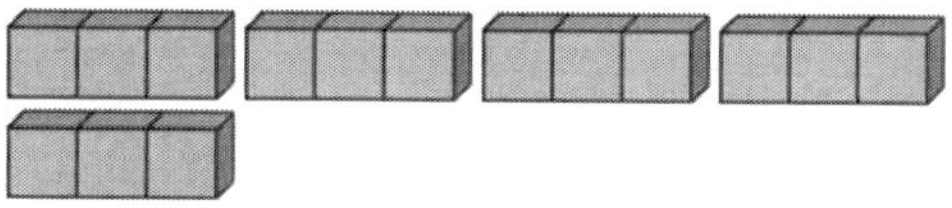

Five sets of cubes represents five packages of socks.

2. **Step 1**
Determine the unknown.

Multiplication can be explained as *total = number of groups × number in each group.* The variable *C* represents the number in each group.

Step 2
Create a problem.

Write a problem where there is a total of 200 items. The items are placed in five groups with an unknown number of items in each group.

Hannah was helping the librarian sort 200 books. She had five equal piles of books on the table. How many books were in each pile?

Step 3
Solve the problem.

$$200 = 5 \times C$$
$$200 \div 5 = C$$
$$200 \div 5 = 40$$
$$C = 40$$

There were 40 books in each pile.

3. **Step 1**
Determine the equation.

The equation that represents the problem is $r + b = 4$.

Step 2
Explain how the values can change.

The value of b will change depending on the value of r.

- If Dylan has one red balloon, then he has three blue balloons.
- If Dylan has two red balloons, then he has two blue balloons.
- If Dylan has three red balloons, then he has one blue balloon.

PRACTICE EXERCISE ANSWERS AND SOLUTIONS

1. **Step 1**
Identify the unknown.

The unknown is Ronald's weight. This number is represented by the variable k.

Step 2
Express the problem as an equation.

Since the combined weight of the boys is 130 kg, the problem can be expressed as $70 + k = 130$.

Step 3
Solve the problem symbolically.

$$70 + k = 130$$
$$k = 130 - 70$$
$$k = 60$$

Ronald's weight is 60 kg.

Step 4
Solve the problem concretely.

Use base ten blocks to help solve the problem.

Represent Bernie's weight of 70 kg with seven ten rods.

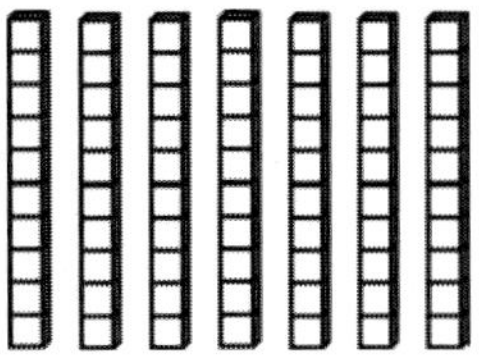

$70 + k = 130$

Represent the total weight of 130 kg with a hundred flat and three ten rods.

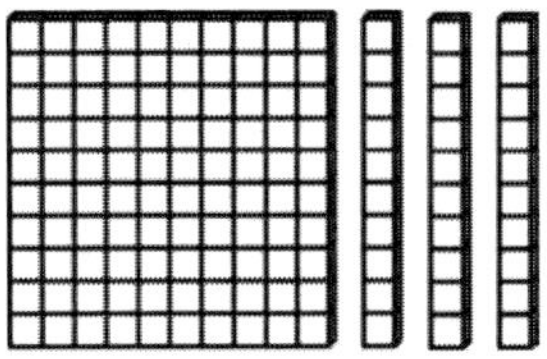

Think of how many ten rods must be added to the seven rods to equal the value of one hundred flat and three ten rods (130).

Start at 70 and count up by 10s as each ten rod is added up to 130.

70:

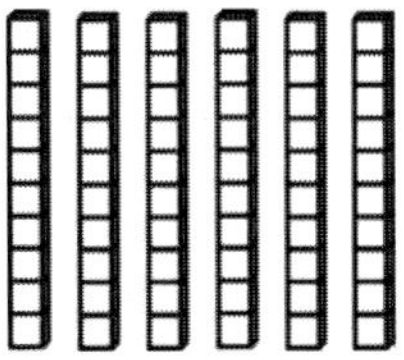

80, 90, 100, 110, 120, 130

Since six ten rods were added, Ronald's weight is 60 kg.

3. The letters l and w are the variables. They represent changing values.

The value of l will change depending on the value of w. This is shown in the following examples:

$A = l \times w$
$18 = 3 \times 6$
$18 = 9 \times 2$

5. In this problem, the unknown variable is the number of rows. Each row in the bus has four seats (except for the front row, which has three seats).

The number of seats on the bus can be represented by the expression $3 + r \times 4$. Since there are 51 seats in total, the equation $3 + r \times 4 = 51$ can be used to express this problem.

Lesson 4—Preserving Equality

TIME TO TRY ANSWERS AND SOLUTIONS

1. Divide the two blocks on the left side by 2, and there is one block left. Divide the eight balls on the right side by 2, and there are four balls left.

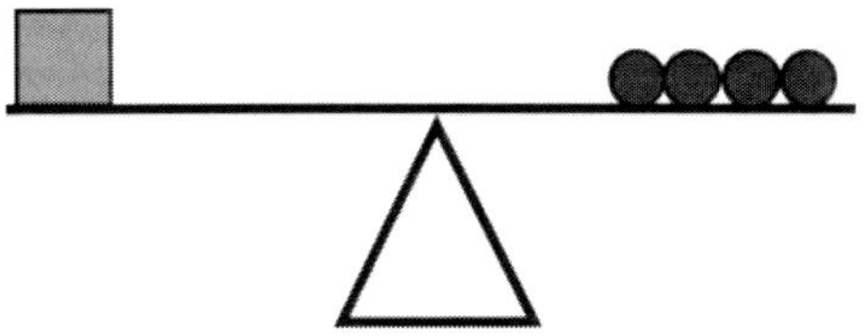

$x = 4$

If two blocks ($2x$) balance eight balls, then one block (x) balances four balls.

$x = 4$
$2x = 8$
$x - 8 : 2$
$x = 4$

Substitute 4 into the equation to show that both sides of the equation are equal.

$2x = 8$
$2(4) = 8$
$8 = 8$

PRACTICE EXERCISE ANSWERS AND SOLUTIONS

1. **Step 1**
Apply the property of division to scale A.

Divide the number of cylinders in half on the left side, and also divide the number of blocks in half on the right side.

Therefore, two cylinders will balance three blocks, as shown in the following diagram. For every two cylinders, three blocks are needed.

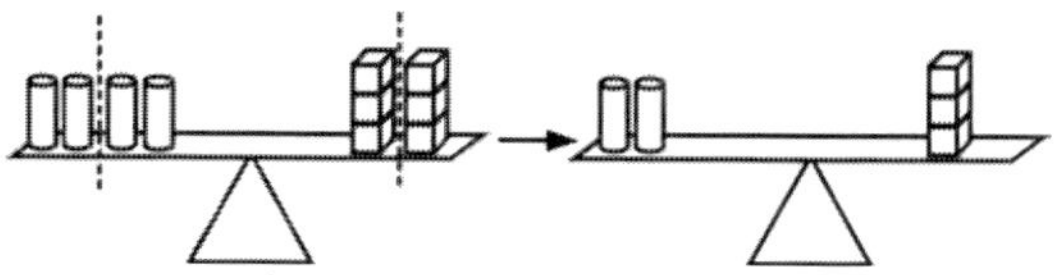

Scale A Scale A

Step 2
Calculate the number of blocks needed to balance six cylinders.

Since there are three groups of two cylinders on the left side of scale B, three groups of three blocks are needed on the right side of scale B to show equality.

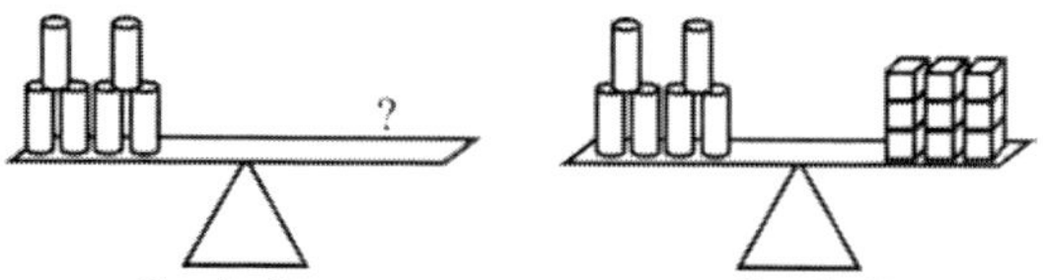

Scale B Scale B

Nine blocks must be placed on the right side of scale B to balance the six cylinders.

3. **Step 1**
Determine the mass of the cube and three cylinders on the right side of the scale.
$5+4+4+4=17$ g

Step 2
Determine the mass of the rectangular block.

Since the scale is balanced, that means that the mass of the rectangular block is equal to the mass of the cube and three cylinders. Equality is preserved.
17 g = 17 g

The mass of the rectangular block is 17 g.

5. **Step 1**
Determine the mass of the six cubes.

Since the scale is balanced, the mass of the six cubes is the same as the mass of one rectangular block.
30 g = 30 g

Step 2
Determine the mass of each cube.

Since there are six cubes on the left side, divide 6 by 6 and 30 by 6 to determine the mass of one cube.

$$6x=30$$
$$6x \div 6=30 \div 6$$
$$x=5$$

Each cube has a mass of 5 g.

7. **Step 1**
Determine the number of cones needed to balance four cubes.

Since three cones = two cubes,
six cones = four cubes.

The number of cones needs to be doubled because the number of cubes is doubled.
$3+3=6$ cones
$2+2=4$ cubes

Six cones are needed to balance four cubes.

Step 2
Determine the number of cones needed to balance one cylinder.

Since six cones = four cubes and
four cubes = one cylinder,
six cones = one cylinder.

Step 3
Determine the number of cones needed to balance two cylinders.

Since six cones = one cylinder,
12 cones = two cylinders.

The number of cones needs to be doubled because the number of cylinders is doubled.
$1+1=2$ cylinders
$6+6=12$ cones

To balance two cylinders, you would need to use 12 cones.

Practice Test

ANSWERS AND SOLUTIONS

1. **Step 1**
Think multiplication.

Since three hearts = two triangles,
six hearts = four triangles.

The number of hearts needs to be doubled because the number of triangles is doubled.
3 triangles $\times 2=4$ triangles
3 hearts $\times 2=6$ hearts

Step 2
Determine the number of hearts needed.

Since six hearts = four triangles, and
four triangles = one square, then
six hearts = one square.

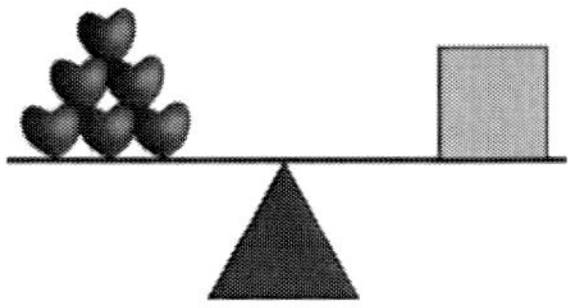

3. $4 \times 10 \div 5 + 2 = 10$

First operation	Multiplication: $4 \times 10 = 40$	$40 \div 5 + 2$
Second operation	Division: $40 \div 5 = 8$	$8 + 2$
Third operation	Addition: $8 + 2 = 10$	$4 \times 10 \div 5 + 2 = 10$

5. Step 1
Determine the equation.

The equation that represents the problem is $p + g = 5$.

Step 2
Explain how the values can change.

The value of *g* will change depending on the value of *p*.

- If Daria has one pink hair band, then she has four green hair bands.
- If Daria has two pink hair bands, then she has three green hair bands.
- If Daria has three pink hair bands, then she has two green hair bands.
- If Daria has four pink hair bands, then she has one green hair band.

7. Step 1
Write an equation to express the problem.

The following is one way to express the problem:
$60 + x = 150$

Step 2
Solve the problem symbolically.
$60 + x = 150$
$x = 150 - 60$
$x = 90$

The older brother weighs 90 kg.

Step 3
Solve the problem concretely.

Use base ten blocks to help solve the problem.

Represent Carl's weight of 60 kg with six ten rods.

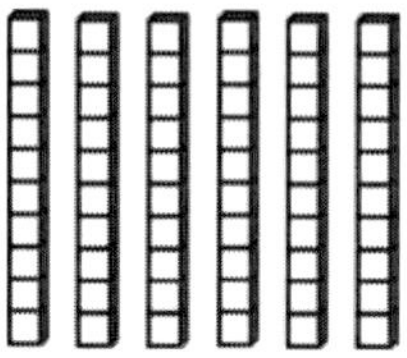

$60 + x = 150$

Represent the total weight of 150 kg with a hundred flat and five ten rods.

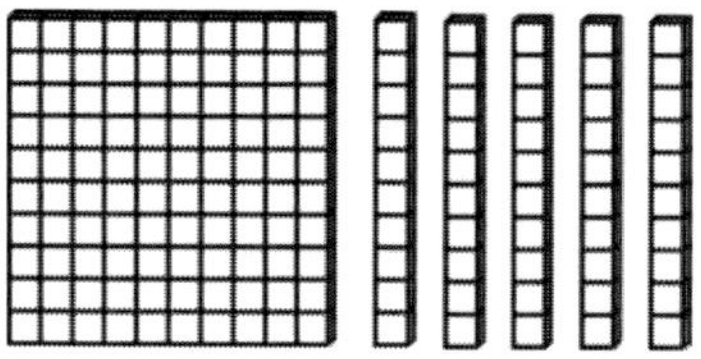

Think of how many ten rods need to be added to the six rods to equal the value of one hundred flat and five ten rods (150).

Start at 70 and count up by 10s as you add each ten rod, until you reach 150.

70:

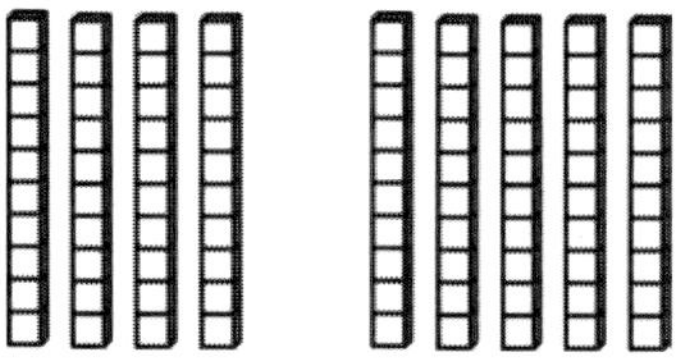

70, 80, 90,… 130, 140, 150

Since you added nine ten rods, the older brother's weight is 90 kg.

9. Each term number is three times more than the term number before it. The relationship between the term numbers can be expressed as $a \times 3$.

11. **Step 1**
Apply the property of subtraction.

Remove five squares from each side of the scale.

The rectangle on the left side and five squares on the right side are left.

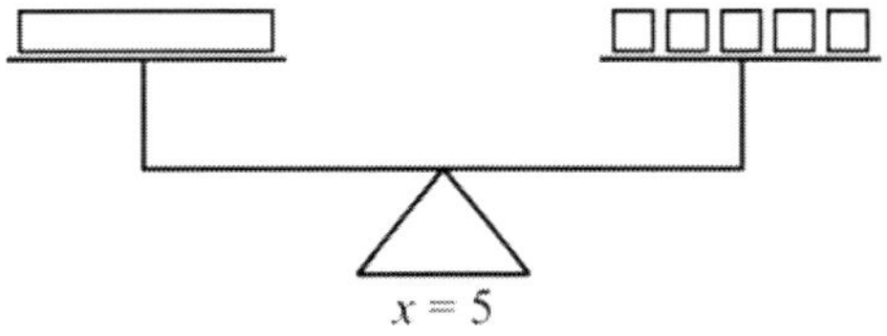

$$x + 5 = 10$$
$$x + 5 - 5 = 10 - 5$$
$$x = 5$$

Step 2
Check the answer by substituting the value of x into the original equation.

$$x + 5 = 10$$
$$5 + 5 = 10$$
$$10 = 10$$

Equality has been preserved because both sides of the equation are equal.

DATA ANALYSIS

Lesson 1—Line Graphs

TIME TO TRY ANSWERS AND SOLUTIONS

1. The number of correct answers on a student's quiz is an example of discrete data. The answers are either correct or wrong. There are no values between the scores.

2. There are two attributes missing from the graph. It does not have a title and the numbers on the y-axis are not labelled. As a result, we have no idea what the numbers represent or what the graph is about. The graph is confusing.

PRACTICE EXERCISES ANSWERS AND SOLUTION

1. Continuous data: The height of the tree will continuously change.

3. The data displayed is continuous data. As the hours go by, the speed changes. You can track the speeds in between the given values (hours).

5. The graph is missing a title and a label for the x-axis. Without these two attributes, the graph is confusing. We have no idea what the graph is about.

Lesson 2—Creating and Interpreting Line Graphs

TIME TO TRY ANSWERS AND SOLUTIONS

1. The graph could look like this:

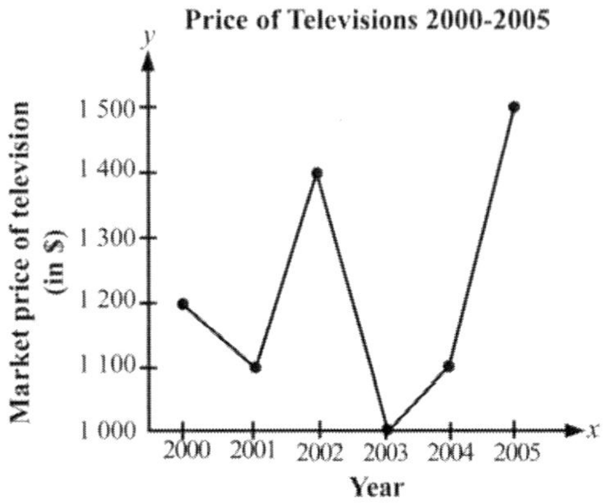

2. There are several conclusions that can be drawn from the data. Here are examples of two conclusions:

- Chuck makes $9 for every hour he works, up to eight hours a day.
- After eight hours a day, Chuck's wage doubles for each hour worked.

PRACTICE EXERCISES ANSWERS AND SOLUTIONS

1. The graph could look like this:

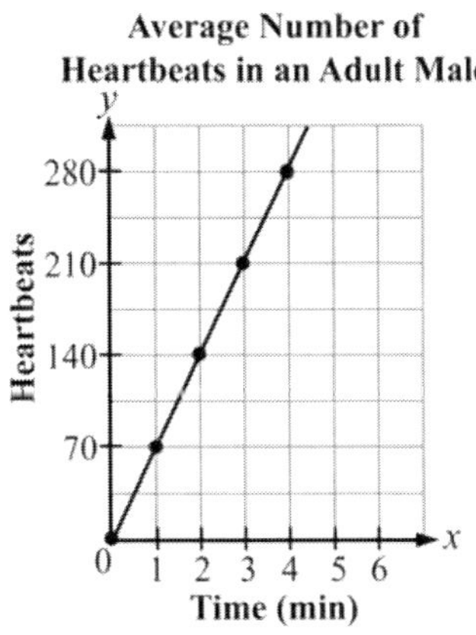

3. The graph could look like this:

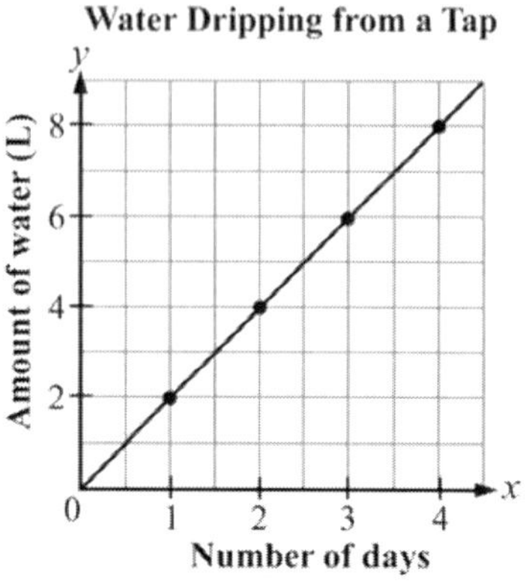

5. It can be concluded that the price of the ticket stays the same no matter how many tickets are purchased. There is no reduced price or reward for buying more tickets.

Lesson 3—Selecting, Justifying, and Using Appropriate Methods of Collecting Data

TIME TO TRY ANSWERS AND SOLUTIONS

1. **Step 1**
Offer choices.

Jared could offer choices such as playing sports, reading books, watching television, playing video games, spending time with friends, or other.

Step 2
Determine how to collect the data.

Depending on how many Grade 5 and 6 students are in his school, Jared could ask for a show of hands in each class or he could give the students a written questionnaire that they could fill out and return to him.

Step 3
Organize a method for recording results.

If Jared used a written questionnaire, he could use a table to record the responses.

Favourite Weekend Activity	Number of Students
Playing sports	25
Reading books	15
Watching television	8
Playing video games	11
Spending time with friends	17
Other	4

2. **Step 1**
Determine the number of times the action will be repeated.

Jayden can repeat the action of kicking the soccer ball until it lands 5 times.

Step 2
Determine how to record the observations.

Jayden can use a tally chart to record the number of kicks for each of the five tries.

Try	Number of Kicks
First	卌 卌 |||
Second	卌 卌 卌 |
Third	卌 卌 卌
Fourth	卌 卌 卌
Fifth	卌 卌 |||

Step 3
Draw a conclusion from the results.

Based on the results (13, 16, 15, 15, and 13), Jayden can conclude that he can keep the ball in the air with about 15 kicks before it lands on the ground.

3. The following search terms could be used:

Average temperature in Canada over the past 100 years.

Average temperature changes in Canada over the past 100 years

4. **Step 1**
Select a method of collecting data.

Trace should administer a questionnaire to answer his question.

Step 2
Justify your choice of method.

A questionnaire is a research instrument consisting of one or more questions, often with choices given, that are asked of a group of people in order to discover their opinions.

Trace could formulate a question with choices like, "Which cereal do you prefer for breakfast: Peanut Butter Delight, Rainbow Clusters, Cinnamon Squares, or something else?"

Trace could use a written questionnaire where his classmates check off their choices, or he could ask students to raise their hands as he reads each option. Trace could then record the number of students who chose each option.

PRACTICE EXERCISES ANSWERS AND SOLUTIONS

1. A database would be appropriate as a source of data. Using accurate search terms, you should be able to locate this information in an online database quite quickly.

3. A database would not be appropriate as a source of data. This information is too current to be in a database. You would be better off doing an Internet search for this information.

5. **Step 1**
Determine the most appropriate method of collecting the data.

This type of data would best be collected by performing an experiment.

Step 2
Justify your answer.

An experiment is an activity where you carry out an action once or several times to gather results, which then become the data.
The action to be carried out is to repeatedly perform jumping jacks. The results will be the number of times you can perform jumping jacks in a row without stopping.

Lesson 4—Selecting and Justifying an Appropriate Graph to Display Data

TIME TO TRY ANSWERS AND SOLUTIONS

1. The data would best be represented by a line graph because line graphs show changes over time, and temperature changes over time.

PRACTICE EXERCISES ANSWERS AND SOLUTIONS

1. The data would best be represented on a bar graph. Each of the four activities is countable and the relationships among the four choices would be easily seen so that comparisons could be made.

3. A double bar graph is the most appropriate because it is designed to make comparisons between and within sets of data. In this case, you are comparing the prediction with the actual amount for each of the five students.

Lesson 5—Graphing Collected Data and Interpreting the Resulting Graph

TIME TO TRY ANSWERS AND SOLUTIONS

1. **Step 1**
Graph the collected data.

The double bar graph could look like this.

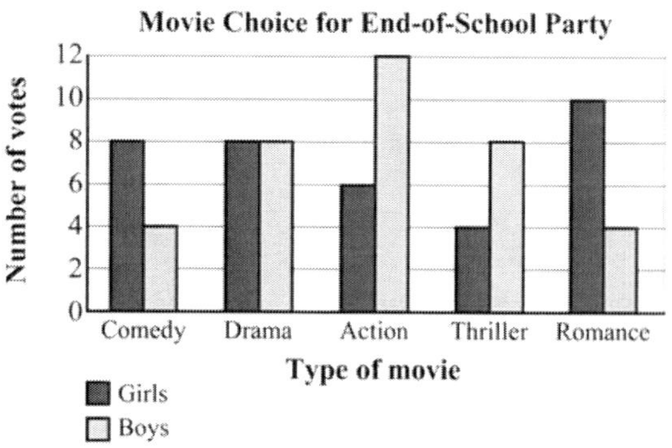

Step 2
Solve the problem.

Because the boys and girls in this class have completely different tastes in movies, except for drama, you can draw the conclusion that a drama movie should be shown at the year-end party. Both boys and girls would most likely be satisfied as they like drama equally.

PRACTICE EXERCISES ANSWERS AND SOLUTIONS

1. **Step 1**
Graph the collected data.

The graph could look like this:

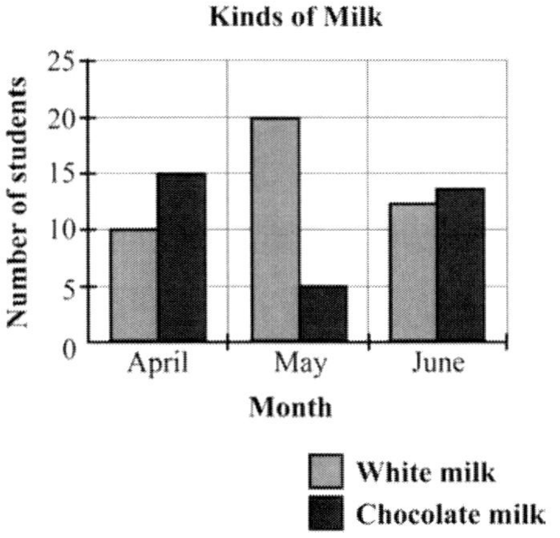

Step 2
Solve the problem.

From this graph, it can be concluded that overall, more students chose to drink white milk $(10+20+12=42)$ than chocolate milk $(15+5+13=33)$.

3. **Step 1**
Graph the collected data.

The displayed data could look like this.

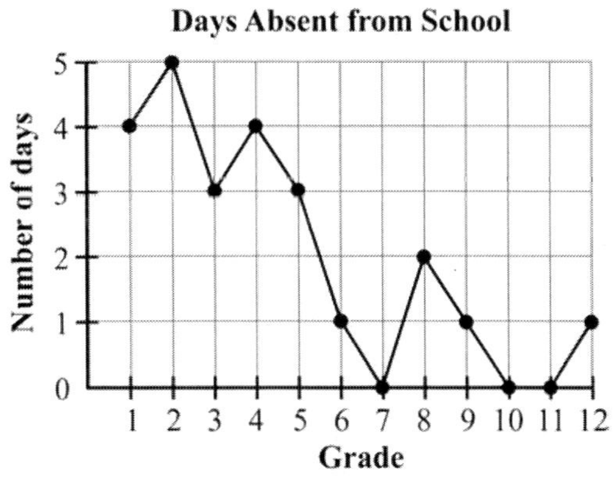

Step 2
Solve the problem.

The biggest drop in absences was 2 days. It happened twice:

- In Grade 2, Grayson was absent for 5 days and in Grade 3, he was only absent for 3 days.
$5-3=2$
- In Grade 5, Grayson was absent for 3 days, and in Grade 6, he was only absent for 1 day.
$3-1=2$

Practice Test

ANSWERS AND SOLUTIONS

1. In order for data to be presented clearly on a graph, the following attributes need to be present:
   - A graph needs a title to show what the graph is about.
   - An appropriate scale (interval) should be used to clearly display the data.
   - Both the x- and y-axes should be labelled to show what data is being presented.

3. Continuous data: The time will change continuously as the runner runs. There will be values between the given distances.

5. Based on the fact that at the end of week five, Brent had \$0, it can be concluded that sometime during week five, Brent spent all of the money he had saved over the four-week period.

7. **Step 1**
Determine an appropriate graph for displaying the data.

A line graph would be most appropriate.

Step 2
Justify your answer.

A line graph is best used when you want to show a change over time. In this case, the temperature is changing over a period of time during one particular day. A line graph would help you see the temperatures between the four given times.

9. A database would not be appropriate as a source of data. This information is too current to be in a database. Tom would be better off doing an Internet search for this information.

PROBABILITY

Lesson 1—Theoretical Probability

TIME TO TRY ANSWERS AND SOLUTIONS

1. **Step 1**
Find the number of possible outcomes.
There are two possible outcomes.

Step 2
Explain your answer.

A coin has two sides: heads on one side and tails on the other side. When flipped, the coin can land on either heads or tails.

2. **Step 1**
Find the favourable outcome.

Each regular ten-sided die is numbered from one to ten, so the die has 1 two on it. Therefore, there is 1 favourable outcome.

Step 2
Find the possible outcomes.

The die has 10 sides, so it could land on any number from 1 to 10. Therefore, the total number of possible outcomes is 10.

Step 3
Write the theoretical probability.

$$\frac{\text{number of favourable outcomes}}{\text{number of possible outcomes}}$$

The theoretical probability of rolling a two is 1 out of 10, or $\frac{1}{10}$.

3. **Step 1**
Find the favourable outcome.

The 12-sided die has 2 fives on it. Therefore, the number of favourable outcomes is 2.

Step 2
Find the possible outcomes.

The die has 12 sides, so it could land on any one of the 12 sides, regardless of what number is on it.
2, 2, 2, 3, 3, 4, 4, 4, 5, 5, 6, 6

Therefore, the total number of possible outcomes is 12.

Step 3
Write the theoretical probability.

$$\frac{\text{number of favourable outcomes}}{\text{number of possible outcomes}}$$

The theoretical probability of rolling a five is 2 out of 12, or $\frac{2}{12}$.

Step 4
Simplify the fraction.

$$\frac{2}{12}=\frac{2\div 2}{12\div 2}=\frac{1}{6}$$

The probability fraction $\frac{2}{12}$ can also be expressed as $\frac{1}{6}$.

4. **Step 1**
Find the theoretical probability.

Since there is only 1 five, there is 1 favourable outcome. Since there are six sides with numbers on a six-sided die, there are 6 possible outcomes.

The theoretical probability is 1 out of 6, or $\frac{1}{6}$.

Step 2
Make a prediction.

Since there will be 30 rolls of the die, write a fraction equivalent to $\frac{1}{6}$ with a denominator of 30.

$$\frac{1}{6}=\frac{1\times 5}{6\times 5}=\frac{5}{30}$$

A good prediction would be that Audrey may roll a five, 5 times out of 30 rolls.

PRACTICE EXERCISES ANSWERS AND SOLUTIONS

1. The die has eight sides, so there are 8 possible outcomes.

Even though there are only three different numbers used on the sides (6, 8, or 9), the eight-sided die can land on any one of the eight sides, even if the numbers are the same.

3. Out of 12 parts, there are 3 nines: a red nine, an orange nine, and a black nine.
The theoretical probability of the spinner stopping on a part with a nine is 3out of 12, or $\frac{3}{12}$.

The probability can also be expressed as 1out of 4, or $\frac{1}{4}$.

$$\frac{3}{12}=\frac{3\div 3}{12\div 3}=\frac{1}{4}$$

5. Out of 12 parts, there are 4 black parts: 1 with a seven, 1 with a four, 1 with an eight, and 1 with a nine.

The theoretical probability of the spinner stopping on a part that is black is 4 out of 12, or $\frac{4}{12}$.

The probability can also be expressed as one out of three, or $\frac{1}{3}$.

$$\frac{4}{12} = \frac{4 \div 4}{12 \div 4} = \frac{1}{3}$$

7. **Step 1**
Find the theoretical probability.

Since there are 7 numbered parts containing even numbers (72, 2, 32, 18, 22, 20, 12), there are 7 favourable outcomes.

Since there are 10 numbered parts in all, there are 10 possible outcomes.

The theoretical probability is $\frac{7}{10}$.

Step 2
Make a prediction.

Since there will be 30 spins, write a fraction equivalent to $\frac{7}{10}$ with a denominator of 30.

$$\frac{7}{10} = \frac{7 \times 3}{10 \times 3} = \frac{21}{30}$$

A good prediction is that the spinner will stop on an even number 21 times out of 30 spins.

Lesson 2—Experimental Probability

TIME TO TRY ANSWERS AND SOLUTIONS

1. **Step 1**
Identify the number of times the outcome happened.

Jenn rolled a three 17 times.

Step 2
Identify the number of times the event happened.

Jenn rolled the die 60 times.

Step 3
Find the experimental probability.

The experimental probability of rolling a three was $\frac{17}{60}$.

2. **Step 1**
Find the theoretical probability.

Since there is 1 favourable outcome (landing on heads) out of 2 possible outcomes (heads or tails), the theoretical probability is $\frac{1}{2}$.

Step 2
Find the experimental probability.

Since the coin landed on heads 9 times out of 14 tosses, the experimental probability is $\frac{9}{14}$.

Step 3
Compare the experimental results with the theoretical probability.

Write an equivalent fraction for the theoretical probability $\left(\frac{1}{2}\right)$. Use a denominator of 14 since the coin was tossed 14 times.
$\frac{1}{2}=\frac{1\times 7}{2\times 7}=\frac{7}{14}$

The experimental probability was greater than the theoretical probability.
$\frac{9}{14}>\frac{7}{14}\text{ or }\frac{1}{2}$

PRACTICE EXERCISES ANSWERS AND SOLUTIONS

1. Write the experimental probability.
$\frac{\text{number of times outcome occurs}}{\text{number of times event tested}}$

The experimental probability of landing on the red section is $\frac{19}{100}$.

3. **Step 1**
Find the theoretical probability.

Since there is 1 favourable outcome (landing on tails) out of 2 possible outcomes (heads or tails), the theoretical probability is $\frac{1}{2}$.

Step 2
Find the experimental probability.

Since the coin landed on tails 16 times out of 30 tosses, the experimental probability is $\frac{16}{30}$.

Step 3
Compare the experimental results with the theoretical probability.

Write an equivalent fraction for the theoretical probability of $\frac{1}{2}$. Use a denominator of 30 because the coin was tossed 30 times.
$\frac{1}{2}=\frac{15}{30}$

In the experiment, the coin landed on tails 16 times, so the experimental probability was just a little greater than the theoretical probability.
$\frac{16}{30}>\frac{15}{30}$

Practice Test

ANSWERS AND SOLUTIONS

1. **Step 1**
Identify the probability.
The event is based on theoretical probability.

Step 2
Explain your answer.

Rhonda is not performing an experiment. She is calculating a ratio fraction to show that there is 1 favourable outcome (rolling a 9) out of 12 possible outcomes (a 12-sided die).

3. **Step 1**
Find the favourable outcomes.

The favourable outcomes are four, five, and six. There are 3 favourable outcomes.

Step 2
Find the total number of possible outcomes.

There are six numbers on a six-sided die, so there are 6 possible outcomes.

Step 3
Write the theoretical probability.
$$\frac{\text{total number of favourable outcomes}}{\text{total number of possible outcomes}}$$

The theoretical probability of rolling a number greater than three is 3 out of 6, or $\frac{3}{6}$ (simplified to $\frac{1}{2}$).

5. Write the experimental probability.
$$\frac{\text{number of times outcome occurs}}{\text{number of times event tested}}$$

The experimental probability of rolling a six was 4 out of 12, or $\frac{4}{12}$.

7. The theoretical probability of rolling a one is 1 out of 4, or $\frac{1}{4}$.

The experimental probability of rolling a one is 3 out of 12, or $\frac{3}{12}$.

Since an equivalent fraction for $\frac{1}{4}$ (theoretical probability) is $\frac{3}{12}$ (experimental probability), the two probabilities are the same.

9. Write the experimental probability.
$$\frac{\text{number of times outcome occurs}}{\text{number of times event tested}}$$

The experimental probability of the spinner landing on the number three is $\frac{5}{18}$.

DEVELOPING AND APPLYING FORMULAS FOR PERIMETER, AREA, AND VOLUME

Lesson 1—Determining the Perimeter of Any Polygon

TIME TO TRY ANSWERS AND SOLUTIONS

1. Count the number of units for each side, and then add the four sets of units.

There are 5 units in the top length and 5 units in the bottom length.

There are 5 units on the right side of the shape.

There are 3 units on the left side of the shape.

Recall that you can add the numbers in any order.
$5+5+5+3=18$

The perimeter of the shape is 18 units.

2. **Step 1**
Develop a formula.

Since there are four sides to the figure, the formula you can use is $P=s+s+s+s$.

Step 2
Apply the formula.

Substitute the given lengths into the formula.
$$\begin{aligned}P&=24+24+30+20\\&=98\end{aligned}$$

The perimeter of the trapezoid is 98 m.

3. One method is to count the number of units for a side length. There are 4 units in each side length.

You can then add four to itself four times ($4+4+4+4=16$) or multiply 4 by 4 ($4\times4=16$).

The perimeter of the square is 16 units.

4. Since all the side lengths are the same for a square, you can use a formula like $P=l+l+l+l$ or $P=4\times l$.

Substitute the length of 10 m into the formula.
$P=10+10+10+10=40$ or $P=10\times4=40$

The perimeter of the frame is 40 m.

PRACTICE EXERCISES ANSWERS AND SOLUTIONS

1. To find the perimeter of the rectangle, count the number of centimetre units in each of the four sides and add them together. The numbers can be added in any order.
$4+4+2+2=12$

The perimeter of the rectangle is 12 cm.

3. Since the shape is a square, all four sides have the same length.

One method of finding the perimeter is to start by counting the number of units in one side length. Since there are 6 units in each side length, add four 6s to find the perimeter.
$6+6+6+6=24$

Another method of finding the perimeter of the square is to multiply the number of units in one side length (6) by 4.
$6\times4=24$

The perimeter of the square is 24 units.

Lesson 2—Determining the Area of Any Rectangle

TIME TO TRY ANSWERS AND SOLUTIONS

1. One method that you can use to find the area is to count the number of units that cover one length (6 units). Then, repeatedly add that number (6) for the number of rows that make up the width (6 units).
$6+6+6+6+6+6=36$

Since there are a total of 36 units that cover the surface of the square, the area of the square is 36 units^2.

2. Substitute the length (6 m) and the width (4 m) of the garden into the area formula.

$$\begin{aligned}A&=l\times w\\&=6\times4\\&=24\end{aligned}$$

Tom needs to buy enough peat moss to cover 24 m^2.

PRACTICE EXERCISES ANSWERS AND SOLUTIONS

1. **Step 1**
Explain one method.

One method is to use repeated addition.

The length of the rectangle is made out of five square units.

Since there are three rows, add 5 three times.
$5+5+5=15$ units^2

Step 2
Explain another method.

Another method is to use multiplication.

Multiply the number of square units in the length (5) by the number of square units in the width (3).
$5 \times 3 = 15 \text{ units}^2$

3. Count the number of square units for the lengths and widths of each rectangle and substitute the numbers into the area formula.

Rectangle A: $A = 2 \times 10$
$= 20 \text{ units}^2$
Rectangle B: $A = 8 \times 4$
$= 32 \text{ units}^2$
Rectangle C: $A = 6 \times 5$
$= 30 \text{ units}^2$
Rectangle D: $A = 5 \times 4$
$= 20 \text{ units}^2$

Rectangles A and D have an area of 20 units^2.

Lesson 3—Determining the Volume of Any Right Rectangular Prism

TIME TO TRY ANSWERS AND SOLUTIONS

1. **Step 1**
Count the number of cubic units.

Length of base: 6 cubic units
Width of base: 2 cubic units
Height of prism: 3 cubic units

Step 2
Calculate the area of the base.
$A = l \times w$
$= 6 \times 2$
$= 12 \text{ units}^2$

Step 3
Multiply the area of the base by the height of the prism.
$12 \times 3 = 36$

The volume of the right rectangular prism is 36 units^3.

2. **Step 1**
Select a formula for volume.

Since you know the area of the base of the prism, you can use a formula like $V = A \times h$, in which you multiply the area of the base by the height of the prism.

Step 2
Apply the formula.

Substitute the numbers for the area of the base (25) and the height of the prism (6) into the formula.
$A = 25 \times 6$
$= 150 \text{ cm}^3$

The volume of the prism is 150 cm^3.

PRACTICE EXERCISES ANSWERS AND SOLUTIONS

1. **Step 1**
Count the number of cubic units.

Length of base: 4 cubic units
Width of base: 2 cubic units
Height of prism: 3 cubic units

Step 2
Calculate the area of the base.
$A = l \times w$
$= 4 \times 2$
$= 8 \text{ units}^2$

Step 3
Multiply the area of the base by the height of the prism.
$8 \times 3 = 24$

The volume of the right rectangular prism is 24 units^3.

3. Step 1
Determine the height of the prism.

To find the height, count the number of layers that make up the prism. Since there are 5 layers of cubes, the height of the prism is 5 cubes.

Step 2
Apply a formula for volume.

Substitute the numbers for the length (12), the width (10), and the height (5) into a formula like $V = l \times w \times h$.
$V = 12 \times 10 \times 5$
$= 600 \text{ cm}^3$

The volume of the rectangular prism is 600 cm^3.

Practice Test

ANSWERS AND SOLUTIONS

1. To find the perimeter of the rectangle, count the number of centimetre units in each of the four sides and add them together. The numbers can be added in any order.
$5 + 5 + 3 + 3 = 16$

The perimeter of the rectangle is 16 cm.

3. Step 1
Find the area of the base.

The base is covered with 5 rows of cubes. Each row is made up of 11 cubes.

Use the area formula to calculate the total number of cubes used to cover the base.
$A = l \times w$
$= 11 \times 5$
$= 55 \text{ units}^2$

Step 2
Find the height.

Five layers of cubes are needed to reach the top of the prism.

Use a volume formula to calculate the total number of cubes needed to fill the prism.
$V = A \times h$
$= 55 \times 5$
$= 275 \text{ units}^3$

5. Substitute the numbers for the length (9) and the width (6) into the formula for area.
$A = l \times w$
$= 9 \times 6$
$= 54 \text{ units}^2$

Since it takes 54 square units to cover the rectangular shape of the game board, the area of the board is 54 units^2.

7. Since you need to add the lengths of the three sides to find the perimeter, you can use the formula $P = s + s + s$.

Substitute a side length for each letter s in the formula.
$P = 11 + 9 + 5$
$= 25 \text{ cm}$

9. Substitute the numbers for the length (5), the width (4), and the height (3) into a formula for volume.
$V = l \times w \times h$
$= 5 \times 4 \times 3$
$= 60 \text{ m}^3$

The volume of the tank is 60 m^3.

PROBLEM SOLVING

Lesson 1—Steps for Problem Solving

TIME TO TRY ANSWERS AND SOLUTIONS

1. **Step 1**
   Determine the operation.

   The problem does not involve equal groups. The keyword *left* means that subtraction can be used.

   Step 2
   Write an expression to solve the problem.

   Rachel started with 24 cookies and gave away 10 cookies, then 6 cookies.

   The amount of cookies that are left is given by 24 – 10 – 6.

   Step 3
   Solve the problem.

$$\begin{array}{r} 24 \\ 10 \\ -\ 6 \\ \hline 8 \end{array}$$

   Rachel has 8 cookies left.

PRACTICE EXERCISES ANSWERS AND SOLUTIONS

1. **Step 1**
   Determine the operation.

   The boxes all have the same mass.
   The keyword *total* means that multiplication can be used.

   Step 2
   Write an expression to solve the problem.

   There are 6 boxes, each with a mass of 7.48 grams. The total mass is given by 6×7.48 or 7.48×6.

   Step 3
   Solve the problem.

$$\begin{array}{r} 7.48 \\ \times\ \ 6 \\ \hline 44.88 \end{array}$$

   The total mass of the 6 boxes in 44.88 g.

3. **Step 1**
   Determine the operation.

   The 7 pieces of cloth are of equal length. The keyword *into* means that division can be used.

   Step 2
   Write an expression to solve the problem.

   Ms. Carlson had 9.8 m of cloth that she cut into 7 equal parts. The length of each piece is given by 9.
   8 ÷ 7

   Step 3
   Solve the problem.

$$\begin{array}{r} 1.4 \\ 7\overline{)9.8} \\ \underline{7} \\ 2\ 8 \\ \underline{2\ 8} \\ 0 \end{array}$$

   Each piece of cloth is 1.4 m long.

Lesson 2—Estimating Solutions

TIME TO TRY ANSWERS AND SOLUTIONS

1. a) Step 1
Determine the operation.

The problem does not involve equal groups. The keyword *total* means that addition can be used.

Step 2
Write an expression to solve the problem using estimation.

Drop the decimal point and the digits to the right of the decimal point.
$12.45 \rightarrow 12$
$11.90 \rightarrow 11$

The approximate cost of both items is given by $12 + 11$.

Step 3
Solve the problem using estimation.

$$\begin{array}{r} 12 \\ +11 \\ \hline 23 \end{array}$$

The approximate cost for the two items is \$23.

b) Step 1
Write an expression to solve the problem without estimation.

The total cost of both items is given by $12.45 + 11.90$.

Step 2
Solve the problem without estimation.

$$\begin{array}{r} 12.45 \\ +11.90 \\ \hline 24.35 \end{array}$$

The actual cost of both items is \$24.35.

Step 3
Compare the estimated solution to the actual solution.

Subtract the estimated cost from the actual cost to find the difference between the two solutions.

$$\begin{array}{r} 24.35 \\ -23 \\ \hline 1.35 \end{array}$$

The estimated cost is \$1.35 lower than the actual cost.
If Evelyn relies on the estimate, she will think that she can afford both items, which she cannot.

Because the problem involves money, this is not a reasonable estimate.

PRACTICE EXERCISES ANSWERS AND SOLUTIONS

1. Step 1
Determine the operation.

Each lap took an equal amount of time to complete. The keyword *total* means that multiplication can be used.

Step 2
Write an expression to solve the problem using estimation.

Use the front-end digits.
$2.61 \rightarrow 2$

The approximate time to complete the laps is given by 2×9.

Step 3
Solve the problem using estimation.

$$\begin{array}{r} 2 \\ \times 9 \\ \hline 18 \end{array}$$

It took Jill approximately 18 minutes to complete all nine laps.

3. Step 1
Determine the **operation**.

The problem does not involve equal groups. The keyword *left* means that subtraction can be used.

Step 2
Write an expression to solve the problem using estimation.

Use the front-end digit of each number.
$218 \rightarrow 200$
$37 \rightarrow 30$
$103 \rightarrow 100$
The approximate number of blocks left in the basket is given by 200 – 30 – 100.

Step 3
Solve the problem using estimation.

$$\begin{array}{r} 200 \\ 30 \\ -100 \\ \hline 70 \end{array}$$

There are approximately 70 blocks left in the basket.

Lesson 3—Problem Solving with Large Numbers

TIME TO TRY ANSWERS AND SOLUTIONS

1. Step 1
Determine the operation.

The **problem** does not involve equal groups. The keyword *combined* means that addition can be used.

Step 2
Write an expression to solve the problem.

The total score is given by
35 295 218 + 51 021 754 + 29 717 698.

Step 3
Solve the problem.

Because the problem involves several very large numbers, use a calculator to perform the additions.

[3] [5] [2] [9] [5] [2] [1] [8] [+]
[5] [1] [0] [2] [1] [7] [5] [4] [+]
[2] [9] [7] [1] [7] [6] [9] [8] [=]

The combined score of the team was 116 034 670 points.

PRACTICE EXERCISES ANSWERS AND SOLUTIONS

1. Step 1
Determine the operation.

The problem does not involve equal groups. The keyword *left* means that subtraction can be used.
Step 2
Write an expression to solve the problem.

The amount Daniela still needs to collect is given by 10 000 – 2 922.76 – 3 391.21.

Step 3
Solve the problem.

Because the problem involves several very large numbers, use a calculator to perform the subtractions.
[1] [0] [0] [0] [0] [–]
[2] [9] [2] [2] [.] [7] [6] [–]
[3] [3] [9] [1] [.] [2] [1] [=]

Daniela still needs to collect $3 686.03 to reach her goal.

3. **Step 1**
Determine the operation.

Each truck will carry an equal number of chocolate bars. The keyword *split* means that division can be **used**.

Step 2
Write an expression to solve the problem.

The number of chocolate bars in each truck is given by 1 423 037 ÷ 11.

Step 3
Solve the problem.

Because the problem involves a large number, use a calculator to perform the division.
[1][4][2][3][0][3][7][÷][1][1][=]

Each truck will carry 129 367 chocolate bars.

Practice Test

ANSWERS AND SOLUTIONS

1. **Step 1**
Determine the operation.

The pages all contain an equal number of stickers. The keyword *total* means that multiplication can be used.

Step 2
Write an expression to solve the problem.

There are 25 pages, each with 11 stickers. The total number of stickers is given by 25 × 11.

Step 3
Solve the problem.

$$\begin{array}{r} 25 \\ \times 11 \\ \hline 25 \\ \underline{250} \\ 275 \end{array}$$

Beth has 275 stickers in total.

3. **Step 1**
Determine the operation.

The problem does not involve equal groups. The keyword *total* means that addition can be used.

Step 2
Write an expression to solve the problem.

Marissa used 1.42 L, then 1.29 L, and finally 1.61 L of water. The total amount of water is given by 1.42 + 1.29 + 1.61.

Step 3
Solve the problem.

$$\begin{array}{r} 1.42 \\ 1.29 \\ +1.61 \\ \hline 4.32 \end{array}$$

Marissa used 4.32 L of water in total.

5. **Step 1**
Determine the operation.

The problem does not involve equal groups. The keyword *total* means that addition can be used.

Step 2
Write an expression to solve the problem using estimation.

Use the front-end digits.
25.8 → 20
21.3 → 20

The approximate amount of fabric is given by 20 + 20.

Step 3
Solve the problem using estimation.

$$\begin{array}{r} 20 \\ +20 \\ \hline 40 \end{array}$$

Lorraine has approximately 40 yards of fabric.

7. **Step 1**
Determine the operation.

The problem does not involve equal groups. The keyword *left* means that subtraction can be used.

Step 2
Write an expression to solve the problem.

The distance left to fly is given by 13 288.9 – 5 519.6.

Step 3
Solve the problem.

Because the problem involves large numbers, use a calculator to perform the subtraction.

[1] [3] [2] [8] [8] [.] [9]
[–] [5] [5] [1] [9] [.] [6] [=]

The airplane still needs to fly 7 769.3 km before reaching Sydney.

Credits

ORDERING INFORMATION

SCHOOL ORDERS

Please contact the Learning Resource Centre (LRC) for school discount and order information.

***THE KEY* Study Guides** are specifically designed to assist students in preparing for unit tests, final exams, and provincial examinations.

***THE KEY* Study Guides** – $29.95 each plus G.S.T.

SENIOR HIGH		JUNIOR HIGH	ELEMENTARY
Biology 30	Biology 20	English Language Arts 9	English Language Arts 6
Chemistry 30	Chemistry 20	Math 9	Math 6
English 30-1	English 20-1	Science 9	Science 6
English 30-2	Mathematics 20-1	Social Studies 9	Social Studies 6
Applied Math 30	Physics 20	Math 8	Math 4
Pure Math 30	Social Studies 20-1		
Physics 30	English 10-1	Math 7	English Language Arts 3
Social Studies 30-1	Math 10 Combined		Math 3
Social Studies 30-2	Science 10		
	Social Studies 10-1		

Student Notes and Problems (SNAP) Workbooks contain complete explanations of curriculum concepts, examples, and exercise questions.

SNAP Workbooks – $29.95 each plus G.S.T.

SENIOR HIGH		JUNIOR HIGH	ELEMENTARY
Biology 30	Biology 20	Math 9	Math 6
Chemistry 30	Chemistry 20	Science 9	Math 5
Applied Math 30	Mathematics 20-1	Math 8	Math 4
Pure Math 30	Physics 20	Science 8	
Math 31	Math 10 Combined		Math 3
Physics 30	Science 10	Math 7	
		Science 7	

Visit our website for a tour of resource content and features or order resources online at

<u>www.castlerockresearch.com</u>

#2340, 10180 – 101 Street
Edmonton, AB Canada T5J 3S4
e-mail: learn@castlerockresearch.com

Phone: 780.448.9619
Toll-free: 1.800.840.6224
Fax: 780.426.3917

CASTLE ROCK
RESEARCH CORP

ORDER FORM

THE KEY	QUANTITY
Biology 30	
Chemistry 30	
English 30-1	
English 30-2	
Applied Math 30	
Pure Math 30	
Physics 30	
Social Studies 30-1	
Social Studies 30-2	
Biology 20	
Chemistry 20	
English 20-1	
Mathematics 20-1	
Physics 20	
Social Studies 20-1	
English 10-1	
Math 10 Combined	
Science 10	
Social Studies 10-1	
English Language Arts 9	
Math 9	
Science 9	
Social Studies 9	
Math 8	
Math 7	
English Language Arts 6	
Math 6	
Science 6	
Social Studies 6	
Math 4	
English Language Arts 3	
Math 3	

Student Notes and Problems Workbooks	QUANTITY SNAP Workbooks	QUANTITY Solution Manuals
Math 31		
Biology 30		
Chemistry 30		
Applied Math 30		
Pure Math 30		
Physics 30		
Biology 20		
Chemistry 20		
Mathematics 20-1		
Physics 20		
Math 10 Combined		
Science 10		
Math 9		
Science 9		
Math 8		
Science 8		
Math 7		
Science 7		
Math 6		
Math 5		
Math 4		
Math 3		

TOTALS

KEYS	
SNAP WORKBOOKS	
SOLUTION MANUALS	
SOLUTION MANUALS	

Learning Resources Centre

Castle Rock Research is pleased to announce an exclusive distribution arrangement with the Learning Resources Centre (LRC). Under this agreement, schools can now place all their orders with LRC for order fulfillment. As well, these resources are eligible for applying the Learning Resource Credit Allocation (LRCA), which gives schools a 25% discount off LRC's selling price. Call LRC for details.

Orders may be placed with LRC by
Telephone: 780.427.2767
Fax: 780.422.9750
Internet: www.lrc.education.gov.ab.ca
Or mail: 12360 – 142 Street NW
Edmonton, AB T5L 4X9

Learning Resources Centre

PAYMENT AND SHIPPING INFORMATION

Name: ____________________
School Telephone: ____________________
SHIP TO
School: ____________________
Address: ____________________
City: __________ Postal Code: __________
PAYMENT
☐ by credit card
VISA/MC Number: ____________________
Expiry Date: ____________________
Name on card: ____________________
☐ enclosed cheque
☐ invoice school P.O. number: __________

CASTLE ROCK RESEARCH CORP

#2340, 10180 – 101 Street, Edmonton, AB T5J 3S4 **Phone:** 780.448.9619 Fax: 780.426.3917
Email: learn@castlerockresearch.com **Toll-free:** 1.800.840.6224
www.castlerockresearch.com

Made in United States
North Haven, CT
12 November 2026

22200105R00267